KEEPING
FOOD FRESH

Books by Janet Bailey

CHICAGO HOUSES
KEEPING FOOD FRESH

KEEPING FOOD FRESH

Revised Edition

———————————◆⟨⟩◆———————————

JANET BAILEY

Kitchen Testing by
Sue Spitler

PERENNIAL LIBRARY

Harper & Row, Publishers, New York
Grand·Rapids, Philadelphia, St. Louis, San Francisco
London, Singapore, Sydney, Tokyo, Toronto

The first edition of this book was published in 1985 by the Dial Press.

First PERENNIAL LIBRARY edition published 1989.

LIBRARY OF CONGRESS CATALOG CARD NUMBER 87-45085
ISBN 0-06-272503-3

93   94   95   96   97   RRD   10   9   8   7   6   5   4   3   2   1

For my parents, Edward and LaVerne Parrott;
for my sister, Katherine Caruso;
for my mentor, Laurence Gonzales.

ACKNOWLEDGMENTS

For their help in sifting through hundreds of books and towering stacks of technical publications, I thank my researchers, Carolyn Lorence, Marijo Racciati, and Susan Zaczek. I am grateful to Sue Spitler, whose wide experience and bright, bustling test kitchen have been my best sources. And special thanks also to my most faithful reader, Leo Barliant.

CONTENTS

Introduction   xiii

1 Vegetables   1

2 Fruits and Nuts   67

3 Milk, Cheese, and Eggs   129

4 Meat, Poultry, and Fish   173

5 Grains and Grain Products   243

6 Staples   273

7 Snacks and Sweets   315

8 Kitchen Systems   325

9 Kitchen Crises   359

Index   379

INTRODUCTION

One day several years ago I brought home ten pounds of white rice from a warehouse supermarket. The price was cheap and the bag was a quaint cream-colored cotton. For about six months the rice was settled comfortably in the back of the pantry and I dipped out what I needed, but soon there were only a few cupfuls left. The sack disappeared behind cans of tuna and a stack of cookbooks and I lost count of the time.

When the bag of rice came back into view it looked okay; it smelled okay; but was it okay? I didn't know and I couldn't seem to find out without making long-distance phone calls.

A couple of hundred years of technology have taken the task of storing food almost entirely out of our hands, leaving us ignorant of a basic craft. We rely on a giant, sprawling network of growers, harvesters, packers, processors, canners, dairies, shippers, government inspectors, wholesalers, and retailers to supply us with food that is clean and fresh.

There is nothing at all wrong with depending on the food industry: subsistence farming is only attractive seen through mists of nostalgia. But most of us have *no* idea where food comes from, how fresh it is, or how it has been treated before we get it. We usually have only a sketchy idea of how to store it once it is in our kitchens. Should I wash the berries first? Can I just scrape this mold off and eat the Cheddar anyway? Labels sometimes give us dates or cryptic instructions like "Refrigerate after opening." (Fine, but for how long? And what about this brown stuff on the rim of the mayonnaise jar?) Too many questions go unanswered.

So, standing there rice in hand, thinking of all the things I didn't know about food and couldn't easily find out, I decided to write Keeping Food Fresh.

I found some surprises. While my rice would keep indefinitely, a certain protein in wheat flour deteriorates, usually in twelve months. The flavor of regular soft drinks will go off after a year; natural soft drinks (with no preservatives) can ferment and should not be stored for more than a few months. Onions and potatoes do not make good shelf mates: each gives off a gas that spoils the other.

Even though the mainspring of this book is telling you how to store all the foodstuffs in your kitchen—what is the right packaging, the right temperature, the right length of time?—I have also made lots of suggestions about how to choose fruits, vegetables, meats, dairy products, canned goods, breads, flours, and everything else you buy. Obviously, fresh, wholesome, high-quality foods are going to last longer and be a pleasure to eat.

The specific details about how to handle food and exactly how long to store it have been culled, bit by bit, from hundreds of books and technical publications, from conversations with growers, processors, and wholesalers, and from scores of kitchen tests. I have considered storage life at an end when a food is noticeably less palatable than it was at its peak. Examples would be wilted lettuce or overripe nectarines. Another standard for storage times relates to nutrients: a food that has lost a considerable number of vitamins, proteins, or minerals is no longer acceptable. Whenever a food is potentially dangerous after storage time has elapsed, the problem is specifically explained in the text.

Because a freezer is available to almost everyone, I have also described how to freeze and properly defrost many foods so that they will keep for months and come out almost as good as they went in.

Chapter 8, "Kitchen Systems," explains some of the fundamentals of food science. When you understand what exactly is going on inside your eggplant or your canned tuna to make it spoil, and when you know what it takes to slow the spoilage, you can make some judgments for yourself about how to manage your food supply.

The final chapter of KEEPING FOOD FRESH deals with crises in the storage life of your food. You can find out what to do about troubles like mice in the pantry and malfunctioning freezers full of half-thawed hamburger patties. The section on food poisoning tells you how to make sure food is safe.

With all this information at hand, you should be able to keep your food at its peak—colorful, succulent, and with nutrients intact—until the moment you begin to cook and the recipe books take over.

1

Vegetables

VEGETABLES

Plants are generous with themselves, yielding up edible leaves, fruits, seeds, stalks, tubers, and roots. And this bounty has a startling characteristic: even plucked from the field, boxed, stored, shipped, and supermarketed, vegetables continue to live. The celery in the refrigerator and the sweet potato on the shelf both breathe, taking in oxygen and giving off carbon dioxide. As long as a fresh vegetable is stored, its oxygen-fueled internal chemistry labors along, altering the way a vegetable looks, tastes, and feels. Only the deep slice of your kitchen knife or the heat of cooking finally stops the process.

No matter which particular part of the plant is harvested, the fresh vegetable breathes. Complex proteins called enzymes form within each cell, triggering chemical reactions soon visible to the cook. Many types of vegetables, such as onions and potatoes, sprout, sending out small, pale shoots that can become full-blown plants, which, as they grow, consume the parent vegetable for nourishment. Other enzymes cause leafy and stem vegetables to yellow and toughen after they are picked. When a vegetable rots, it is at least in part because of the action of its own enzymes.

Enzymes alter the flavor of vegetables in storage, as well as their texture and appearance. Compared to fruits, vegetables have less sugar, less acid, more starch, and therefore less intense flavor. They become even blander during storage as the already low sugar is gradually converted to starch. One of the reasons most root, tuber, and bulb vegetables last longer in storage than seeds, fruits, leaves, and other types is that they have fully matured on the plant before being picked. Because they are mature, these vegetables begin with a higher proportion of starch and their loss of sugar is slower and less noticeable. At the other end of the scale are corn and peas, which are very high in sugar at harvest. Enzymes can reduce their sugar content to virtually zero within a few hours of picking, until the garden taste of the vegetables is totally lost.

The only way to keep a vegetable fresh for very long is to suppress its enzymes. Cooling a vegetable to the right temperature slows down the exchange of oxygen and carbon dioxide, which in turn curbs the enzymes that cause unwanted changes.

Vegetables also tend to dry out in storage. Deep within the tissue structure of every newly picked vegetable are vacuoles like tiny balloons plumped up with nutrient-rich water. The pressure these vacuoles exert on cell walls makes vegetables crisp, firm, and erect. As this internal moisture evaporates into the outside air, a vegetable droops, losing not only its pleasing texture but also many vitamins. Because of their very high water content, leafy vegetables such as lettuce and stem vegetables such as celery suffer most dramatically from the effects of dry air, but they are not alone in needing high humidity during storage. Almost all vegetables do.

A fresh-picked vegetable is immediately assailed by microbes. Potatoes are attacked by blackheart and slimy soft rot. Onions suffer from smudge, tomatoes from sour rot, and green beans from cottony leak. Invading bacteria, parasites, and molds can corrupt plant tissue to the point where the vegetable becomes soggy, discolored, and totally unpalatable. Because it is alive, a healthy vegetable is able, to some degree, to resist these destroyers. Any cut or bruise sustained during handling, however, is a breach through which microbes may infiltrate. Warm temperatures, of course, encourage microbes to multiply; cool temperatures inhibit them.

Growers and shippers apply a complicated technology to harvested vegetables to control all the factors of decay. Different gases and chemicals applied to the vegetables alter the function of the enzymes and discourage microbes (reason enough to wash vegetables before you serve them). The cold temperatures in refrigerated warehouses retard the metabolism of the plant tissues stored there and so slow aging. The high humidity maintained in warehouses prevents drying and wilting. Many vegetables are cradled from the field to the warehouse to the supermarket in specially designed packing materials that protect them from damage. Almost all of these same controls can be applied in the kitchen to prolong the useful life of the vegetables you buy.

Time, heat, enzyme activity, dry air, microbes, and injury—the same forces that spoil vegetables also destroy vitamins. When you store vegetables with care, you are not only protecting their vitality and flavor; you are maintaining the potency of the food's nutrients.

Buying Fresh Vegetables

Whether you are choosing ripe tomatoes or papery-skinned onions, there are some commonsense observations that cover all vegetable buying.

- Choose beautiful vegetables. Rich, bright color and firm texture are definitive measures of both freshness and vitamin content in a vegetable. A crisp, deep orange carrot has more nutrients—namely, vitamins A, C, and folacin, a B vitamin—than one that is limp and faded. The connection between color and certain vitamins is so direct that even on the same head of lettuce the darker outer leaves have more vitamin A than the paler ones close to the heart.
- Bruises, cuts, and tears are forerunners of rapid decay. Avoid damaged vegetables and those that already show obvious signs of spoilage such as discolored patches, spots, and bleached surfaces.
- Many vegetables as they mature develop a tough, woody substance, lignin, that even long cooking cannot soften. Common sites for lignin are the core of carrots, beets, and parsnips; the stalks of broccoli and asparagus; and the ribs of chard, kale, spinach, and other greens. Choose smaller vegetables to get the youngest, most tender-textured.
- Take into account the grocer's display methods. For example, lettuce stacked up for sale at room temperature is losing storage life by the hour. It may be cheap, but it may not last more than one or two days once you get it home.

Storing Fresh Vegetables

A refrigerator is, in most cases, exactly the right instrument for storing fresh vegetables. Most vegetables last longest at low temperatures and high humidity. The closer asparagus, celery, peas, and spinach, to name a few, get to 32° F. and 90 percent humidity, the better they do. An efficient refrigerator can hover around the correct temperature, but the air inside is too dry. The vegetable crisper, that drawer in the bottom of the refrigerator, works because it closes the food off from circulating air that draws up moisture.

You can humidify the vegetables in your refrigerator most effectively with the help of plastic bags or other closed containers. Put the fresh, *unwashed* vegetable in the bag, close it up, and you create a small but very damp environment, humidified by the vegetable's own natural moisture. A vegetable that is actually wet on its surface is too receptive to bacterial growth, so while the air around it must be moist the vegetable itself should feel dry to the touch.

A number of vegetables, notably potatoes and tomatoes, don't do well at all in the refrigerator. They thrive at a higher temperature and slightly lower humidity, namely 50° F. and 80 percent. This is a tough combination to achieve in the average kitchen. An unheated porch or cellar, if you have one, may be the answer. Sometimes the best you can do is to keep these vegetables in the coolest spot in the cupboard.

If you have been careful to buy undamaged vegetables, it only makes sense to treat them gently once you've got them home. Don't wash, peel, or cut them until the last minute before cooking or serving. Don't soak them in water. Cutting and soaking vegetables leaches out water-soluble vitamins and minerals. To minimize the chances of bacteria and mold growth, store vegetables in clean containers and keep refrigerator surfaces wiped free of decaying food particles.

Many supermarkets sell produce preselected, preweighed, and prewrapped in plastic. Inside the package, it is not unusual to find a few brown and scrawny brussels sprouts or pea pods buried beneath their healthy-looking companions. As soon as you get them home, sort through your vegetables and remove any decayed or damaged individuals that might transfer contaminants to healthy ones during storage.

For specific storage advice, see under name of particular vegetable.

Buying and Storing Canned and Frozen Vegetables

Both fresh and processed vegetables are subject to United States Department of Agriculture grading. The grade is much more likely, however, to appear on the label of canned or frozen food than on fresh vegetables in the produce department. The grading system evaluates size, color, and shape of each type of vegetable. Grade A (also called Fancy) green beans may be prettier and tenderer than Grade B (Extra Standard) or Grade C (Standard), but all the grades have equivalent food value and storage life.

A frozen vegetable is often preferable to a canned one. Both frozen and canned vegetables are processed almost immediately after harvest, when they are at a peak of freshness and nutritional vitality, but canning alters the character of a vegetable and reduces its vitamin and mineral content. Of course cupboard space is easier to come by than freezer space, so canned foods are sometimes the most practical choice. And there are a few vegetables—tomatoes and sweet potatoes, for instance—that are more appealing canned than frozen. (Canned tomatoes are actually superior to the dull, pithy cue balls passed off as fresh tomatoes during the winter months.) But, in general, frozen vegetables are best because they are more like fresh, more nutritious, and they keep just as long as canned.

When you do buy canned vegetables, choose cans that are clean and rust-free. Many cans are dented by the time they reach the grocer's shelf and, though most of them are perfectly safe, don't buy any can that is so misshapen that it can't be stacked. Bulging ends and breaks in the seams are signs of dangerous spoilage. Glass containers should be very tightly sealed; avoid any jars with signs of leakage.

After opening, canned vegetables must be refrigerated. Transfer them to tightly lidded containers and store them for no more than three to five days.

When you pick a frozen vegetable from the grocer's case, give it a squeeze and a shake before you drop it in your cart. It should feel very hard. A mushy frozen vegetable just isn't cold enough to hold on to its flavor and texture. If the box rattles because each individual pea, bean, or corn kernel is still separate and frozen hard, you have found a rare package indeed, one that has survived shipping and handling without thawing even a little. The least bit of melting will fuse frozen vegetables into a solid block, an event which, while not a disaster, is a compromise.

Canned and frozen vegetables are not immune to mishandling. If a can of food goes through any temperature extreme—accidental freezing or heat above normal room temperature—the contents will lose desirable flavor, color and texture. On a cool, dry shelf, canned vegetables will keep well for up to one year. After that, they will be safe to eat but not as tasty or nutritious.

Frozen vegetables must be kept constantly at or below 0° F. to protect their taste and texture. They will, under the right conditions, keep well for ten to twelve months.

Both canned and frozen vegetables may be cooked and then frozen for further storage. They will be safe to eat after thawing, but very soft in texture and far less nutritious.

Freezing Fresh Vegetables

Most fresh vegetables can be home-frozen with very good results, particularly if they are just picked. Here is the procedure that works for all but a few types of vegetables:

Step 1. Wash or scrub vegetables meticulously in water.

Step 2. Prepare vegetables as you would for cooking—peel, trim, and cut into uniform pieces as necessary.

Step 3. Blanch the vegetables by boiling or steaming them for a few minutes, depending on their size and texture. This heat treatment destroys enzymes that would otherwise continue to age the vegetable even at freezing temperatures. Vegetables frozen without blanching soften and lose flavor and color after several months of storage.

You must time the blanching process precisely. Underblanching is worse than no blanching at all. Partial blanching breaks down a vegetable's tissues but does not destroy all the enzymes. These leftover enzymes age the softened vegetable tissues very rapidly in the freezer.

Blanching for too long is no more desirable. Overblanching discolors vegetables, softens their texture, and leaches out more vitamins than you want to lose.

To blanch in boiling water, use a large kettle full of water and a basket that fits inside it. (You can buy a pot made for blanching or rig your own.) Ideal proportions are 1 pint of vegetables to 1 gallon of water. The vegetables must be completely immersed in the water during blanching. As soon as the water is at a full boil, lower in the basket of vegetables and start timing. (Optimum blanching times for different vegetables are recommended in the individual sections that follow.) Jiggle the basket to make sure the vegetables get evenly heated. When the recommended time is up, pull out the basket and immediately plunge it into cold water. The best cooling method is to submerge the vegetables in a sinkful of ice water for the same number of minutes they boiled.

You can also blanch by steaming, a method that preserves more nutrients. The vegetables must be in a basket suspended above a couple of inches of boiling water. Steam no more than one pint of vegetables at a time, spread in a single layer in the basket. Start counting minutes as soon as the lid goes on the pot. When the recommended time is up, quickly cool the vegetables as described above.

A microwave oven can blanch vegetables perfectly, too, but you will have to experiment with timing and quantities or follow the individual instructions in your oven's manual for good results. It can be worth the effort, though, since the process is less cumbersome than boiling or steaming.

Note: If you are at 3000 feet or higher, add one minute to all the recommended blanching times.

Step 4. Drain the vegetables thoroughly on paper toweling. Pack them in plastic freezer bags, forcing out all the air and sealing the bag snugly (heat-sealed bags are most secure). Or you can use freezer containers with tight-fitting lids. (See pp. 349–54 for more information on freezer containers.) Certain vegetables, such as green beans and peas, may be spread in a single layer on a tray and quick-frozen. When they are frozen solid, seal them in plastic freezer bags or other containers. Tray-frozen vegetables won't stick together, so you can take out as many as you need when it is time to cook them.

Vegetables that are blanched, sealed in airtight containers, and kept at 0° F. or lower will keep for ten to twelve months.

Cooking frozen vegetables is a snap. Put unthawed vegetables into a little boiling salted water, gently separate the pieces, and simmer, covered, to desired degree of doneness. The best proportion is 1/4 cup water to 1 cup vegetables—the less water and the less cooking time, the more vitamins and minerals you save.

You can also thaw vegetables just enough to separate them, add butter, and bake them in a covered casserole at 350° F. They will cook in about 30 minutes.

You can microwave a cup of frozen vegetables quickly with little or no water in a covered dish (except for corn-on-the-cob and lima beans: these will need more water for cooking). Follow the oven's manual; this method will give you the best results.

Storing and Freezing Cooked Vegetables

Leftover cooked vegetables may be kept in a covered container in the refrigerator for three to five days.

Many leftover cooked vegetables can also be frozen. Seal the cooked vegetable airtight in a plastic bag or rigid container and freeze. The cooked vegetables can later be added to soups, stews, or the stockpot. Most thawed cooked vegetables will be too soft to be particularly good simply reheated as is.

Fully cooked vegetables also keep ten to twelve months at 0° F.

LETTUCE

Lettuce is the most guileless of vegetables. When fresh, the leaves are bright green and crisply erect. Not-so-fresh lettuce is an entirely different matter. The head withers, browns, and sags, leaving no doubt as to its condition.

Lettuce is also a vegetable that frankly resists all forms of preservation. Try to freeze, can, or pickle it and it fades into a bland pulp.

The ancestor of lettuce is thought to be a still flourishing spiny weed called the compass plant, whose leaves twist and turn to follow the sun across the sky. This association with sun worship seems to lend weight to lettuce's reputation as a most healthful food.

Buying Lettuce. There are three common types of lettuce: iceberg, romaine (or cos), and butterhead.

Iceberg is the most popular, though it probably doesn't deserve to be. It has less flavor and less vitamin A than its competitors but does store longer and holds its extra-crisp texture extremely well. A smooth, round head of iceberg should be light green, not too pale, and firm but not hard. The stem end may be slightly browned but this is just oxidation and not spoilage. Look for crisp leaves that show no signs of browning at the edges.

Romaine lettuce has an elongated shape and darker leaves. The leaves should be almost as crisp as iceberg's, but the head will feel looser, not so firm.

Boston and Bibb are varieties of butterhead lettuce that share soft, fragile leaves. The leaves are not stiffly crisp, but they should not be wilted or limp.

There is a fourth type, leaf lettuce, more common in the West than elsewhere. Leaf lettuce is more of a loose fan of leaves than a head, and it may be green or red-tinged. Treat leaf lettuce as you would romaine.

Hydroponic lettuce, grown in water and free of chemical sprays, makes increasingly frequent appearances in the market. The roots may be attached at the stem end, but the head can be handled just like field lettuce.

A good grocer keeps lettuce thoroughly chilled. Lettuce treated any other way has an unpredictable storage life.

Storing Lettuce. Lettuce loves the cold. The closer it gets to 32° F. the longer it lasts. One degree colder than 32° F. will damage the leaves, however, so the temperature in your refrigerator should range around 35° F. to be on the safe side. Once the head is chilled, all it needs is humidity to keep it crisp and green.

Put the lettuce, uncut and unwashed, in a sealed plastic bag and store it in the coldest part of the refrigerator. Keep it away from fresh fruits, especially apples. The ethylene gas given off by many ripe fruits will turn the lettuce brown.

This advice may sound very different from what you have heard before about lettuce but here are the reasons it works: uncut and uncored whole leaves retain vitamins and minerals best. Washing the lettuce adds surface water that promotes bacterial growth. Finally, plastic bags help retain the natural moisture.

Stored this way, iceberg lettuce lasts one to two weeks, romaine seven to ten days, and butterhead three to four days.

Prepare the lettuce for the table by rinsing it gently but thoroughly in cold water. You can core iceberg lettuce by using a knife (sharp knives cause less tearing and bruising) or by thumping the core against the edge of a counter and pulling it out with your fingers. Once cored, run cold water through the opened end, then drain the head core end down. Other types of lettuce can be rinsed leaf by leaf. Dunk and swish lettuce in a sinkful of cold water for a really good wash—dirt settles into the sink rather than onto the leaves—but never soak the head. Soaking extracts nutrients.

Dry the lettuce completely with paper towels (a painstaking job) or in a salad spinner.

Washed and dried, the lettuce is ready for the salad bowl. When you have more than you need, put the extra lettuce in a clean plastic bag, fill it with air, seal it, and refrigerate it. If the washed lettuce is completely dry when you store it, and the plastic bag is filled with air before you seal it, the lettuce may keep two or three weeks in the refrigerator. This method keeps lettuce crisp and tasty but is not preferred because lots of nutrients are lost in the process.

To crisp lackluster lettuce, put the washed and dried leaves in the freezer for 2 to 3 minutes. Don't leave them in a minute longer or the lettuce will be badly damaged by the cold.

Don't put salad dressing on the lettuce until the last moment before serving. Leftover lettuce coated in dressing cannot be stored for more than a few hours before it wilts.

There is no need to toss out lettuce at the end of its term in the refrigerator. You can braise chopped lettuce and serve it as a side dish or in soup.

Do not freeze lettuce.

ENDIVE, ESCAROLE, AND CHICORY

The names "endive," "escarole," and "chicory" are all applied haphazardly to the leafy tops of the plant whose Latin name is *Cicorium endivia.*

A similar but more versatile species is *Cicorium intybus. C. intybus* produces a smooth green salad leaf always called "chicory." Once these leaves are harvested, the root of *C. intybus* can be transplanted indoors where it will sprout a compact, self-blanching head called French or

Belgian endive. As if that weren't enough, the roots of this hardworking individual can be ground into a coffee substitute, also called chicory.

Radicchio, known to home gardeners as red-leafed chicory, is flown in from fields in northern Italy to restaurants and specialty stores. Its deeply colored leaf may be rose red or maroon.

All of these pungent greens add variety to the salad bowl and adapt to cooking in much the same way as lettuce does. Their flavors can be quite distinct. If the leaves are overmature at harvest, their taste may be entirely too harsh.

Buying Endive, Escarole, and Chicory. Curly endive (a variety of *C. endivia*) has a full, lacy head of leaves, deep green at the top going to yellow toward the center. The straight-leafed type is also bushy and richly green. Heads of endive and radicchio should be crisp but delicate-looking. Thick, tough leaves indicate bitter flavor.

Brown or yellow patches on the leaves are a sign of decay. Avoid prepackaged heads.

The tightly wrapped leaves of Belgian or French endive form a solid bullet about 6 inches long. It is best young and newly picked. The pale leaves may be lightly tipped with yellow, but the whiter and firmer the head, the fresher it is. And a small endive is a young one.

Belgian endive has a slightly acrid flavor that will turn nasty unless the vegetable is coddled. To keep it mild, you must store it in the dark and never soak it in water. Therefore, buy heads that are clean enough to require little rinsing and that have been wrapped in tissue or otherwise protected from light in the store.

Storing Endive, Escarole, and Chicory. The leafy versions of these vegetables should be treated like lettuce. Refrigerate them, unwashed, in plastic bags. They should keep three to five days.

Wrap Belgian endive in a damp cloth or paper towel inside a plastic bag and refrigerate it. Use Belgian endive the same day you buy it or the next day at the latest.

Do not freeze endive, escarole, or chicory.

SPINACH

Like many of our most familiar vegetables, spinach was originally cultivated in the Middle East. Migrating eastward, spinach was eventually so highly regarded by the King of Nepal that he considered it a fit tribute to the Emperor of China. Spinach captivated this warrior-king, T'ai Tsung, and it was called the "Persian herb," the name still used in China.

During its more recent history, spinach acquired an altogether different label. For many decades it was decidedly not popular in the United

States. Look at a nineteenth-century cookbook: instructions were to boil it for 25 minutes, and certainly no flavor could survive this torture. Somewhere along the way it was discovered that spinach's tangy, lush green leaves make a delectable salad, a dish so successful here that Europeans sometimes call it "American salad."

Bright, full bunches of freshly picked spinach are plentiful every month of the year.

Buying Spinach. The two varieties of spinach grocers regularly offer are equally good and interchangeable in cooking. One has a smooth, spade-shaped leaf; the other, the Savoy type, is deeply crinkled.

Fresh spinach often comes prepackaged. While the wrapping may mean the bunch has been pampered from the moment it was picked, you can't judge the freshness of the vegetable through a layer of plastic. Choose loose spinach when it is available.

Fresh spinach should have broad, crisp leaves with deep green color. There may be a bit of grit on the leaves, but bunches with too much dirt should be left at the store. Don't settle for spinach with any signs of decay: yellowing, wilting, or slime. Spinach harvested late has leathery leaves and elongated stalks with perhaps a seed-bearing stem in the center and even if fresh will be tough.

One pound of spinach, when cooked, serves 2. As a raw salad, one pound serves 4 to 6.

Storing Spinach. Spinach keeps two to three days in a plastic bag in the refrigerator. Wash and cut it right before you serve it, no sooner. If you have bought the spinach prepackaged, open it up and pick out any wilted or yellowing leaves, then rewrap it in a plastic bag.

Leftover cooked spinach keeps three to five days, well covered, in the refrigerator.

Freezing Spinach. Start by stripping off the stems: with one hand, fold the sides of each leaf downward, leaving the stem side facing up; with the other hand, grasp the end of the stem and pull upward away from the leaf. The stem should separate easily, leaving the leaf whole. Discard wilted and yellowing leaves. Dunk the spinach in a sinkful of cold, salted water to rinse off all the grit. Take care with this step—a Savoy spinach leaf is a labyrinth of dirt traps—and repeat the washing in fresh water until no more sand appears at the bottom of the sink.

Blanch the spinach 2 minutes in boiling water or 1½ minutes in steam. Cool in ice water and drain on paper towels. *Very gently* pat off extra water with another towel.

Place the spinach in a plastic bag, expel all the air in the bag, seal, and label it. Freeze immediately at 0° F. or lower and it will keep at that temperature ten to twelve months.

Cook the unthawed spinach for 3 to 5 minutes in just a thimbleful of water or in a little butter.

OTHER GREENS

Kale, collard, beet, mustard, turnip, and dandelion greens, sorrel, and fiddlehead ferns are nutritional powerhouses. These humble native greens are now regularly shipped to every part of the country and displayed with pride at the most serious food stores. This has not always been the case. To the flamboyant Louisiana politician Huey Long belongs some credit for elevating these vegetables above the category of coarse, poorhouse fare. During his campaigns he praised the vitamin-drenched stock called "pot likker," brewed from greens, that was a staple in the Southern household. Commerce in these worthy crops has prospered ever since.

Buying Greens. All of these greens are best when very young and tender. Older, larger leaves are likely to be too bitter. These fragile vegetables are not as quick-selling as the more familiar leafy greens, so examine them carefully.

• KALE'S Latin name translates as "headless cabbage," because it develops into a loose cluster rather than a compact sphere like its cabbage cousins. Kale has thickly ruffled leaves that are usually green but may also be found in brilliant tints of purple and blue or cream. Avoid kale with woody stalks or yellowing edges. Brown edges can be cut off but yellowing means the leaves are over the hill.

• COLLARD GREENS have broad, flat, dark green leaves. Check them for insects as well as for signs of bruising or decay.

Kale and collards are most abundant in midwinter and are hard to find during the summer.

• BEET GREENS turn red instead of yellow as they wither, so buy them bright green. Summer is the peak season for beet greens.

• MUSTARD AND TURNIP GREENS are not necessarily crisp, but they shouldn't be limp or bruised looking. Color may vary but brown or yellow patches are a bad sign. They are available year round.

• DANDELION GREENS should be bought (and stored) still attached to their roots. Look for the best dandelion greens in spring: buy them in the store or dig them up in the yard.

• SORREL'S sour, arrow-shaped leaves should be pale green and unblemished. Large, dark leaves will be stringy and unpleasantly tart.

• NOPALITOS or CACTUS PADS are the crisp, green leaves of prickly pear cactus. Diced and cooked, they taste like firm, tart green beans.

• FIDDLEHEAD FERNS are the young shoots of the ostrich fern. Because the fronds are still tightly coiled at harvest, they resemble a violin's scroll.

Fiddleheads are expensive, so choose moist, bright specimens with no browning or bruises and use them the same day.

• RAPE, also called raby or *broccoli di rape,* has slightly bitter, deep green leaves. Clusters of rape may sprout tiny broccoli-like flowers that indicate maturity, but in themselves are perfectly tasty. Don't worry so much about the flowers, but do pay attention to woody stems or yellowed, wilting leaves.

Storing Greens. These greens are very short-lived. They wilt and fade so quickly that it is worth your while to literally rush them from market to table. Nopalitos are an exception: they are going to last one to two weeks in a plastic bag in the refrigerator. Kale is pretty tough for a leaf vegetable. Treat it like spinach.

Put unwashed greens into plastic bags in the coldest part of the refrigerator. Serve them the same day they're bought or the next day at the latest.

Freezing Greens. All of these greens, except fiddlehead ferns and nopalitos, will freeze successfully. Trim off thick stems and ribs; cut large leaves into uniform pieces; wash thoroughly; and follow the freezing directions for spinach above.

WATERCRESS

Watercress belongs to a family of edible plants that counts among its members pepper root, cuckoo flower, lamb's cress, and lady's smock. They grow wild in temperate climates where they are gathered for food in the spring. Most share watercress's intense, tangy flavor.

Fortunately, watercress is also commercially grown in sufficient quantities to keep it in stores every month of the year.

Buying and Storing Watercress. Watercress should be bright green and crisp with no yellowing or wilted leaves. It keeps two to three days in a plastic bag in the refrigerator.

Do not freeze watercress.

ARUGULA

Arugula, also called rocket, is an everyday salad green around the Mediterranean but a rarity elsewhere. You are most likely to find this peppery-tasting vegetable in summer.

Buying and Storing Arugula. The smaller leaves have the mildest flavor. The bunch should be nicely green and not too sandy.

Store unwashed arugula up to three days sealed in a plastic bag. Do not freeze.

POTATOES

The potato is a tuber—that is, not a root but a fattened underground stem that stores surplus carbohydrates for the leafy green plant sprouting above the soil. Left alone, the green plant eventually bears fruit resembling small green tomatoes which are emphatically not good eating.

Until well into the eighteenth century, Europeans in general paid no attention to the potato as a food. Potato plants, with their abundant ivory, pink, and violet star-shaped blossoms, grew in flower gardens. Their underground tubers were accused of causing leprosy and other frightful diseases.

In Ireland it was a different story. Originally brought from South America, potatoes grew so well in the poor soil that they became the major food crop. Returning across the Atlantic to the New World, the vegetable became known as the Irish potato to set it apart from what was then the exceedingly popular sweet potato.

During the last two hundred years the potato has acquired unparalleled importance in the vegetable world. It is produced in vastly greater quantities than any other vegetable.

The most common varieties of potatoes fall into one of four categories. Round Red potatoes are waxy, smooth skinned, and good for boiling. The Round Whites, also boiling potatoes, have tannish skin. Long Whites are all-purpose potatoes, oblong in shape with tan skin and few eyes. The Russet or Idaho potato is a long, slightly flattened cylinder with brown leathery skin that has the look of netting. The Russet is the perfect potato for baking. These cooking recommendations are not fool-proof, so if your potato recipe fails the fault may well lie with the personality of the potato. Experiment with different varieties within each broad category. For potatoes with more flair, pick up either the Finnish variety—they are sweet, creamy, and yellow—or purple potatoes, also sweet, and somewhat softer. Purple potatoes, when cooked, turn the color of red cabbage. There is one suggestion that won't fail: the baked Idaho potato is a glorious vegetable that would have won over even the fussiest seventeenth-century courtier. Furthermore, all types of potatoes grown today are superior in every way to the small, starchy, tasteless potatoes of history.

When any potato is harvested young, it is called a "new" potato. The skin is thin and fragile and apt to tear; low in starch and high in moisture, new potatoes are what cookbooks mean when "waxy" potatoes are called for. Mature potatoes have been left longer in the ground, where their skins have grown thicker and more durable. The majority of mature potatoes you eat are harvested between September and November. You see them in markets year round because potatoes store very well under the right conditions. (New potatoes skip storage and go directly to grocers.) Potatoes store so successfully because that is the botanical task they are assigned by nature.

Not too many years ago, growers and packers dug holes in the ground, put in the potatoes, and covered them with dirt and straw. The potatoes would be good as new in spring. Not surprisingly, a good deal more effort now goes into storing them until spring. The temperature in potato warehouses is controlled, first to cure the vegetable, then to store it.

When growers dig up potatoes, the tubers inevitably suffer a few cuts and bruises. Within two weeks, stored at moderately low temperatures (45° F. to 60° F.), the potatoes' own vital processes heal these wounds. Once cured, the potatoes can stay dormant and relatively free of decay for five to eight months at temperatures of 40° F. to 45° F. Warmer temperatures, aside from encouraging the ever present rot organisms, eventually awaken the sprouts.

To extend storage time further, warehousers spray potatoes with chemicals that inhibit sprouting. When the eyes send out these tiny shoots, a potato begins, very gradually, to shrivel and turn rubbery. A sprouted potato is not a bad potato but it is at the beginning of the end.

Another precaution must be taken to protect potatoes in the warehouse. Light slowly turns the surface of the vegetable green, which causes it to have a bitter flavor. Worse than that, the chemical reaction between light and potato may create solanin, a chemical toxic to man in very large doses. Potatoes must be kept in the dark to avoid greening.

Buying Fresh Potatoes. Don't just grab the top bag of potatoes and run. Look through the plastic or the netting and check each potato in the sack. Like apples, one bad one in the group can contaminate the lot of them. If you want your potatoes to last as long as possible, buy them loose and choose each one carefully.

Select potatoes that are smooth and well shaped with no bumps and gullies to hide dirt and decay. Gently squeeze the potatoes—they should be firm and have tight-fitting skin. They should also be clean, since clean vegetables are less likely to be under attack by microorganisms.

The skin of a mature potato should be free of breaks and bruises. More delicate, new potatoes will show some superficial peeling; this is acceptable and unavoidable.

Look for signs of rot such as soft spots, stickiness, and deep cracks. Blackened eyes suggest that the potato has been accidentally frozen. A green tinge to the skin means a bitter taste and the presence of solanin. Sprouting potatoes are edible, but avoid buying them if you can.

Recipes specifying medium potatoes call for potatoes that weigh about 1/2 pound each. Large potatoes can weigh between 3/4 and 1 full pound. You can plan on between 1/2 and 1 pound of potatoes per serving.

Never handle potatoes roughly—tossing a 10-pound sack around as if it were a bag of rocks is a costly mistake. A bruise on a potato is an open door for mold and other microbes.

Storing Fresh Potatoes. The best place for potatoes is a spot not easy to find in most American homes. Potatoes last longest at 45° F. to 50° F. In a cool environment, such as a cellar or unheated porch, mature potatoes can last for several weeks, even as long as two months; new potatoes for one to two weeks. At normal room temperature, potatoes begin to sprout and wither quickly. They will stay fresh a week to ten days at the most. New potatoes can be counted on for only four to seven days.

The refrigerator is not the best answer. Chilling causes the starch in a potato to convert to sugar, changing its flavor. You can stretch a potato's storage life in the refrigerator, then leave it out at room temperature for a day or two before cooking—the sugar will reconvert to starch.

If you have bought prebagged potatoes, sort through them before you put them in storage. Remove any individuals with cracks, bruises, or soft spots.

Keep potatoes in a basket, burlap bag, or even a paper bag—anything that allows air circulation. It is important to store the potatoes in the dark since every little bit of light contributes to greening. A buildup of moisture encourages decay, so don't wash potatoes until you're ready to prepare them.

No matter what you do, some sprouts are liable to spring up on the potatoes. Just dig them out and cook the potatoes anyway. Use heavily sprouted potatoes in soups or stews rather than in quick-cooking dishes.

Dry onions and potatoes store best under the same conditions, but don't store them side by side. Each exudes a different gas that shortens the storage life of the other.

Leftover cooked potatoes can be refrigerated for three to five days. Whole, unpeeled potatoes retain many more nutrients than pared ones. It is not a good idea to try to make salad out of cold cooked potatoes because they simply will not absorb dressing well.

Buying and Storing Commercially Prepared Potatoes. Canned potatoes are available, but they are greatly inferior to the fresh. Since fresh potatoes are always inexpensive and available, there are few reasons to buy the canned. Canned potatoes keep up to one year on a cool, dry shelf.

Commercially frozen potatoes can be very convenient, since many are partially prepared as french fries and other specialties that need only brief heating. However, these potato products, when thawed, all have a less pleasing, softer texture than their freshly cooked counterparts, and much less flavor. Frozen potatoes can be stored up to one year at 0° F.

Dry, boxed, instant mashed potato mixes, which contain many additives and preservatives, keep up to one year on a cool, dry shelf. Seal the box inside a plastic bag after opening for extra protection against bugs and humidity.

Freezing Fresh Potatoes. Potatoes are not good candidates for freezing. Raw or cooked potatoes that have been frozen at home are watery and simply disintegrate when thawed. Commercially frozen potatoes do better because they have been processed more quickly than is possible at home.

There is one way to save aging potatoes from the garbage can: cook them, mash them, and then freeze them. Pack the mashed potatoes in a rigid container, leaving 1/2 inch headspace. Sealed airtight, mashed potatoes keep ten to twelve months at 0° F. To reheat, put a tablespoon or two of milk or water in a saucepan, add the frozen potatoes, and warm them at a medium low setting, stirring every couple of minutes so they don't stick.

SWEET POTATOES

Two types of fresh sweet potatoes show up in supermarkets. One has dry, mealy flesh; the other is moister and has more sugar. Both are sweet potatoes, even though the moist variety is usually labeled "yam." (A true yam is a foreign-born tuber that looks like a sweet potato but can grow to 100 pounds.)

The dry sweet potato has lighter skin—yellowish tan—and pale orange flesh. The flesh of the moist variety is deep yellow or orange red and the skin color ranges from light tan to brownish red. In fact, it is fairly difficult to distinguish between them in the grocery store unless you ask. In winter, look for the long, thin, deep red sweet potatoes from California, so rich they need no butter or seasoning.

As for home storage, both varieties are the same—short-lived and injury-prone. It looks tough but the sweet potato, once bruised or invaded by decay organisms, deteriorates very quickly. Furthermore, a bad spot on the potato affects the flavor of the whole vegetable, so it does no good to just cut away the damaged area.

Having selected close to perfect specimens, bag and carry them as carefully as you would a carton of eggs.

Buying Fresh Sweet Potatoes. It is best to buy sweet potatoes that are loose rather than prepackaged so you can examine each one for defects. Look for a smooth, unbroken skin and reject any potato with shriveled, soft, or sunken patches. The skin should be dry and the flesh firm.

A pound of sweet potatoes yields 3 to 4 servings.

Storing Fresh Sweet Potatoes. Put the dry potatoes in a perforated plastic bag. This keeps the air around them humid but doesn't allow too much moisture to build up.

If you have an unheated area in your home that stays around 45° F. to 50° F., it is ideal for storing sweet potatoes. Healthy specimens can last for as long as two months at these temperatures. At room temperature, plan to keep sweet potatoes for no more than a week.

You can refrigerate sweet potatoes if your kitchen is very warm, but you must make sure they do not accidentally freeze. Freezing, even in one spot, darkens the flesh and imparts a bitter flavor to the whole potato. Sweet potatoes stored in a plastic bag keep seven to ten days in the refrigerator. If you bought more sweet potatoes than you needed for cooking, slice them and serve them raw as a snack or with a dip.

Cooked sweet potatoes can be refrigerated for up to five days. Use leftover sweet potatoes to make potato pancakes or combine them in a puree with leftover white potatoes. Cooked sweet potatoes are particularly delicious in baked goods.

Freezing Sweet Potatoes. Sweet potatoes freeze well when they are fully cooked. Sliced, boiled, or mashed, they all benefit from the addition of a little lemon juice, which will prevent discoloration. Pack the cooked sweet potatoes in a rigid container with an airtight lid. Leave ½ inch headroom for the food to expand. Label the container. Sweet potatoes keep ten to twelve months at 0° F.

CARROTS

The similarity between the feathery green tops of carrots and Queen Anne's lace is no coincidence. The two plants are botanical twins, close enough to cross breed. Some food historians theorize that Queen Anne's lace is the descendant of cultivated carrots that escaped from colonial American gardens. Luckily, the fat orange, sugary carrots farmed today taste nothing at all like the thin, bitter roots of the wildflower.

Growers harvest the youngest carrots to sell fresh. Some carrots are "topped" (cut away from their bushy, green stems) and bagged. Some keep their tops all the way to market. Topped carrots may be held in storage for a while before they are shipped to stores.

Among the carrot's winning qualities is the fact that its generous vitamin A content actually increases during the first twenty weeks of commercial storage.

Buying Carrots. Carrots are never more appealing than when you see them stacked up in glossy bunches with their brilliant green tops intact. As far as a carrot plant is concerned, however, the greenery has priority over the orange root below. The top, which is sustained by the root while it is still in the ground, continues to draw moisture even after the plant is harvested. As the top draws up moisture, the carrot begins to shrivel.

The choice between topped, prepackaged carrots and ones with tops intact you will have to make for yourself. Bagged carrots may be higher in nutrition; bunched carrots are much easier to judge for freshness. If you cut off the greenery immediately, carrots bought with tops on will keep as long as the prepackaged kind.

Look for firm, well-shaped carrots with no cracks. A soft or shriveled carrot has been stored improperly. Any mushy spots indicate decay. Thin hairy rootlets sprouting from the skin are a sign of age. A distinct green tinge at the stem end means the carrot has been left in the sun too long. The green part is bitter and should be discarded.

Very pale carrots are not as nutritious as those that are vivid orange, so beware of plastic bags tinted to exaggerate the color of the vegetable.

Some grocers carry tiny, very sweet "finger" carrots. These are almost always bagged. They are just as sturdy as the larger variety and should live up to the same standards of appearance.

One pound of carrots yields 3 to 4 servings.

Storing Carrots. Carrots like cold temperatures and high humidity. Leave prepackaged carrots in their bags and refrigerate in the vegetable crisper. Cut off the tops of bunched carrots and store them in a plastic bag. They will keep very nicely up to two weeks.

Apples manufacture a gas that causes a bitter flavor in carrots, so keep them apart during storage.

Wilted carrots are not a total loss. They still have plenty of flavor and lots of vitamin A. Use them in soups, stews, purees, and stock, where texture is not important. You can recrisp slightly wilted carrots by soaking them in ice water for 30 minutes.

Carrot juice is very susceptible to bacterial growth. Keep fresh-made and opened cans chilled and store no more than one or two days.

Freezing Carrots. Wash the carrots, and peel them if desired. Leave tiny ones whole, slice larger ones 1/4 inch thick. Blanch in boiling water: whole carrots for 5 minutes, sliced carrots for 2 minutes. For steaming, add 1 minute to the boiling times. Cool and drain. Pack in plastic bags, expel air from the bag, seal, and label. At 0° F. carrots will keep ten to twelve months.

BEETS

A row of beets is customary in the home garden. Since the attached household is likely to have beets in abundance all at once, it is fortunate that this ruby-red vegetable adapts to all kinds of treatments in the kitchen. A beet is perfect for pickling; it can be baked whole in its skin; as soup, it becomes any of a dozen varieties of borscht; it is crisp and colorful served raw.

A fresh beet leaks nearly indelible purple juice all over cooks and kitchens after the slightest break in its thin skin or long, tapering tail. Once in the pan, however, the inky liquid is an asset because, unlike most vegetable pigments, it is unaffected by prolonged cooking. A beet stays deep violet-red whether it is stewed, canned, or otherwise processed.

Fresh beets are most plentiful in late summer but good varieties are shipped from warm climates every month of the year.

Buying Fresh Beets. Beets come bunched, with a full head of greens, or packaged with only a short length of stem. Buying beets with their tops on has two advantages: the greens indicate the freshness of the vegetables, and they are a tasty vegetable in their own right (see p. 14). Smaller beets are best. They are less likely to have unusable, woody cores.

Beets should be clean and free of cuts and soft spots. Wilted greens with red or brown edges mean the beet has already been stored too long.

Beets with trimmed tops should still have at least ½ inch of stem showing; otherwise the vegetables will leak their juices. The root tip should be at least 2 inches long.

Buy about ¼ pound of beets per serving.

Storing Beets. Cut off full tops immediately—they draw moisture up from the root. Leave at least 2 inches of stem.

Store unwashed beets in a plastic bag in the refrigerator; they keep for seven to ten days. (Dry air and warmer temperatures cause beets to shrivel.)

Cooked beets can be stored in the refrigerator three to five days.

Canned Beets. Canned beets are either plain or pickled. The flavor and texture of beets survive processing remarkably well, but they do not store as well as other vegetables. Keep unopened cans and jars on a cool, dry, dark shelf no more than six months, and refrigerate after opening. Opened, plain beets keep up to one week in the refrigerator; pickled beets up to two weeks.

Freezing Beets. Leftover beets can be sealed in plastic bags and frozen for long-term storage. If they are coated in sauce, they keep six to eight months.

To prepare fresh beets for freezing you must cook them until they are fork tender—blanching is not sufficient. Rinse the beets very carefully and boil them whole, with root tips and 2 inches of stem left on. Cool them and slide the skins off. Pack them sliced or whole in plastic bags or rigid containers, seal airtight, and label. They will keep ten to twelve months at 0° F.

TURNIPS AND RUTABAGAS

Turnips and rutabagas are smooth-skinned root vegetables, related in taste and texture. The round, baseball-sized turnip has sweet but peppery white flesh. The larger rutabaga's yellow flesh has a stronger flavor, but the two vegetables are interchangeable in most recipes.

Turnips and rutabagas belong to the cabbage family and share their relatives' sulphurous character. Stored in quantity, in a root cellar for instance, they give off an odor that makes them most undesirable bedfellows. Fortunately, keeping a few in the kitchen won't cause this kind of trouble.

Buying Turnips and Rutabagas. Turnips may come with their green tops, which can be used separately (see p. 14). Clean, unwilted greens that have no browning mean the turnip is fresh. Turnips alone should be firm and without a lot of roots.

Rutabagas should feel firm and heavy for their size. Avoid any with punctures or other damage. Glossy rutabagas have been waxed with clear paraffin to hold in their moisture. Peeling will remove the wax.

Variations in the shape and color of these vegetables reflect their different varieties but not their quality.

Buy 1/4 pound of turnips or rutabagas per serving.

Storing Turnips and Rutabagas. Like most vegetables, turnips and rutabagas thrive on high humidity. Store them in plastic bags in the refrigerator. Turnips keep up to seven days; rutabagas can be stored up to two weeks. You can also leave rutabagas at room temperature for up to one week. Don't wash the vegetables until you are ready to use them.

Freezing Turnips and Rutabagas. Either blanched or fully cooked, these vegetables are suitable for freezing.

To blanch, first wash, peel, and cut into 1/2-inch cubes. Boil for 2 minutes or steam-blanch for 3 minutes, cool rapidly, and drain. Seal in plastic bags, label, and freeze. If you use rigid containers, leave 1/2 inch headroom.

Fully cooked turnips and rutabagas freeze well after pureeing. Another possibility is to grate the vegetable and sauté it in butter, then pack it, juice and all, for the freezer. Whether blanched or fully cooked, you can keep them eight to ten months at 0° F.

PARSNIPS

The parsnip, a root vegetable that looks like an ivory carrot, was once as ubiquitous as the potato is today. There is really no good reason for it to have fallen so far out of favor. It is inexpensive, nutritious, and has an appealing nutty flavor.

Parsnips are durable enough to be left in the ground through a whole winter of frosts—the cold temperatures improve the flavor by converting starch to sugar until the parsnips are at their sweetest in the early spring.

Parsnips are on the market every month of the year. Good ones have a very sweet, rich flavor however they are prepared. If the parsnips you buy are uneven in quality, cook them in their skins, peel them, and scoop out the tough inner cores before serving.

Buying Parsnips. A wilted, shriveled parsnip has lost its juices. Look for firm flesh with no soft spots or cuts. Very large parsnips are apt to have woody centers. If the tops are on, they should be green and fresh-looking.

A pound of parsnips serves 2.

Storing Parsnips. Remove the green tops before storing. Keep the parsnips in a plastic bag in the refrigerator seven to ten days.

The leftover cooked vegetable will keep in the refrigerator for three or four days.

Freezing Parsnips. Parsnips can be treated just like turnips for the freezer (see pp. 23–24). They will keep ten to twelve months at 0° F.

EXOTIC ROOTS AND TUBERS

The root and tuber vegetables listed below have very little in common, but there is one rule of thumb that applies to all—they should feel very firm, not at all flabby. Look for soft spots—any mushiness is evidence of rot, whether you can see it or not.

• CASSAVA. This yucca root is also called manioc. It looks like a small tree branch with white flesh and brown, barklike skin. It contains prussic acid and is poisonous unless cooked. Cassava, which is very bland and starchy, is used in Caribbean and Latin American cooking and is also the source for tapioca. Refrigerated in a plastic bag, fresh cassava will keep two days.

• CELERY ROOT OR CELERIAC. In the store it looks like a gnarled clump of brown bulbs and roots. (The roots are not used, so pick the specimen with the fewest.) Miniature green stalks, distant cousins of common celery, may be emerging from the top of the root, which does taste very much like celery, but slightly nutty, too. One pound of whole celery root yields 1/2 pound peeled. Trim off the greenery and the stringy roots before storing. It will keep for several days in a plastic bag in the refrigerator.

• GINGER. The sharp flavor of ginger is essential in oriental cooking. A single ginger root can have any number of bulging tan knobs growing at odd angles. The root should be plump and have unbroken, somewhat pliable skin. Young ginger has a pinkish, soft skin and a brighter taste than mature, but is available only in oriental and Indian markets in spring.

Ginger is very durable in storage, and there are at least three satisfactory methods for keeping it. You can simply slice off what you need and store what's left, tightly wrapped, in the refrigerator. It will last two to three weeks. Seal unpeeled ginger airtight in a plastic bag and it will keep for four weeks in the freezer. During that time you can take out the ginger, slice off the amount you need, and return the unused portion to the freezer. Ginger won't lose much zest in storage, but it may dry out and a musty scent is a sure sign that it is too old.

The most intriguing storage method is to steep the ginger root in wine. Cover it with white wine or sherry in a covered jar and refrigerate. It will last for at least six months. Periodically pour off some of the wine for salad dressings and for stir-fry dishes; replace what you have used with fresh wine to cover the ginger.

Pickled ginger can be refrigerated in its own brine one to two months.

• GOBO ROOT. A rough, brown-skinned root, also called burdock, it is used in Japanese and Chinese dishes. Refrigerate up to two weeks.

• JÍCAMA. The jícama is a large, bland, bulbous root with the texture of a hard, crisp apple. It is common in Latin American cooking. Look for firm flesh and unblemished skin. Refrigerate it up to two weeks.

• HORSERADISH. Its fiery flavor is best when the vegetable is fresh. Keep the woody-looking root refrigerated in a plastic bag for up to three weeks. Freezing is not recommended.

Prepared horseradish is mixed with vinegar and packed in jars; the red type has beets in it too. You can store prepared horseradish in the refrigerator for three to four months, but as it ages it loses pungency.

• JERUSALEM ARTICHOKES. These bumpy little tubers have tan skin and crunchy white flesh. Purveyors may label them "sunchokes." Look for smooth unblemished ones, with no cracks. They have a strong but pleasant flavor. Jerusalem artichokes will store for two weeks in a plastic bag in the refrigerator. Do not freeze them.

• SALSIFY. Another long, tapering root, salsify looks like a shaggy beige carrot. The very subtle flavor has earned salsify the alternate name of oyster plant, though it tastes nothing at all like the shellfish. If it is crisp and undamaged when you buy it, you can keep salsify one to two weeks in a sealed plastic bag in the refrigerator. Do not freeze salsify.

• TARO. This root is also known by an equally lyrical name, dasheen. In Hawaii it is eaten as poi; Latin Americans treat it like a potato. Refrigerated in a plastic bag, a firm taro root lasts two to three days.

RADISHES

The most familiar radish is the round red "button." White icicle radishes, usually available only in summer and fall, are about 5 inches long and are usually sold in supermarkets sealed in a plastic bag. During the winter ethnic markets sell black radishes the size and shape of turnips. These are the most intensely flavored radishes. Daikon, a long, thin white variety also known as Japanese radish, is sold year round in specialty markets. The daikon is often cooked like turnip. All varieties have crisp white flesh with flavor ranging from pleasantly piquant to fiery.

Buying Radishes. Radishes should feel hard; sponginess is most undesirable. If greens are attached, they should be uniformly green and unwilted. Avoid radishes with cracks, scrapes, or soft spots.

Storing Radishes. Remove tops before you store radishes. Keep them in a sealed plastic bag in the refrigerator. They should last two weeks.

A brief soak in ice water will crisp up a sagging radish.

Do not freeze radishes.

TOMATOES

The tomato is, botanically speaking, a fruit, but in a United States Supreme Court ruling of 1893 it was designated a vegetable for purposes of commerce. The tomato is no stranger to debate, legal, moral, or otherwise.

Imported to Europe from Peru in the sixteenth century in the form of a tiny, yellow, unattractive globe, the tomato met with little success and some frenzied opposition. For instance, it was accused of being poisonous, an allegation that stuck to some extent until as recently as a hundred years ago. The habit used to be, when eating tomatoes at all, to douse them in vinegar and spices and cook them for hours, to neutralize the poison.

For most of their history, tomatoes were grown only at home—most enthusiastically in Italy, where horticulturists bred plump red tomatoes very similar to those seen today. Too fragile and too short-lived to transport to market successfully, those tomatoes were never an important commercial crop. Their useful storage life spanned the distance from garden to kitchen and not much more.

As soon as the general public embraced the tomato as a favorite vegetable trouble lay ahead. In order to bring tomatoes to the market day to day, growers must pick them green and keep them cooled off. Cold temperatures discombobulate a tomato, putting an end to the process that brings about full flavor. The resulting tomatoes are too often bland and rubbery, and, like their ancestors, the object of suspicion and hard feelings. Nobody believes tomatoes taste as good as they used to.

But not every tomato in every market is as unappetizing as a sponge If you shop with cunning, you can bring home a delicious tomato.

Always keep in mind that, federal judges notwithstanding, a tomato is a fruit. A mature tomato can ripen into a sweet, juicy one; an immature tomato, picked before its time, can turn red but will be tasteless and pulpy.

Look for tomatoes labeled vine-ripened. A vine-ripened tomato, if accurately labeled, developed a hint of red before it was picked.

Most tomatoes are harvested when they are still all green. Growers pick green tomatoes because they are more crush-resistant and therefore more likely to arrive whole at the retailer's than tomatoes that have begun to turn red. A green tomato may be a mature one, but very possibly not. Green tomatoes are often treated with ethylene gas to force reddening.

You are more likely to find good vine-ripened tomatoes at those grocery stores that emphasize high-quality produce. And the very best tomatoes will usually be on the market only in the warmest months of the year.

Although many people find them pale imitators of the field-grown tomato, greenhouse tomatoes are on the market year round and they are left to mature on the vine. Greenhouse tomatoes usually have about half the vitamin C that tomatoes grown in the sun contain.

Hydroponic tomatoes come on the market in spring. Their attraction is that they have been grown without chemical sprays. They are shipped pink and therefore should ripen normally.

If you keep coming up with tasteless tomatoes, try different varieties to find what you like. There are fresh Italian plum or cherry tomatoes, for instance, and small, sugary, yellow or red teardrop tomatoes. Or wait until summer and hope for locally-grown supplies to come to the store.

Buying Tomatoes. Unless you need it right away, buy a mature, preferably vine-ripened tomato before it is fully ripe. You will get the longest storage life from a tomato that is purchased still firm and then further ripened at home.

Look for a tomato heavy for its size—it will be the juiciest. It should be smooth and round with no cracks in the surface, although a few superficial scars around the stem end are not an indication of trouble. Avoid any tomato with bruises or mold. These problems are not easy to spot unless the tomatoes are displayed loose rather than prepackaged.

Greenhouse tomatoes still have stems and a cap of green leaves attached. Fresh leaves mean the tomato has not been stored long. A fully ripe greenhouse tomato with bright fresh leaves was undoubtedly left on the vine until the last possible moment and will be sweet and juicy.

A red, ripe tomato should be soft and yielding to the touch but not squishy. And a tomato should *smell* like a tomato.

These guidelines apply to all fresh red tomatoes, including beefsteak, plum, and cherry varieties.

Storing Tomatoes. The refrigerator is no place for a tomato. Temperatures below 50° F. interfere with ripening. A refrigerated tomato will turn red but it will not become sweet and juicy at the same time. On the other hand, don't set tomatoes on the window sill to ripen. The sun will heat the tomato and cause it to ripen unevenly.

Keep all tomatoes at room temperature until they are nearly overripe. Even a ripe tomato will keep one or two days at room temperature.

There is one very small space in most households that stays at the ideal temperature for tomatoes, between 50° F. and 60° F. In that cool range, they will easily ripen and fend off rapid decay. Test the butter compartment in your refrigerator with a thermometer. If it is higher than 50° F., you can fit two large or four small tomatoes in there and they will keep up to twice as long.

When you need a succulent red tomato sooner than you thought, put one inside a perforated paper bag or in a covered bowl with an apple. The apple gives off ethylene gas that speeds up the ripening process.

Sort through your supply of tomatoes frequently. Even if they were identical when you bought them, they will ripen at individual rates. Also, if one of the batch becomes moldy, the rest are more likely to succumb.

Go ahead and refrigerate a very ripe tomato if you can't use it right away, but take it out about an hour before you serve it so that it loses its chill. A room-temperature tomato tastes far better than a cold one.

If you have a cellar, attic, or porch that stays between 55° F. and 70° F., you can store mature green (not vine-ripened) tomatoes there from four to six weeks. Pack them in shredded newspaper, no more than two layers deep. Once a week sort out any that have begun to turn red and store them separately.

Keep sliced fresh tomatoes in the refrigerator covered with plastic wrap for two to three days. You can still set them out for a while before serving for better flavor.

Canned Tomato Products. Canned tomatoes, whole, sliced, stewed, in sauce and in paste, should be regulars on the pantry shelf. These products are usually sweet and well colored. For cooking, canned tomatoes are likely to be an improvement over poor-quality fresh tomatoes. The canned imported Italian varieties, especially those from the San Marzano region, have the springiest texture and the fullest flavor.

Store the unopened cans a maximum of six months on a cool, dry shelf. Canned tomato products do not keep as well as other canned vegetables.

After opening, refrigerate canned tomato products in clean, covered glass containers. They tend to take on a metallic flavor if left in their cans. You can keep them, tightly lidded, for a week. You can freeze leftover tomato paste and tomato sauce and keep it for up to two months in airtight containers.

Freezing Tomatoes. Raw tomatoes, whole or in pieces, can go right into the freezer without any ado; just seal them in an airtight plastic bag. They will get mushy but are flavorful for soups, sauces, and stews. Before you freeze tomatoes in large quantities, you should test a small batch. Thawed tomatoes may not appeal to you as a substitute for fresh-cooked or even canned tomatoes.

Sun-dried Tomatoes. You can save a tomato harvest by cutting the tomatoes in half and drying them in the sun on screens or in baskets. The process takes several days, until they turn dark. If it rains, put them in the oven at the lowest setting for a few hours. These are particularly delicious marinated in olive oil for several months. Store both dry and marinated tomatoes in tightly sealed containers in the refrigerator for six to nine months.

PEPPERS

The green pepper is totally unrelated to black pepper, salt's perennial companion. But, just as Columbus assumed he had found the East Indies when he landed in the West Indies, he concluded that the pungent green vegetable used in local cooking was simply another version of black pepper, one of the valuable spices he was in search of, and so named it accordingly.

Sweet green peppers are also called bell peppers because of their plump, cylindrical shape. Italian peppers, a less common variety, are paler and narrower. Bell peppers are bright, glossy green at maturity and turn rich red as they ripen. You will find both red and green sweet peppers for sale, but it is unlikely that you can ripen your own green ones at home—the process requires controlled temperature and humidity. Another variation, yellow bell peppers, are only occasionally available. Their mild flavor is an asset in salads or sautés and the glowing color lights up a plate.

Hot (chili) peppers—jalapeños, poblanos, serranos, New Mexican, among others—are long and tapered. Hot peppers can be green, red, yellow, and even brown at different stages.

Chili peppers can be found fresh in most markets, but their nomenclature varies so from region to region, you may need to ask an expert before you buy chilis for cooking. Any type of chili can range from mild to very hot depending on its particular subvariety and cultivation. The most common fresh chilis are: the jalapeño, a bright green pepper about 3 inches long that ranges from fairly hot to fiery; the New Mexico, 6 inches long, medium green, mild to slightly hot; the mild Anaheim, a bright green pepper 6 to 8 inches long; the serrano, very hot, which grows to only 2 inches long and is very skinny; the black poblano chili, a plump 5 inches long, which has a mild, full flavor; and the guero, a very hot yellow pepper the size of a jalapeño. All these chilis turn redder, darker, and sweeter as they ripen.

Buying Peppers. Look for a firm, crisp pepper with smooth skin. Because it withers as it spoils, soft, limp flesh and wrinkled skin mean the pepper is on its last legs. The pepper should have thick, resilient skin that is glossy and unblemished.

Some sweet peppers may come to market waxed. These are the least desirable, because the waxing can accelerate bacterial growth in the flesh.

Storing Peppers. Peppers are another vegetable, like tomatoes, that have no natural resting place in the kitchen. They do best at around 50° F., with high humidity. You can't keep peppers at room temperature, though—they will rot much too quickly.

Put them in a paper bag in the vegetable crisper and plan to keep them no longer than four to five days. You may eke out another day or two if you keep a pepper in the butter compartment.

Wrap a sliced pepper in plastic and refrigerate it no more than two days.

Canned Peppers. Store canned, pickled peppers and pimientos no more than six months on a cool, dry shelf. After opening, pickled peppers can be stored, covered, in the refrigerator one to two months. Opened pimientos keep in the refrigerator no more than ten days.

Dried Peppers. Ethnic markets carry many different types of whole, dried hot peppers, each with a distinct character. Examine each closely for insect infestation and mold. Store the peppers loose and airy: hang them up or toss them in a basket. If it's cool and dry, they keep a year or more. For information on other dried pepper products, see pp. 290–92.

Freezing Peppers. Paradoxically, the same pepper that is ill at ease in the refrigerator takes to the freezer like a duck to water. You can dice a raw pepper, put it in an airtight plastic bag, and stick it right in the freezer with no blanching required. Small, hot peppers can be left whole, then seeded after a brief thawing. Peppers will lose crispness after freezing but can still be used for cooking.

Peppers keep at 0° F. for six to eight months, though hot peppers lose a lot of zing.

TAMARILLO

A tamarillo, or tree tomato, is a gorgeous plum-red oval with smooth, waxy skin. Its flavor can go either way: sweeten it to make it a fruit, or drop it into a stew as a vegetable.

Skin tamarillos, seed them, and eat them raw or cooked. A ripe tamarillo is one with a deep purplish cast. Keep them refrigerated one week to ten days.

TOMATILLOS

Also known as tomates verdes, tomatillos are small green fruits—from 1/2 to 2 inches in diameter—widely available in Mexican markets. They grow their own loose, papery wrapping. Tomatillos yellow as they ripen, but most recipes call for firm, acidic green ones. They keep up to two weeks on the shelf. To freeze, peel off the husk, cover in water, and simmer very slowly until tender. Pour both fruit and cooking water into a tightly lidded container and freeze them ten to twelve months at 0° F.

CUCUMBERS

Most grocers carry at least two types of fresh cucumbers. One has smooth, shiny green skin while the other, usually labeled "pickling," is smaller and has whitish-green skin with bumps. Less common are the long English seedless cucumbers sealed in plastic.

Cucumbers are on the market and equally good in every season. Most cucumbers you buy will be sweet, mild, and easy on your digestion: the burp has been bred out. Occasionally a bitter one turns up—the unpleasant flavor is an accident of growing conditions—but it will look just like the others, so getting one is a matter of bad luck.

Buying Cucumbers. The smooth-skinned variety should have a dark green color. The cucumber is a vegetable fruit that, like a tomato, ripens, turning yellow, even after harvest. Contrary to what you might expect, you don't want a ripe cucumber because its seeds are hard. The dark green variety usually has a coating of edible wax on the skin and you must peel the cucumber to remove it completely, but peeling removes the better part of the vegetable's vitamin A. The wax also interferes with the pickling process. The bumpy cucumbers, meant for pickling, may naturally have some yellow streaks in the skin, so are harder to judge for ripeness.

Buy uniformly solid cucumbers—they turn limp and shrivel as they spoil.

Storing Cucumbers. Put them, unwashed, in a perforated plastic bag in the refrigerator, or unwrapped in the vegetable crisper. A cucumber should keep four to five days. If you happen to have a warm zone in the refrigerator, perhaps in the door, where the temperature is above 40° F., cucumbers will store best there.

Sliced, raw cucumber will hold in the refrigerator one to two days, but be sure to wrap it very securely to keep the odor from penetrating other foods. Sliced cucumbers soaked in dressing can also be stored two days.

Do not freeze cucumbers.

For advice on storing pickles, see p. 62.

EGGPLANT

At one time women in the Orient used a black dye made from eggplant to stain their teeth a gunmetal gray. The dye probably came from the same type of dark purple globes we recognize as eggplant in the market now.

The eggplant does not seem out of place as a tool of fashion. It is certainly ornamental: glossy, rounded, deeply colored. But it is a very workmanlike vegetable, rich and solid enough to serve as a main dish.

In addition to the familiar purple eggplant, which is available every month of the year, there are other varieties: black, yellow, white, or red ones for instance, also small, spherical, or long, thin ones. These are very rarely available commercially.

Buying Eggplant. The smaller the eggplant the more likely it is to be sweet and tender. The skin should be a deep purple, smooth and totally free of cuts and scars. Tan patches are a sign of rot. The eggplant should feel heavy for its size and fairly firm.

If you plan to use it as a side dish, buy about 1/3 pound per person. Plan for more—1/2 to 3/4 pound per serving—when eggplant is the main dish.

Cushion the vegetable against bumps and bruises as much as you can while bringing it home from the market.

Storing Eggplant. It is difficult to keep an eggplant firm and fresh beyond three or four days. Keep it in a plastic bag in the refrigerator if you want to hold on to it that long. If you plan to cook it the same day you buy it, leave it out at room temperature.

Large eggplants tend to be bitter. You can offset this by salting slices and letting them drain 20 to 30 minutes before cooking.

Cooked eggplant can be refrigerated for two or three days but gets mushy when it is reheated.

Freezing Eggplant. Wash and peel one eggplant at a time—the flesh begins to discolor after a few minutes' exposure to the air. Blanch 1/4-inch-thick slices in salted water for 4 1/2 minutes or steam for 5 minutes. After a quick chilling, dip each slice in a solution of 1 1/2 tablespoons lemon juice to 1 cup water—the dip helps keep the eggplant from darkening. Drain well, seal in plastic bags, label, and freeze.

You can freeze fully cooked eggplant in the form of a puree—a little lemon juice is called for here too.

Eggplant keeps for six to eight months at 0° F.

OKRA

Okra is a green, rough-skinned pod tapering to a narrow tip. When cut it exudes a viscous juice that traditionally thickens the Southern stew called gumbo. Its pronounced flavor mixes well with other vegetables, particularly tomatoes, peppers, and onions.

Fresh okra is always plentiful in the South. In other parts of the country supply is greatest in July and August.

Okra is common in the cooking of India, Greece, Africa, the Middle East, and the Caribbean. Ethnic markets often carry dried okra pods and leaves. The vegetable is all but unheard of in most European cuisines.

Buying Okra. Fresh pods should be crisp and small, no more than 3 to 4 inches long. Test the tips with your finger—they should be flexible. Older pods become tough and woody.

A good green color is important. Bruising shows up as black discoloration, and warm temperatures during storage cause bleaching. Avoid okra that looks wilted or has pitted skin.

There is practically no waste in preparing okra, so you can allow as little as 1/4 pound per serving when you buy.

Storing Okra. Because fresh okra is very perishable, plan to keep it no more than two or three days in the refrigerator. Store it in a plastic bag but keep the pods very dry. Any water sprinkled on the okra will soon turn the surface slimy.

Many people are put off by okra's slick, gluey juices. You can minimize this effect by sauteing whole okra briefly in butter before adding it to soups or stews.

Leftover cooked okra stands up to reheating fairly well. You can store the cooked vegetable in a well-covered container in the refrigerator for three to four days.

Store dried okra in a tightly closed container. On a cool, dry shelf it should keep two to three months; in the refrigerator, three to six months.

Freezing Okra. Wash the fresh okra and trim off stems but leave the caps intact. Sort them for size. Blanch small pods 3 minutes in boiling water, large pods 4 minutes. Add 1 minute for steam blanching. Chill and drain. You can pack okra whole or slice it after the draining. Store in airtight plastic bags, sealed and labeled. (You may want to tray-freeze pods or slices first.) Okra will keep ten to twelve months at 0° F.

BITTER MELON

Bitter melons are vegetables used in Asian cooking. They may be four to ten inches long with the general appearance of a warty cucumber. The bitterness comes from quinine, and diminishes somewhat as the melon ripens. You want one that is firm and turning yellow, but not yet altogether yellow or orange. Blanch the flesh first if you use it in stir-frying and always discard the seeds. Refrigerate bitter melons up to five days in a plastic bag.

CABBAGE

The cabbage has at one time or another had a place on every rung of the social ladder. The Greeks ate a lot of it but paid it no particular mind. The Romans held cabbage dear, and their emperors paid whatever the cost to get it. During the Middle Ages, Europeans, who rarely ate any vegetables, made no exception for cabbage. Eventually it was grown in such abundance that the poorest citizen could afford it, and it was consequently regarded as peasant food.

The pale, bald heads of cabbage common in stores year round have come out of cold storage and perhaps deserve their plebeian reputation on looks alone. The same cabbage fresh from the field in summer, still wearing its lush outer leaves, is like a vast green rose. Another elegant cabbage is the crinkly-leaved Savoy variety. The red cabbage is interchangeable with the green in recipes and has the allure of its intense color.

Cabbage and all of its botanic family, including cauliflower, brussels sprouts, kale, mustard, broccoli, turnips, and rutabagas, contain sulfur compounds. When the plant tissue is cut, one of these compounds, sinigrin, meets with an enzyme that changes the sinigrin into mustard oil. It is mustard oil that gives raw cabbage its biting flavor.

When cabbage cooks, the number of its sulfur compounds and other volatile substances increases. Cooking cabbage and its relatives too long eventually produces unappetizing flavors and odors.

Buying Cabbage. New cabbage, either green, red, or Savoy, will have loose outer leaves, called wrapper leaves, around a firm, tightly packed head. The wrapper leaves should be well colored, with no wilting or signs of decay. Some spring green cabbage will feel soft by comparison, which is fine as long as the leaves seem fresh and crisp.

The cabbage held in storage has been stripped of wrapper leaves, leaving the firm center head that has almost no green color. This head should feel hard and heavy for its size and should look moist and fresh. Cabbages stored too long begin to separate from the stem, so check the base of the vegetable.

A medium-size head of cabbage weighs about 2 pounds. For cooked cabbage, you will need about 1/4 pound per serving. A 2-pound head equals about 10 cups shredded.

Storing Cabbage. Store whole, unwashed heads in a plastic bag in the refrigerator and the cabbage should keep one to two weeks. Don't slice or shred cabbage and expect to store it for any length of time—it will wilt noticeably within several hours and lose a lot of nutrients at the same time. When you use less than a whole cabbage, leave the remainder of the head in one piece, wrap it tightly in plastic, and store it another day or two.

Freezing Cabbage. Cut cabbage into coarse shreds. Blanch in steam for 1 1/2 minutes. Drain, chill, pack into rigid container, leaving 1/2 inch head space. Label, seal, and freeze. Thawed cabbage is suitable for cooking only. It will keep ten to twelve months at 0° F.

Sauerkraut. Sauerkraut is cabbage that has been salted and fermented in its own juices. Canned sauerkraut tastes best when used within six months (rather than the usual year for canned vegetables). Fresh sauerkraut, packed in jars or plastic bags and sold out of refrigerator cases at delis and supermarkets, needs cold temperatures. You can store it one week in the refrigerator.

CHINESE CABBAGE

Also called celery cabbage or Nappa cabbage, Chinese cabbage has an elongated, crinkly leaf with a wide, pearly white stalk running up the center. It has a mellower flavor than other cabbages, but with a faint peppery undertone to make it interesting. The texture is light and crisp.

Buying Chinese Cabbage. The leaves should have a bleached look with more yellow and white than green. The leaves should be crisp and un-damaged.

Storing Chinese Cabbage. Seal Chinese cabbage in a plastic bag and store it in the refrigerator four to five days.

To freeze, follow instructions for cabbage, see above.

CAULIFLOWER

Cauliflower is truly a flower, one that grows from a type of cabbage plant. Instead of opening outward, the flower forms a compact, edible mass of underdeveloped florets, called the curd.

Heavy, cream-colored heads of cauliflower are most abundant in the autumn months and scarcest in summer. Cauliflower is usually sold with all but a narrow collar of its green outer leaves trimmed away.

Buying Cauliflower. The cauliflower's white florets should be dense and firmly packed together. A head that has begun to loosen and spread is too old. Look at whatever greenery remains attached to the head—if it looks fresh and crisp, the florets are fresh too. If the florets look bristly rather than smooth, or if there are a few green leaves coming through the white florets, the head is still perfectly good.

Brown areas on the florets are not desirable, but if the browning is not too widespread the patches can be trimmed off and the rest of the cauli-flower will be fine.

Some stores sell packaged florets that have been trimmed away from the head. These should be uniformly white and free of bruises. Buy the precut cauliflower to cook within one day, and don't plan to store it longer.

A medium-size head of cauliflower weighs around 2 pounds and will serve 4 to 6 people after trimming.

Storing Cauliflower. Handle the head gently to avoid bumps that will crush and discolor the florets. Store the cauliflower covered loosely in plastic wrap in the vegetable crisper. It should keep for four to seven days. Don't wash the cauliflower before you store it but rinse it thor-oughly in running water before serving it, since some heads are treated with chemicals to preserve their freshness.

You can keep cooked leftovers two to three days in the refrigerator.

If a full head of cooked cauliflower is bound to be too much for one meal, trim off the florets you won't need and serve them raw in salads.

Freezing Cauliflower. Cut the head into pieces about 1 inch in diameter. Wash thoroughly in salted water. Blanch the cauliflower in salted water for 3 minutes or steam it for 5 minutes. Chill, drain, and pack in plastic bags. If you are using a rigid container for cauliflower, it is not necessary to leave headspace for expansion.

Cauliflower tends to be watery after thawing, so you get better results from cooking and pureeing it for freezing.

Cauliflower, in pieces or pureed, keeps ten to twelve months at 0° F.

BROCCOLI

At a point early in its career a broccoli plant looks identical to the cauliflower plant. After a time the similarity recedes, as the loosely packed green buds and thick stalk of the broccoli flower emerge. For a thousand years cooks and gardeners alike made no distinction at all between the two vegetables, which are both only elaborate variations of a simple cabbage.

Broccoli is less temperamental about growing conditions than cauliflower, but a bit more troublesome in storage. It tends to grow woody with age even after harvesting. When you shop for broccoli you should be on the lookout not only for decay but also for tough stems.

Buying Broccoli. The buds at the top of the broccoli should be rich green (some varieties have a dark bluish or purplish cast). Any yellowing means overaging and therefore a much less palatable vegetable. The buds should be firm and hug tightly together. Decay shows up there as soft, slippery spots. Any leaves still attached to the stems should be bright and unwilted. Check the stems: the youngest, tenderest broccoli has slender stalks; older broccoli has thick stalks that show a hollow core at the cut end.

One pound of broccoli yields 4 servings.

Storing Broccoli. Fresh broccoli keeps well for only three to five days in a plastic bag in the refrigerator.

Overcooked broccoli is not at all successful as a leftover because the delicate buds usually disintegrate when reheated. If the broccoli is first cooked only until barely tender, it can be rewarmed gently over low heat.

Freezing Broccoli. Remove leaves and, if you like, peel the stalks. Wash the broccoli thoroughly and cut lengthwise into pieces about 1½ inches across. Blanch in steam for 5 minutes. Chill, drain, and pack into a sealed plastic bag. You can also freeze broccoli in rigid containers—there is no need to leave headroom. You can save a little space when you pack the broccoli by laying some pieces in upside down, alternating the broad bud clusters with the narrow stalks. Broccoli will keep ten to twelve months at 0° F.

BRUSSELS SPROUTS

The brussels sprout is something of an oddity among its cabbage relatives. Instead of growing as a single head or flower in a nest of leaves at the top of the plant stem, brussels sprouts grow in clusters up and down an elongated stalk, shaded by umbrella-like leaves above.

The same cool, coastline counties of California that produce artichokes also produce almost all the brussels sprouts cultivated in the United States, with a nod to upstate New York, which grows most of the rest. Great Britain supports six times more acreage of the crop than all the U.S. farms together. Brussels sprouts are dearly loved by the British.

Buying Brussels Sprouts. A brussels sprout fades from green to yellow quite readily if it has been mishandled, so above all buy brussels sprouts that are deep green.

Each sprout should be firm, the leaves fitting close together.

Brussels sprouts are often prepacked and wrapped in cellophane so that telltale signs of insect damage are very difficult to see. Examine the sprouts as closely as you can for tiny holes or dirty smudges on the outer leaves.

Brussels sprouts often come packed in quart or pint cartons. One quart (approximately 1 pound) is enough for 4 to 6 servings.

Storing Brussels Sprouts. You can keep brussels sprouts for three to five days in a plastic bag in the refrigerator.

Cooked brussels sprouts can be refrigerated for three to four days. Reheat them over very low heat or they will turn gray and bitter.

Freezing Brussels Sprouts. Wash thoroughly and sort the sprouts for size. Blanch smaller heads in boiling water for 4 minutes, larger heads for 5 minutes. Add 1 minute for steam blanching. Chill, drain, and pack in plastic bags or tray-freeze first. Seal, label, and freeze at 0° F. Brussels sprouts will keep ten to twelve months in the freezer.

KOHLRABI

Kohlrabi is simply another kind of cabbage, not a flower like broccoli nor thickly layered leaves like brussels sprouts, but a bulging length of stem: its light green bulb looks like a root, but it actually grows above the ground. Long stalks with leaves at the top grow upward from the bulb.

Sometimes kohlrabi is sold with these edible leaves attached and sometimes it is topped before marketing. It is in stores almost exclusively in
June and July.

Buying Kohlrabi. The best kohlrabi is small, under 3 inches in diameter,
and firm, with no soft or yellowing spots. The bulb should have a fresh
scent and attached leaves should be green and crisp.

Figure on serving 1 medium or 2 small kohlrabies per person.

Storing Kohlrabi. You can keep kohlrabi in a plastic bag in the refrigerator four to five days.

Leftover cooked kohlrabi can be mashed and reheated as a puree. Store
it, well wrapped, in the refrigerator, up to four days.

Freezing Kohlrabi. Wash and peel the bulb. You can leave it whole or
dice it. Blanch whole kohlrabi 3 minutes in boiling water or 4 minutes in
steam. Diced, it should be boiled 1 minute or steamed 2 minutes. Chill,
drain, and seal in plastic bags.

Kohlrabi keeps ten to twelve months at 0° F.

ASPARAGUS

Asparagus is a prima donna of a vegetable—it has an unsurpassed delicacy of flavor; it is temperamental when it comes to storage. As soon as
asparagus is harvested it begins to lose its sweetness and to develop
tough stringy fibers along its stalk. Small differences in the appearance of
asparagus denote big differences in eating quality.

There are basically two types of asparagus: those that turn dark green
in sunlight, among them varieties named Martha Washington, Reading
Giant, and Palmetto; and those that stay pale green or white, such as
Conover's Colossal. These light-colored spears should not be confused
with blanched asparagus, which is grown in darkness, buried in earth so
that chlorophyll cannot develop.

Buying Asparagus. Look first at the tip of the asparagus. The tiny buds
should fit together tightly and smoothly. The tip should be firm and free
of grit, mold, and other damage.

The stalk should be green down at least two thirds of its length. The
blunt white end of each spear is desirable, however, because it retains
moisture better than a cut green end. The stalk should be well rounded;
ridges are a sign of age.

The whole spear should be crisp and taut enough to be easily punctured. Wilted asparagus will not be tender when cooked.

When buying fresh asparagus, allow 1/3 pound per serving.

Canned Asparagus. Canned asparagus comes in a number of different forms: whole spears, tips (about half the spear), and points (just the tender heads). It may be either green or blanched.

Because canned asparagus may lose its shape during shipping and handling, you can't be sure you are buying whole, undamaged spears. Also canned green asparagus fades after long storage, so plan to keep canned asparagus on a cool, dry shelf no more than six months.

Storing Asparagus. You can keep fresh asparagus four to six days in a plastic bag in the refrigerator, but it will steadily lose its natural sugars and turn tougher with each day of storage. Take extra care to keep the tips dry.

It makes much more sense to cook asparagus the same day you buy it, serve it hot the first time, and chill the leftovers for salads. Or you can eat fresh asparagus raw and still get the benefit of its special flavor.

Freezing Asparagus. Wash asparagus and snap off the bottom 2 inches of stalk. Sort spears according to thickness, then blanch in boiling water 2 minutes for small stalks, 3 minutes for medium, and 4 minutes for large. Add 1 minute for steam blanching. Chill and drain. Tray-freeze asparagus before packing so you won't have to poke at the delicate flesh to separate the spears when you cook them.

Asparagus will keep eight to ten months at 0° F.

ARTICHOKES

The artichoke can be a Chinese box to the noninitiate, a puzzle whose prize is perhaps not worth the trouble of unlocking. The vegetable is, after all, a thistle with stickers and bristles designed to discourage invasion. But anyone who has dipped the succulent leaves in melted butter or delved into the artichoke's meaty base can tell you there are *two* treasures worth the winning.

The artichoke is not easy to grow—it thrives only in the evenhanded climate of the California seacoast. It travels well after harvest, though, withstanding most of the adversities of shipping and storing.

Buying Artichokes. Spring artichokes should be bright green. During colder months bronze-colored artichokes come on the market. These have actually been frostbitten but have suffered no loss of flavor or tenderness.

An artichoke should be heavy for its size and have thick, compact leaves. The leaves of aging artichokes begin to spread outward like petals. They may wilt or turn black from bruises and decay. If the stem end of an artichoke has tiny worm holes there may be a lot more damage on the inside of the vegetable.

The size of an artichoke is not an indicator of its quality. Buy 1 medium-size artichoke per person. One large artichoke can serve 2. On occasion, you may see very tiny fresh artichokes for sale. These are immature and can be eaten whole.

Canned and Frozen Artichokes. Most grocers carry a variety of canned artichoke products, which come three ways: marinated, packed in oil, and packed in water. Canned artichoke "hearts" are baby artichokes with tender leaves and a bottom whose bristly "choke" is not developed and is therefore edible. Artichoke bottoms are also canned and are surprisingly tasty. Unopened jars and cans will keep for a year on a cool, dry pantry shelf. Once opened, store canned artichokes in the refrigerator up to four days.

Storing Artichokes. You can keep fresh artichokes in the refrigerator up to one week. They should be stored in a plastic bag. Even when they begin to wilt, they are still quite palatable.

"Leftover" seems the wrong word to apply to chilled, cooked artichokes because they are just as delectable served cold as they are hot, leaves and all. Try sprinkling them with chopped fresh mint before they go into the refrigerator. You can store them, wrapped, in the refrigerator for three to four days.

Freezing Artichokes. You can freeze cooked artichoke bottoms, but coat them with lemon juice to prevent darkening. Even a whole steamed artichoke can be frozen. Cook it through and wrap it tightly in foil or plastic wrap, then seal it inside a plastic bag. Frozen artichokes keep six to eight months at 0° F. Thawed artichoke bottoms are reasonably close to fresh in texture, but thawed leaves are less so.

CELERY

It is difficult to picture a refrigerator without a perky green bundle of this indispensable ingredient in the crisper. And, though celery provides the underpinning for many a great recipe, it is also capable of being cooked with flair and served as a perfectly respectable vegetable by itself. Even omnivorous Greek and Roman gourmets made the mistake of regarding celery as no more than a seasoning, but those ancient strains were probably more bitter than the sweet stalks grown now.

Buying Celery. Start by looking at the leaves. They should be green and no more than slightly wilted. The ribs, which may be light, medium, or dark green, depending on variety, should be rigid and crisp, as if they would make a resounding snap when broken in two. The celery may be light, medium, or dark green, depending on the variety.

Obviously, avoid celery with brown or gray discoloration. Peer into the inside surface of the stalks to check for hollow, pithy-looking patches.

Celery should be on display in the refrigerator case only, and if it is kept moist with a daily sprinkling of water, so much the better. Grocers sometimes slice bruised areas off celery to make it more attractive. Stalks showing this kind of handling will not store as long as completely undamaged celery.

Storing Celery. Celery can keep up to two weeks in the refrigerator as long as it is stored unwashed in a plastic bag. If it has begun to wilt, you can try reviving it by setting the bottom end of the stalk in water, still in the refrigerator. To serve raw, you can recrisp sliced celery by soaking it in ice water for 2 or 3 hours.

Freezing Celery. Celery is inexpensive and always available, but if you have reason to store it long term you can freeze it. The thawed celery will not be as crisp as fresh-cooked. Wash the celery and trim away leaves. Cut the stalk into 1-inch sections. Blanch for 3 minutes in boiling water or for 4 minutes in steam. Chill, drain, and pack into plastic bags. Celery will keep ten to twelve months at 0° F.

SWISS CHARD

Swiss chard is a beet that has been coaxed by breeders to develop its stems and leaves more than its root.

It must be treated as two vegetables in one. Swiss chard's delicate greenery can be cooked like spinach; its heavy stalk is similar to celery. Only the youngest chard is eaten raw, since the older, larger vegetable is somewhat bitter until cooked.

Buying Swiss Chard. The leaves should be crisp and evenly green with no yellowing. (One variety has a red tinge to the leaf tips.) The stalk should be crisp. The smaller the specimen the sweeter it will be.

You will need 2 pounds of fresh chard to serve 4 as a cooked side dish.

Storing Swiss Chard. You can keep chard in the refrigerator two or three days in a plastic bag. Plan to use it as soon as possible, because at the first sign of yellowing the chard may have already become bitter.

Freezing Swiss Chard. Separate the leaf from the stalk. Blanch the washed leaves 2 minutes in boiling water. Chill, drain, and pack into plastic bags. Cut the stalk into 1-inch pieces, wash, and blanch for 3 minutes in boiling water. Chill, drain, and pack into plastic bags. Both parts of the chard will keep ten to twelve months at 0° F.

FENNEL

Fennel is bulbous at the bottom, has narrow stalks in the middle, and sprouts feathery leaves at the top. Its taste is reminiscent of licorice and it is sometimes labeled anise in markets. All parts of the plant are edible, including the seed.

Fennel turns up in French, Italian, and Chinese cooking and Shakespeare mentioned it as a flavoring for fish. Nevertheless, it is not a particularly common vegetable anywhere.

Buying Fennel. The bulb should be firm and white to light green in color. The stalks should be darker green and the leaves darker still. Avoid wilted or browning fennel. The light snappy texture of a fresh bulb is one of its most irresistible pleasures.

If you are planning to serve fennel as a separate vegetable, buy 1 medium-size bulb per serving.

It is tasty raw, so cook only what you need and save both bulb and stalks to serve cold. Use a little fennel as a seasoning in tomato-based sauces and in fish or chicken salads.

Storing and Freezing Fennel. With the leaves removed, fennel keeps for seven to ten days in a plastic bag in the refrigerator.

Freeze fennel cooked and pureed, sealed in an airtight container with 1/2 inch headspace. It keeps ten to twelve months at 0° F.

BOK CHOY

Bok choy has footlong white stalks that are fringed with dark green leaves at the top. It cooks quickly and is a frequent addition to Chinese stir-fry recipes.

Buying Bok Choy. The vegetable should be crisp, both stalks and leaves unwilted. The stalks should be pearly smooth, without browning or marks of any kind.

Storing and Freezing Bok Choy. Bok choy will keep three to four days in a plastic bag in the refrigerator. To freeze, follow the instructions for Swiss chard on pp. 42–43.

PEAS

Canned or frozen, peas are an everyday item, pleasing to be sure, but not likely to inspire rapture. Most peas that are farmed end up canned or frozen for the same reason they are a rare sight in the fresh produce section—they are very perishable and, because of the pod, time-consuming to prepare.

Fresh peas, on the market from January to June, are a particular treat. Their taste is brighter, their texture bouncier, than the processed kind offers. But even fresh, store-bought peas do not reveal the vegetable's many possibilities. Only the avid home gardener knows all that a pea can be. The type of pea most often grown commercially is smooth, round, and sweet, but there are many varieties, often less pretty, that have fuller, more distinctive flavors.

A serious greengrocer will offer you three kinds of fresh peas. The English pea is the same variety that is canned and frozen. (*Petits pois* are baby English peas.) Snow, also called sugar, peas are used in Chinese cooking. The pod is smaller, flatter, and peas are tiny compared to the English variety. Snow peas are usually eaten pod and all. Sugar snap peas are a cross between English and snow peas.

Because commercial-variety English peas are a bit lackluster to start with, and because they quickly become chalky after harvest, they can often disappoint you. Much more reliable are delicate snow and sugar snap peas.

Buying Peas. Buy all fresh peas in the pod. English peas should be crisp, fat, and grass-green. If you can peek inside a pod, the peas should be glossy and fresh-smelling. Buy 3/4 to 1 full pound of pods per serving.

Snow pea pods should be green and thin-skinned so you can see the shape of the peas inside. Absolutely avoid snow peas that have begun to wilt. Sugar snap peas should be green and well rounded. Buy 1/4 pound snow or sugar snap peas per serving, since you eat them pod and all.

Beware of yellowing or speckles on the pod, signs of aging. An over-the-hill pea pod is not worth the price.

Storing Peas. You should eat fresh peas the same day you buy them. Otherwise, you can store them for three to four days in a plastic bag in the refrigerator. Don't shell them until you are ready to cook them.

Cooked peas can be held for three to four days in the refrigerator.

You can serve snow peas and sugar snap peas raw in salads and with dips.

Freezing Fresh Peas. Very fresh peas, from your own garden or a local grower, are worth the trouble of home freezing. They will taste much livelier than commercially frozen peas.

Shell English peas and discard the pod. Blanch 1½ minutes in boiling water or 2½ minutes in steam. Chill, drain, and tray-freeze. Pack in sealed plastic bags.

To freeze snow and sugar snap peas, start by trimming blossom ends and removing the strings that run along one edge of the pod. Blanch the entire pod in steam for 2½ minutes. Chill, drain, and tray-freeze. Pack in plastic bags, seal, and label.

Peas and pods will keep ten to twelve months at 0° F.

SWEET CORN

It was an American Indian, Squanto, who taught the Pilgrims the techniques of cultivating corn. His help was not inconsequential because corn (also called maize and sweet corn), is the only food plant that cannot thrive by itself. It depends on man's intervention to complete its cycle of reproduction.

After maize was introduced into Europe and Africa it eventually became an important subsistence crop (it prospers in almost any climate). But no one adopted corn as quickly as the New World colonists, whose fields of wheat, barley, oats, and rye did not take hold quickly enough to feed the growing population. Following the lead of the Indians, they ate corn fresh, notably in succotash, and they made cornmeal, for bread and puddings.

Tough and adaptable as the plant is, fresh corn itself is a vulnerable vegetable. The very instant it is plucked from the plant an ear of corn begins to convert its considerable store of sugar into starch. The sweet, milky liquid inside the kernels begins to turn bland and pulpy.

There are only two things that arrest this process: heat and cold. You can take the corn directly from the field to the kitchen and boil it. This captures the sugar at its peak but is not a practical plan for the office worker looking forward to fresh corn on the cob after a long commute.

The urban office worker, barring a weekend trip to a country farm stand, must rely on the food industry to bring fresh corn to market with utmost care. Corn must be chilled soon after harvest. Chilled corn stays sweet for a much longer time. During shipping and while it is in the grocery store, every minute the corn spends above 32° F. sugar in the kernels is losing ground to starch.

Buying Fresh Corn. Think twice about buying corn that is displayed anywhere but in a refrigerator unit unless you are sure it has been just picked or just delivered. Corn stored at room temperature loses about half of its total sugar in one day.

Of the more than 200 varieties of sweet corn, most commercially grown corn is yellow and contains more vitamin A than corn with white kernels. To quibble over white vs. yellow corn is really pointless unless you are planting a crop for yourself, since the freshness and age of an ear have much more to do with how it tastes.

The husk of unshucked corn should be green and pliant; the silk should be golden—not brown, moist, and free of decay; the stem end should be damp, pale green, and flexible, not dried and brown.

Some supermarkets partially husk fresh corn and wrap it in plastic for display. If it has been carefully refrigerated, partially husked corn can be as sweet as unshucked corn that has been stored at room temperature.

Freshness is not the only measure of a good ear of corn. Learn to look for signs that the corn was picked at full maturity. The most telling clue to the freshness and maturity of the corn is the condition of the kernels themselves. Be grateful for the merchant who allows you to peel back the husk to view the vegetable you plan to eat. Of course partially husked corn is already open to view. The kernels should be tightly packed together in even rows. Gaps between the rows mean the ear is overmature. If the tip of the ear is bare of kernels, it was picked too soon. The kernels should look plump and juicy. Fresh kernels, when pierced, spurt a milky liquid. As the corn dries out and turns starchy, the center of each kernel sinks inward.

Canned and Frozen Corn. Canned and frozen corn marked U. S. Grade A or Fancy contains the sweetest, most succulent corn. The lower grades, B and C, are assigned to corn that is starchier and chewier. All grades are equally nutritious. Cans of corn keep up to one year on a cool, dry shelf.

Storing Corn. The best advice is, *don't* store fresh corn. Buy it the same day you plan to cook it and keep it in the refrigerator in the meantime.

If the fresh corn was irresistible and you bought more than you could eat in one meal, cook the surplus ears for a minute or two right away, cool them quickly in cold running water, and store them in plastic bags. Use cooked corn within two to three days. You can slice the kernels off the cob with any sharp knife. Stand the ear on end and run the blade downward, slicing a few rows of kernels all at once.

Freezing Corn. Remove husks and silk. Blanch medium-size ears of corn 4 minutes in boiling water, 3 or 4 ears at a time. Cool quickly and drain. Slice the corn off the cob, cutting only to about two thirds the depth of the kernel (this deliberately leaves the milky juices behind). Tray-freeze and seal in plastic bags.

You can prepare cream-style corn by slicing the blanched kernels off at about half their depth. Then, using the back of the knife blade, scrape the pulp and milky juices off the cob. Combine the cut corn with the scrapings and seal the mixture in airtight plastic bags. Label and freeze.

Both regular and cream-style corn will keep ten to twelve months at 0° F.

GREEN BEANS AND WAX BEANS

The slender young pods of both green and wax beans fall into the category of "snap beans," a much more appropriate name than their other label, "string beans." The chewy string that used to grow down the seam of each pod was long ago bred out of all but a few bean plants, but the brisk "snap" you hear when you break off each end of the bean is still characteristic of this popular vegetable.

The two kinds of beans are similar in every way except that the green bean has a bright green color, while the wax bean is yellow or cream-colored.

These beans are sometimes referred to, most unjustly, as French beans. They are, in fact, a product of the Americas, whose name in French, *haricot,* is a rendition of the Aztec *ayacotl.*

Buying Green and Wax Beans. The beans should be firm but pliant. Beans that are too stiff are older and less tender. Beans that are wilted have been stored too long. Slim, smooth beans are best. Beans with bulges are too mature to be really tender.

Look for vivid color, whether green or yellow. Don't buy beans that have serious blemishes. A few brown spots can be cut away—they are damage rather than decay—but avoid badly scarred beans.

One pound of green beans will serve 4 people as a side dish.

Canned Green Beans. Canned green beans should be used within six months of the date of purchase. The beans are likely to lose color and flavor if stored much longer than that.

Storing Green and Wax Beans. Store the unwashed beans in a plastic bag in the refrigerator. They will keep three to five days.

You can keep cooked beans in the refrigerator, well covered, three to four days.

Freezing Green Beans and Wax Beans. Wash beans and trim off ends. Cut larger beans into lengthwise strips and leave smaller beans whole. Blanch in boiling water for 3 minutes. Chill, drain, and pack into an airtight plastic bag. Seal and freeze. Green beans and wax beans keep ten to twelve months at 0° F.

LIMA BEANS AND BROAD BEANS

Lima beans are pale, flat, kidney-shaped beans that come in two varieties, baby limas and Fordhooks. The baby lima is not really a baby, it is a smaller, milder bean. Fordhooks, often called butter beans in the South, are larger and heartier in flavor. You may find fresh limas marketed in pods or shelled and prepackaged in plastic-wrapped trays. For information on dried limas, see p. 51.

Broad beans, also called fava beans, look like limas but are much larger. Broad beans are the type of bean native to Europe. They are sold fresh, still in their very long and bulgy pods.

Buying Lima Beans and Broad Beans. Look for full, green pods with no browning or other marks. They should look moist, feel firm, and show no signs of withering.

Buy 1/2 to 3/4 pound of pods per serving or 1/4 pound of shelled beans.

Canned and Frozen Limas. When you buy canned or frozen limas, look for a USDA grade mark. Grade A will be the tenderest, while B and C are starchier. Canned, they keep up to one year on a cool, dry shelf.

Storing Fresh Lima Beans and Broad Beans. Keep pods in plastic bags in the refrigerator up to two days. Shelled limas should be cooked the same day they are purchased.

Freezing Limas and Broad Beans. Shell limas and broad beans and sort for size. There will be a skin around the broad beans when they come out of the pod. This should be removed before freezing or cooking the beans in any way. Blanch the beans for 2 minutes in boiling water, tray-freeze, and pack into airtight plastic bags. Seal, label, and freeze. They will keep ten to twelve months at 0° F.

LONG BEANS

A common vegetable in Asian cooking, long beans are indeed very long, measuring 1 to 3 feet in length, but otherwise resembling string beans.

The best are thin and dark green, and show no bulges. Treat them just as you would green beans.

SPROUTS

Sprouts are the infant plants that begin to grow out of dried peas, beans, grains, and seeds in warm, moist environments. They have a delicate threadlike greenery with a strong flavor and a gratifying crunch.

Many grocers carry fresh sprouts, and it is also very simple to nurture a crop at home. Health food stores offer a great variety of seeds, grains, and legumes suited to sprouting.

Buying and Storing Seeds for Sprouts. When buying seeds to sprout at home, look for a label that indicates the percentage of germination to expect. Don't buy just any seeds—some seeds have been dried at high temperatures that kill the embryo plant, while some seeds sold for planting have been treated with chemicals poisonous to humans.

Seeds for sprouting can be stored indefinitely in a tightly lidded jar in a cool, dark pantry. Sterilize the jar first with a 30-second bath in boiling water. Make sure both the seeds and the container are perfectly dry before they go into storage.

Buy and store dried beans and grains as described in their respective sections (see pp. 50–52 and pp. 250–52).

Buying and Storing Fresh Sprouts. Look for moist and crisp-looking sprouts with a fresh scent. Bean sprouts are very pale but most other sprouts have a touch of green. Avoid sprouts that look slimy or darkened. The shorter the tendrils the younger and tenderer the sprout.

Fresh sprouts will keep for seven to ten days in a plastic bag in the refrigerator. They should be moist, but don't allow a lot of free water to accumulate on the inside of the bag.

Do not freeze sprouts.

TOFU

Tofu (bean curd) is made from cooked, mashed soybeans. It is cultured, very much like cottage cheese, then pressed into creamy white cakes the consistency of firm custard.

Buying Tofu. Fresh tofu is sold in sealed plastic cartons with see-through tops. It should be in firm, unbroken cakes, completely submerged in water. Because fresh tofu is very perishable, don't buy it unless it is kept refrigerated at the store. Look for a "sell by" date on the label.

Storing Tofu. To keep it soft, store fresh tofu in the refrigerator submerged in lots of water in a covered container. Pour off the liquid once a day and replace it with cold water. Tofu will keep three to five days from the "sell by" date, or up to two weeks if it is very fresh. You can keep tofu stored out of water, tightly wrapped in plastic, in the refrigerator. It will stay fresh as long but become firmer as the days go by.

Freezing Tofu. You can freeze tofu, in water, in its original carton, sealed airtight. It keeps up to two months at 0° F. When thawed, it is spongier and more fragile than fresh and should be added to dishes at the last possible moment.

DRIED BEANS, PEAS, AND LENTILS

Dr⁻ ᵇeans, peas, and lentils are perfectly designed storage units all by themselves. They can sit in a jar on the shelf for months, hanging on to virtually all their nutrients (abundant) and flavor (bland but adaptable).

Any number of regional cuisines include some combination of beans and rice. In the southern United States, they have "Hopping John" and red beans and rice; in Brazil, a mixture of rice, beans, and meat called *feijoada* is the national dish. These recipes are a tradition on which science and folk wisdom can agree. Rice and beans eaten together provide 50 percent more usable protein than beans eaten alone. You get the same nutritional bonus by mixing pork and beans.

Buying Dried Beans, Peas, and Lentils. All the varieties of beans are interchangeable in recipes. Here are some common types:

• BLACK BEANS. Favorite soup beans, they are also used in oriental and Latin American cookery.

• BLACK-EYED PEAS. These are actually beans, small cream-colored ovals with a black spot like a bull's eye on one side. They are an indispensable part of Southern home cooking. If you find these fresh, store and freeze as you would limas, p. 48. They come canned and frozen as well.

• CHICKPEAS. Also called garbanzos, these are tan, round, and meaty. They are often served cold in salads and they are the basis for the pureed Middle Eastern dip, hummus. Fresh chickpeas should be treated like fresh limas, p. 48.

• CRANBERRY BEANS. Glossy white beans speckled with red and green, these are sometimes available fresh in the summertime. Treat fresh cranberry beans as you would English peas, pp. 44–45. Dried, they are well suited to any bean dish.

• GREAT NORTHERN BEANS. These are white, oval beans with a mild flavor perfect for baked bean and soup recipes. They are also available canned.

• KIDNEY BEANS. Large, red, kidney-shaped beans, they are common in chili con carne. You will find two varieties canned, light red and dark red.

• LENTILS. Lentils are neither peas nor beans, but are related to both. They are tiny grayish-brown disks with a most distinctive flavor. They cook more quickly than other dried legumes. Pink lentils (actually orange in color) are used in Middle Eastern cooking. Pink lentils come whole and split.

• LIMA BEANS. Flat, pale, bland beans also available fresh (see p. 48), canned, and frozen.

• MUNG BEANS. Small, sweet beans that may be green, brown, or black. They are a favorite for sprouts. Split golden gram is the split and skinned version of mung beans.

• NAVY BEANS. This name applies to almost any small white beans. Use them for baked beans, soups, and salads.

• PEA BEANS. Small white beans, these are commonly used for baked beans and soups.

• PEAS. There are two types of dried peas, green and yellow. Either kind may be whole or split. You can use yellow and green interchangeably, but whole and split require different cooking times.

• PINTO BEANS. Pinto beans are tan with dark flecks. Used frequently in Southern and Mexican cooking, they are also available canned.

• RED AND PINK BEANS. These are similar to kidney and pinto beans.

• SOYBEANS. A soybean is a powerhouse of protein with an insistent flavor that works best with other foods that can stand up to it—onions, tomatoes, and peppers, for example. Fresh soybeans are sometimes available. They can be stored, cooked, and frozen the same as fresh lima beans.

Buying Dried Beans, Peas, and Lentils. Buy bright dried peas and legumes that you can see. They should be in plastic bags or boxes with cellophane windows.

The main thing you are looking for is uniformity. A mixed bag of small and large beans will be hard to work with because smaller beans cook sooner than larger.

The dried beans, peas, or lentils should all be about the same color. A dried legume fades the longer it is stored, and the longer it is stored the longer it will take to cook. Once again, a mixed bag of bright and dull beans will cook unevenly. Very dry, old beans disintegrate when cooked.

Look for defects such as cracked beans, or debris in the package (twigs or pebbles that can crack a tooth). Tiny pinholes in the beans or peas are evidence of insect damage.

One pound (2⅓ cups) of dried beans cook up to about 6 cups, which will serve 4 to 6 people.

Most of the different types of beans are also sold canned. There is no storage advantage to canned beans—they keep no longer than dried and they take up more space. Canned beans are convenient, but they do lose some nutrients during processing.

Storing Dried Beans, Peas, and Lentils. If they are dry and cool (70° F.), dried beans, peas, and lentils will keep well as long as one year.

The bags in which beans are often sold are fine as storage containers. Once opened, they can be resealed with a paper-wrapped wire twist.

In a closed glass jar, dried beans and peas are quite decorative if you have the space to set them out in view and they keep just as well. Don't pour newly bought beans in with older ones, though. They will cook unevenly.

NOTE: All beans, peas, and lentils, whether fresh or dried, contain lectins, toxins that can cause abdominal pain, nausea, and diarrhea. To destroy the lectins, beans must be cooked at a full boil no less than 10 minutes before heat is reduced for simmering; the toxins in peas and lentils will be destroyed by 2 or 3 minutes of boiling.

Lectins are not, however, the source of flatulence. The current theory is that a certain carbohydrate in beans manufactures gas as it is digested in the stomach. Boiling beans will have no effect on the offending molecule.

Freezing Dried Beans, Peas, and Lentils. Until they are cooked, of course, keep dried peas and legumes on the shelf where they belong. Once they have been made into soup, baked beans, and casseroles, they are perfect candidates for the freezer. See pp. 333–35 for tips on freezing prepared dishes.

SUMMER SQUASH

Squash is one of the triad of vegetables that sustained Indians of North and South America and, in turn, American colonists—corn, beans, and squash. The source of the word is a Massachusetts tribe, who called the vegetable *askutasquash,* meaning "eaten raw."

"Summer" squash are squash that are harvested young, with tender skin and soft seeds. Their designation as summer squash is deceptive, for they are marketed fresh almost every month of the year.

Among the most familiar summer squashes are the long, green zucchini that have a family resemblance to cucumbers. There are two kinds of yellow summer squash—straightneck and crookneck. Their skins may be smooth or bumpy, depending on variety. Both are lemon-colored and shaped something like a bowling pin; the crookneck variety has a cane-handle curve at the top. Sometimes called white squash, the scallop-type summer squash looks like a flying saucer with fluted edges. Scallop squash may be snowy white, pale green, yellow, or striped.

Another soft-skinned squash is the chayote. It is distinct from other summer squash in that its rind and its single large seed are not edible. It is pear-shaped and has deep ridges running from top to bottom. The skin may be light green or white and sometimes has bristles.

Buying Summer Squash. Small squash are the tenderest. They should feel firm and heavy for their size. Squash that are limp and wrinkled have begun to dry up.

Fresh squash will have shiny skin. Dull, pitted skin is a sign of aging and injury. Yellowing on green summer squash, such as zucchini, occurs when the vegetable is past its prime. Check the skin for cuts and bruises, which are forerunners of decay. The squash should be clean—dirt can readily penetrate the thin skin. Summer squash should have about an inch of stem still attached.

One pound of summer squash will serve 4 people.

Storing Summer Squash. Summer squash need cold temperatures and high humidity. You can keep summer squash four to five days in a plastic bag in the refrigerator. Chayote will last longer, up to 2 weeks.

Avoid nicking or bruising the squash, because any injury accelerates spoilage.

Freezing Summer Squash. Wash the squash and slice into 1/2-inch-thick pieces. Blanch 3 minutes in boiling water or 4 minutes in steam. Seal in airtight plastic bags, label, and freeze. Squash will keep ten to twelve months at 0° F.

For freezing chayote, follow directions for winter squash.

WINTER SQUASH

Winter squash have matured on the vine, developing a thick hard rind and tough seeds, both inedible. They are large (some varieties grow to 30 pounds) and they are durable. Like summer squash, they are available almost year round.

An acorn squash has a dark green rind. It is more or less spherical, with evenly spaced ridges running from the stem and to a point at the bottom. An acorn squash weighs from 1 to 3 pounds.

The huge banana squash weighs between 15 and 30 pounds. It looks like a pink watermelon, but you will rarely see it whole. Most grocers cut banana squash into manageable pieces for sale.

A buttercup squash looks like two squashes, one wearing the other as a hat. The bottom half is wide and drum-shaped; the top half is smaller, a smooth bulge centered over the drum. It is dark green and weighs from 3 to 5 pounds.

A butternut squash is a light buff color. It has an elongated pear shape and weighs 2 to 4 pounds.

A hubbard squash has a fat, round middle, a tapering neck at the stem end, and it comes to a point at the bottom. The skin may be dark green, bluish gray, or orange and it is hard and bumpy. Smaller varieties weigh 5 to 8 pounds; larger ones grow to 16 pounds.

The bright orange pumpkin needs no introduction. It is typecast as either a pie or a jack o'lantern, but the pumpkin performs perfectly well in any winter squash recipe.

A spaghetti squash is not a perfect fit in this category, since it falls somewhere between summer and winter squash in character. It is a yellow, elongated oval whose flesh, when cooked, comes out in long thick strings like spaghetti.

Buying Winter Squash. It is hard to judge a winter squash from the outside. First, heft the squash—it should feel heavy for its size. Lightweight squash have lost too much moisture through evaporation.

Second, rap your knuckle on the rind. It should feel thick and hard, and you don't want a squash with any soft spots. A shiny skin means the squash was picked before it was ripe, so look for rinds with a dull finish.

An acorn squash is sure to be fully mature if it shows a little orange on the rind. If more than half the skin is orange, though, the flesh will be stringy.

A butternut squash should be entirely tan—a greenish tint heralds tasteless flesh.

If you buy precut squash, the flesh should look moist and grainy, not stringy.

Buy 1/3 to 1/2 pound fresh winter squash for each serving—you will lose some of the weight in the discarded peel and seeds.

Storing Winter Squash. You can keep winter squash at room temperature up to a week. If you have a cool (50° F.) cellar or porch to store them in, winter squash, particularly the thicker-skinned hubbard and banana types, may keep as long as one month. In the refrigerator, you can store winter squash up to two weeks.

They tend to lose moisture during storage, so dry, overheated air will reduce their shelf life.

Cut squash should be refrigerated, tightly wrapped in plastic, for two to four days.

Winter squash is made to be mashed, so you can reheat cooked squash without worrying about its texture.

Freezing Winter Squash. Cook squash completely before freezing it. You can cut it into pieces and boil the squash or bake it whole in the oven until soft. (Pierce the skin before baking.) Remove seeds and scrape the cooked flesh from the rind. Puree the pulp, or cut it into chunks. Seal in airtight plastic bags, label, and freeze. Winter squash will keep ten to twelve months at 0° F.

MUSHROOMS

A mushroom farm is not a patch of land with sun and rain and weeds and tractors. It is a glass and steel laboratory with rows of test tubes incubating the microscopic spores from which new mushrooms grow. The spores, seed material from the underside of mushroom caps, develop mycelium, a tiny network of strands. The mycelium is next transplanted into a flask filled with grain that nourishes the growing plant. After the strands have enlarged and multiplied, the contents of the flask, grain and all, are moved to a dark and steamy growing room where they are buried under sterile soil. Forty-five days later the mycelium bears its fruit, bulging white mushrooms with domed tops.

The standard issue American mushroom is the plump, white *Agaricus bisporus,* a perfectly pleasant, firm little fungus. But if you have not tried wispy enokis, rich, tan criminis, cèpes, chanterelles, the spongy, dark morels, oyster mushrooms, or shiitakes, don't wait any longer—most markets carry these fresh at least some of the time, and they taste great.

Buying Fresh Mushrooms. Two things happen to an American mushroom as it ages. First, it darkens because of oxidation. Second, its cap, which when fresh hugs close to the stem, begins to flare out like an umbrella opening, exposing the gills.

But the open cap also signals a more robust flavor, for the mushroom gets richer as it ages. The open mushroom will have a shorter storage life, but if you prefer the stronger taste, go ahead and buy mushrooms that have begun to spread open. Of course, you can bring closed mushrooms home and wait for them to mature in your own refrigerator.

The standard American mushroom comes in three shades: off white, cream-colored, or light tan. As long as the color is bright and even and the cap is smooth, the mushroom is fresh.

When mushrooms have been bleached, they should have a label that says so. While you are looking at the label, also make sure they have not been treated with preservatives. These chemicals destroy vitamins.

In any case, avoid mushrooms that are nicked and bruised, excessively dirty, or shriveled. They should be smooth and firm. If the caps are open, check the spokelike gills underneath. Black gills mean the mushroom is too old.

Exotic varieties of mushrooms should be firm, dry, and fragrant.

When you are serving mushrooms as a separate side dish, buy 1/4 pound per person. One pound yields about 5 1/2 cups of sliced mushrooms, which reduce to about 2 cups when they are sautéed.

Storing Fresh Mushrooms. Fresh mushrooms keep best in an open container in the refrigerator. Use a paper bag or the paper tray they often come in. Don't use plastic, because it holds in too much water. To keep them from withering, cover them loosely with a piece of cheesecloth. Don't clean them before you store them. They should remain presentable two to three days. Once they have darkened and begun to dry up, you can still use them chopped up in cooked dishes—their flavor will be fine; only their looks have suffered.

Before you cook mushrooms, clean them by wiping them gently with a damp paper towel or a very soft brush. Don't soak them or peel them.

Recipes often call for just the caps, but for heaven's sake don't throw out the stems. They are every bit as tender and tasty as the caps.

You can add two or three more days to the life of the mushrooms by sautéing them in butter and storing them cooked.

Canned and Dried Mushrooms. Canned mushrooms will keep up to a year on a cool, dry shelf. If you need to store the canned mushrooms after they have been opened, drain off the liquid and keep them in a covered jar in the refrigerator. They should last five to seven days.

The many exotic varieties of dried mushrooms are much more than a substitute for fresh. Keep a supply on the shelf and be ready to add their intense, earthy flavors to your cooking.

Dried mushrooms will last for six months on the shelf as long as they are kept perfectly dry. Seal them in airtight covered jars and keep them in the dark. If your kitchen is very humid, put them in the refrigerator.

Freezing Mushrooms. Wipe the mushrooms clean with a damp paper towel. Sauté them, sliced or whole, in butter with a bit of oil added, until almost done. Pack the cooled mushrooms and buttery juices in an airtight plastic bag, seal, label, and freeze. You can also freeze mushrooms without blanching or sautéing. Tray-freeze them and seal them in plastic bags. Freezing makes them a bit rubbery, but the mushrooms are still good for cooking. Mushrooms will keep ten to twelve months at 0° F.

TRUFFLES

Praise for truffles is exuberant, with pl.rases like "black diamond" and "underground empress" applied freely. The black truffle from Périgord in France is imported fresh in January and February, while Italy's white truffle can be found from October through December. Both varieties are extravagantly expensive—up to $700 per pound—and no one seems to agree on which one is closer to heaven.

A fresh truffle should be plump and well colored, with a powerful perfume. Canned truffle pieces are a less costly alternative to fresh.

A whole, fresh truffle should be submerged in dry rice and refrigerated for up to two weeks. Sliced truffles or those out of a can should be refrigerated submerged in fine, light olive oil or in Madeira in a closed jar and used within a month. Another method is to plunge them into a container of goose fat and store them, covered, up to one month in the refrigerator. The rice, oil, wine, or goose fat will absorb flavor from the truffle and, thus improved, can be used in cooking. Canned truffles may also be transferred to a lidded glass jar after opening and stored in the refrigerator, covered in their own liquid, for up to a week.

GARLIC

Garlic is a pungent cousin of the onion. A *head* of garlic is a cluster of *cloves* doubly sheathed—the entire head has a dry, papery outer skin and each clove has a tough, tight-fitting second skin.

After harvest, garlic is dry-cured to extend its storage life. Each clove, however, stays moist and meaty inside the two protective coverings.

Some markets may carry unusual varieties of garlic in season. As a rule these are larger, sweeter, milder heads grown in warm climates. But the red garlic from Mexico and Argentina is sharper and juicier than the standard variety. Though hard to find, it stores exceptionally well.

Buying Garlic. As time goes by, individual cloves in a head of garlic may dry out or mold and, as a result, virtually turn to dust. The freshest garlic feels firm all around. Avoid heads that have soft or wet spots. Look out for green shoots at the neck, because sprouting shrivels the head.

Storing Garlic. Garlic stores best in a dark, airy spot at that nearly unattainable temperature 50° F. If you put it in the refrigerator it will keep all right, but its fragrance might spice up your milk or banana cream pie. There is no harm in storing garlic at room temperature, but it is an unpredictable arrangement. Put the whole head in an open container on a dark shelf or in a drawer. Take one clove at a time as you need it without cutting into the remaining cloves. Garlic will last anywhere from one week to two months, depending on its age when you bought it and on the heat and humidity in your kitchen. As garlic ages, the center of each clove turns green and bitter, so discard any green parts. Fortunately, garlic is very inexpensive and always available, so losing a few cloves is painless.

You can't buy less than one whole head of garlic. If that is too much for you, try this simple technique for preserving it: peel and chop all the cloves; put the garlic in a small glass jar and cover it with olive or vegetable oil. Tightly lidded, the garlic will keep in the refrigerator for two to three months. You can spoon out small quantities as needed. You can also use the flavored oil for salad dressing and replace it with fresh oil in the jar.

Or you can go to the other extreme and use up all the cloves you have at once in a recipe that features garlic. The longer garlic cooks the milder it tastes, so garlic broth is a perfect base for vegetarian soups and sauces.

To peel garlic easily: lay the clove on a cutting board, cover it with the broad side of a knife blade, and, using the heel of your hand, press down lightly. It is not necessary to smash the clove, for the skin will slip right off even a barely flattened clove. Crushing garlic will produce a slightly stronger flavor than delicately mincing a whole clove. The distinction is only important when the garlic is eaten raw.

If you have a head of garlic that has begun to sprout, hang on to it and let the green shoot grow a little. The cloves in the head will wither but may be usable for a few more days. The garlic sprout itself is delicious as a seasoning—use it as you would a green onion.

Do not freeze garlic.

GREEN ONIONS AND LEEKS

A green onion, also called a scallion, can be any variety of onion that has been picked before the large underground bulb has formed. Its pencil-thin stalk, white at the base, gradually expands into tubular green leaves at the top.

A leek is a special variety of onion with a different, mild flavor, but it looks like a giant scallion. The white base of a leek is a thick cylinder an inch or two across. There is a yearly competition in Britain, first prize going to the largest leek grown, which is usually 3½ to 4 inches in diameter. Giant leeks have no special culinary value but are simply evidence of fervent home gardening. After the awards ceremony, all the vegetable contestants are boiled up into soup. Leeks do well in soup. They are indispensable in vichyssoise and in cock-a-leekie, a traditional Scottish mixture of leeks, beef, chicken, and prunes.

Buying Green Onions and Leeks. The green ends of these vegetables should be brightly colored and crisp. The white ends have a shaggy skirt of roots which should be firmly attached. The white itself should be free of discolored or slick layers of skin. Green onions may rightfully have a round bulbous white end but leeks should be perfectly straight.

Both leeks and green onions can be cooked and served as a vegetable—allow 2 or 3 leeks per serving and a bit more than one average bunch of green onions per serving.

Storing Green Onions and Leeks. In a plastic bag in the refrigerator they should last one to two weeks.

You can eat green onions raw, of course, and use up odds and ends by substituting green onions for regular dry onions in your cooking—they will have a slightly snappier flavor. Use the whole green onion, white and green, no matter how you prepare it.

Leeks require very careful cleaning, because their tight layers trap unbelievable amounts of dirt. To wash, trim off the tops at the point where they turn from dark green to pale. Slice into quarters lengthwise, down to about 1½ inches of the base. Gently spread the layers and dunk the leek, top side down, into a sinkful of water several times. If the leek is still gritty, fan the layers more and soak it.

Freezing Green Onions and Leeks. Slice or chop the vegetable so it takes up less space. Seal in plastic bags without blanching, label, and freeze. Green onions and leeks will keep ten to twelve months at 0° F. These onions soften and lose some pungency in the freezer but can still be used for cooking.

ONIONS

The dual character of the onion—sharp and tonic when raw, mellow and aromatic when cooked—is a result of its distinctive composition, shared by all members of the botanical family, leeks, garlic, and shallots included.

Slicing an onion arouses certain of its enzymes, which then go to work on a passive sulfur compound within the cells, changing it into volatile acids and ammonia. These new chemical combinations are a bit irritating. They are responsible for an onion's biting flavor and they cause the eye to tear. Cooking counteracts the chemicals, soothing their sting.

The many varieties of onions have in common their papery outer skin and their more or less rounded shape.

The globe onion is the most common. It is usually tan but may be white, yellow, or even red. The flavor of any particular onion depends a lot on its breeding and on where and when it was grown, but in general a globe onion is strong. Very small white globe onions are sold as boiling onions. The Bermuda or Spanish types of onion, which may be red, purple, or yellow, are larger and milder, and particularly suitable for eating raw.

Shallots are smaller than the average onion, sometimes no bigger than a garlic bulb. They have orange-tan skin, a slightly out-of-kilter shape, and a keen flavor with undertones of garlic.

In summer you may find some special, very sweet onions at the market. They are usually yellow. Vidalia, Walla Walla, Grano, and Granex are some of the variety names. These onions are a bit softer and shorter-lived than the other types. They are also almost sugary enough to eat like apples.

Soon after harvest, onions are "cured"—left in warm, humid, circulating air. The neck of the onion, the pointed end opposite the roots, dries and closes as it cures, sealing out moisture and decay organisms. An onion is fully cured when its skin is dry enough to rustle. Growers store onions harvested in late summer right through the winter months.

Buying Onions. Globe onions often come in 3-pound open-mesh bags; Spanish and Bermuda onions are sold in bulk; shallots may be boxed or loose. However the onions are displayed, examine each one you plan to purchase since one spoiled onion can spread decay to the whole group.

Start at the neck. It should be tightly closed and completely dry. Avoid onions with thick, woody centers in the neck. The whole onion should be quite firm. Don't buy an onion with soft, water-soaked spots, especially at the neck.

The skin should be bright and shiny. Look for black, powdery patches just under the skin—this is a common mold that eventually spoils the flesh.

Greening occurs in onions that have been cured and stored under too much light. Onions with green areas taste unpleasant, so don't bring one home.

Storing Onions. Store onions in a cool, dry spot, preferably spread out or hung in a basket so air can circulate around each one. Globes should last two or three weeks in the kitchen pantry. Mild, sweet onions last about half that long. Hot, humid weather cuts the shelf life in half.

You can refrigerate onions loose and they will keep as long as two months. They do, however, pass along their aroma to eggs, butter, and milk, so the refrigerator is not the best place for onions unless you've cleaned everything else out and gone on vacation.

Sprouted onions turn mushy, but you may still be able to use them in cooking. The sprout itself is a good substitute for a green onion.

A cut, raw onion should be wrapped tightly in plastic or sealed in a plastic bag and refrigerated. It will keep no more than two or three days.

If you have a batch of onions beginning to soften, use them up in cooking. Store cooked onions in a tightly lidded container up to five days.

Freezing Onions. Peel and chop the onions. Blanching is not necessary. Seal in plastic bags, label, and freeze. Onions keep ten to twelve months at 0° F. but lose flavor gradually during that time. After freezing, the texture of onions changes and they should be used only for cooking.

Often you want just a little onion as a seasoning, but slicing into a whole one seems wasteful. If you chop and freeze every little tag end of raw (or cooked) onion, you will always have a spoonful to perk up your food.

PICKLES AND PICKLED VEGETABLES AND FRUITS

It is the destiny of over half the cucumbers grown in the United States to end their days as pickles. Other vegetables and fruits, as different as chili peppers and watermelon rinds, get pickled too.

There are a number of methods and recipes for pickling vegetables and fruits. What they all have in common is vinegar, which acts as a preservative.

Buying Pickles and Pickled Vegetables and Fruits. The majority of pickled foods are processed, meaning heated and sealed in bottles, and therefore are as stable as other canned foods. These pickled foods lose color when they are stored too long, so avoid jars whose contents look grayish or abnormally pale.

Pickled cucumbers have the honor of being called simply "pickles." They are either dill, sour, or sweet. Dill pickles are flavored with dill, salt, and other spices. Kosher dills have garlic added. Dill pickles may be processed or they may be uncooked. Uncooked dills, called genuine dills, are sold from the refrigerator case in the market and must stay chilled. They are fresh as long as the liquid in the jar is clear, not cloudy. Refrigerated pickles usually have a "use by" date stamped on the label.

Cauliflower, green tomatoes, beets, onions, and chili peppers are other vegetables commonly pickled. Watermelon rinds take well to pickling. Chutney is a fruit-based pickled relish served with curried foods and roasted meat.

Storing Pickles and Pickled Vegetables and Fruits. These products can be stored up to one year on a cool, dry shelf. The exception to this is genuine dills, located in the grocer's refrigerator case. They must be stored in the refrigerator and used before the date stamped on the label.

After opening pickled vegetables and fruits, store them in the refrigerator. They should keep well at least one to two months. Discard them if the liquid becomes cloudy or if any scum develops on the surface. When they start to turn soft and mushy they have been stored too long.

OLIVES

Domestic olives are inexcusably dull compared to imports such as mellow, metallic yellow Calabrese olives from Italy, the wrinkled purple Kalamata from Greece, and Spain's tart green Queen olive.

Fresh-picked olives are bitter and virtually indigestible and must be cured to be edible, so olives of all kinds come to market soaking in oil or brine, packaged in cans or barrels. You can liven up plain old olives by draining off the brine and storing them for a few days covered in oil and garlic in a jar in the refrigerator.

Storing Olives. Unopened cans of olives keep up to one year on a cool, dry shelf. Once you have opened a can, store the olives completely submerged in their own liquid in a covered container in the refrigerator for one to two weeks. Olives bought in bulk and stored steeped in oil keep several months in the refrigerator. Discard olives that have turned soft.

SEAWEED

Nori, hijiki, kombu, Irish moss, grapestone, dulse, wakame, arame, agar, and laver are some of the dozens of kinds of seaweed you may find in oriental and health food stores. Seaweed is an alga which, almost without exception, is sold dried.

Seaweed should be treated like other dried foods. Seal it in a jar or plastic bag and store it in a cool, dry place. On the shelf, dried seaweed should last two to four months; refrigerated, up to six months.

After you have freshened dried seaweed in a bath of warm water, you can store it for another week in a covered container in the refrigerator. Try using the soak water to add flavor to soups or cooked vegetables.

BAMBOO SHOOTS

Bamboo is a giant grass plant that grows towering, woody shafts heavy enough to use in furniture making. But the plant is related to wheat, oats, and other familiar grains, and when bamboo is young and green it is edible.

Only very rarely are fresh bamboo shoots available in American markets. If you find them, refrigerate them in a plastic bag for no more than one week. Only the soft core is edible and it contains hydrocyanic acid when raw. Boil chunks of the fresh core for 5 minutes or until all traces of bitterness are gone.

More likely you will be buying bamboo shoots canned or dried. Store canned shoots up to one year on a cool, dry shelf. Once opened, cover them in water and refrigerate. If you regularly drain the shoots and cover them with fresh water, they will keep one month.

Dried bamboo shoots keep one year on the shelf in a tightly lidded container.

WATER CHESTNUTS

The most common fresh water chestnuts *(maa taai)* are round and woody and about the size of a cherry tomato. *Ling gok* look like little black bats. Only Chinese markets carry them year round, but they are not strictly Asian: water chestnuts grow wild on the banks of the Potomac.

Choose water chestnuts that are quite firm. Refrigerate them, unpeeled, up to one week. Canned water chestnuts are the *maa taai* variety and they can be refrigerated, covered in liquid, for one week after opening.

SUGAR CANE

Anyone with a sweet tooth can skip the frills and go right to the source: a plump chunk of cane dripping with sugary juices. You just pick it up and chew on it. In Louisiana, they sell long stalks of fresh-cut cane in roadside stands, but usually it will be found chopped into short lengths, prepackaged for the produce department of the supermarket. The moister it is the better, so look for cane from California that has been sealed with paraffin wax. Cane from Hawaii has to be sterilized with hot water, a process that leaches out some of the flavor. If you buy it fresh, you can dip the cut ends in paraffin yourself to keep it sweet. Sugar cane, stored in a plastic bag at room temperature will keep three to six months. It is fresh as long as you can see a little liquid bubble up when you squeeze it.

STORAGE TIMETABLE FOR FRESH VEGETABLES

Refrigerated Fresh Vegetables

Type of Vegetable	Days in Refrigerator (in perforated plastic bag unless noted)	Months in Freezer
ARTICHOKES	7	6–8
ARUGULA	3	—
ASPARAGUS	4–6	8–10
BAMBOO SHOOTS	7	N/R
BEANS, BROAD AND LIMA	2	10–12
BEANS, GREEN AND WAX	3–5	10–12
BEETS	7–10	10–12
BITTER MELON	5	N/R
BLACK-EYED PEAS	2	10–12
BOK CHOY	3–4	10–12
BROCCOLI	3–5	10–12
BRUSSELS SPROUTS	3–5	10–12
CABBAGE	7–14	10–12
CARROTS	7–14	10–12
CASSAVAS	2	N/R
CAULIFLOWER	4–7	10–12
CELERY	7–14	10–12
CELERY ROOT	2–3	N/R
CHICKPEAS	2	10–12
CHICORY	3–5	N/R
CHINESE CABBAGE	4–5	10–12
COOKED FRESH VEGETABLES	3–4	See each listing
CORN	1	10–12
CRANBERRY BEANS	1–2	10–12
CUCUMBERS	4–5	N/R
EGGPLANT	3–4	6–8
ENDIVE[1]	3–5	N/R
ESCAROLE	3–5	N/R
FENNEL	7–10	10–12
FIDDLEHEAD FERNS	1–2	N/R
GINGER[2]	14–21	1
GOBO ROOT	7–14	N/R
GREEN ONIONS	7–14	10–12
GREENS: Dandelion, Collard, Beet, Mustard, Turnip	1–2	10–12
HORSERADISH	14–21	N/R
JERUSALEM ARTICHOKES	7–14	N/R

Type of Vegetable	Days in Refrigerator (in perforated plastic bag unless noted)	Months in Freezer
JÍCAMAS	7–14	N/R
KALE	2–3	10–12
KOHLRABI	4–5	10–12
LEEKS	7–14	10–12
LETTUCE, ICEBERG	7–14	N/R
LETTUCE, LEAF	7–10	N/R
LETTUCE, ROMAINE	7–10	N/R
LETTUCE, BUTTERHEAD	3–4	N/R
LONG BEANS	3–5	10–12
MUSHROOMS[3]	4–5	10–12
NOPALITOS	7–14	N/R
OKRA	2–3	10–12
ONIONS: Sliced	2–3	10–12
PARSNIPS	7–10	10–12
PEAS	3–4	10–12
PEPPERS, GREEN AND CHILI	4–5	10–12
PEPPERS, SWEET RED AND YELLOW	2–3	10–12
RADISHES	10–14	N/R
RAPE	2–3	10–12
RUTABAGAS	7–14	8–10
SALSIFY	7–14	N/R
SAUERKRAUT, FRESH	7	N/R
SORREL	1–2	10–12
SOYBEANS	1–2	10–12
SPINACH	2–3	10–12
SPROUTS	7–10	10–12
SQUASH, SUMMER	4–5	10–12
SWISS CHARD	2–3	10–12
TAMARILLO	7–10	N/R
TARO	2–3	N/R
TOMATOES: Overripe or Sliced	1–2	2
TOFU	3–14	2
TRUFFLES[4]	2–14	N/R
TURNIPS	5–7	8–10
WATER CHESTNUTS	7	N/R
WATERCRESS	2–3	N/R

[1] Wrap Belgian endive in a damp paper towel before putting it in the plastic bag.

[2] Ginger submerged in sherry lasts 6 months in the refrigerator.

[3] Keep mushrooms in an open paper bag, loosely covered with a damp paper towel.

[4] Store truffles, covered in sherry, oil, rice, or goose fat, in a closed jar up to one month.

Pantry Vegetables

Type of Vegetable	Time in Pantry	Weeks at 55° F.
BEANS, DRIED	1 year	52
GARLIC	1 week to 2 months	4–12
LENTILS	1 year	52
ONIONS, GLOBE	2–3 weeks	8–12
ONIONS, SWEET	1–2 weeks	4–6
PEAS, DRIED	1 year	52
POTATOES	1–2 weeks	4–8
SEAWEED, DRIED	2–4 months	26
SQUASH, WINTER	1–2 weeks	2–4
SUGAR CANE	3–6 months	—
SWEET POTATOES	5–7 days	4–8
TOMATILLOS	1–2 weeks	2–3
TOMATOES: Ripe[1]	1–2	1

[1] Ideal temperatures for ripening and for storage are 55 to 72°.

STORAGE TIMETABLE FOR CANNED VEGETABLES

Type of Vegetable	Months In Pantry	Unsealed, Weeks in Refrigerator
ASPARAGUS: Canned	6	1
BEANS, GREEN: Canned	6	1
BEETS:		
Canned	6	1
Pickled, canned	6	1–2
HORSERADISH: Canned, prepared	12	12–16
PEPPERS, CHILI:		
Canned	6	1
Pickled	6	4–6
PICKLED VEGETABLES AND FRUITS	12	4–8
SAUERKRAUT: Canned	6	1–2
TOMATOES: Canned whole, sliced, puree, paste, sauce	6	1
ALL OTHER CANNED VEGETABLES	12	1

2

Fruits and Nuts

FRUITS

Perfect fruit is hard to find. The next time you bite into a sweet, succulent melon or peach, give thanks to each member of the relay team that raced it to your table without once dropping the baton. A judicious grower picked that piece of fruit when it was fully mature and not a moment sooner. A nimble-fingered packer tucked it into a box where its tender skin was safe from bumps and jostles. Long before heat and rot could do their worst, a shipper rushed the box to market. Finally, there in the last lap, you chose the right peach or honeydew from the pile, brought it home, and watched it with a motherly eye until it was fully ripe and heavy with juices.

But too many times a promising piece of fresh fruit is bland, woody, bruised, or sour. With so many variables that can go wrong, you can't sidestep every one of these disappointments. What you can do is learn enough about selecting and storing each kind of fruit so that most of the time the fruit you serve will be juicy and delicious.

Maturity. Fruit has a life cycle with an uphill side and a downhill side. On the up side, it goes through all of the growth and the changes that transform it from a blossom into the fruit it was meant to be.

When it has completely organized itself into a fruit and has attained its maximum size, it is mature and can begin to ripen. A fruit picked before this point in its life is stunted—it does not have the ability to develop full sweetness and flavor. Growers must harvest fruit *after* it is mature, otherwise it can never develop the desired taste.

To return to the example of the peach, this is a fruit whose maturity can be measured on the tree. The most important indicator is color. A peach whose overall color has turned from green to yellow is probably ready to pick. Growers may sample peaches using other tests of maturity—how they taste and how firmly the flesh clings to the pit, for instance. Of course even peaches on the same tree, let alone the whole orchard, will be in different stages of growth at any one time. Inevitably some percentage of the fruit is going to be picked before its time. Once the peaches reach the grocer's it is up to you to distinguish between mature and immature specimens.

A fruit that has matured has reached the peak of its life cycle. It immediately departs the uphill slope and starts on its way down the other side: ripening.

Ripening. Ripening is a complex set of chemical activities that happen inside the fruit after maturity. During ripening the color, texture, sweetness, and flavor of a fruit undergo change. A Bartlett pear will go from green, hard, and astringent to yellow, soft, and sweet—with peak flavor—and we call it ripe. The process does not stop there, however. In a day or two the pear will pass right through its desirable stage and continue to change, becoming brown, mushy, and less palatable, in a word, overripe, or, in another word, rotten.

Unripe fruits are green because they contain chlorophyll. The ripening process breaks down the chlorophyll and replaces it with other natural pigments—yellow-orange for apricots, red for cherries, and so on.

In most fruits, starches and acids present in newly mature fruit are converted into sugar as time passes. As a result, ripe fruit is softer and sweeter than unripe. Certain fruits, such as cantaloupes, peaches, and pineapples, have no reserve of starch to make into sugar. These fruits contain all the sugar they are ever going to have by the time they are fully mature, but they do get softer and juicier.

Other by-products of ripening's chemical activities are the organic compounds that give fruits their characteristic flavor. An orange, for instance, contains about 100 different substances that accumulate during ripening, coming together to create the taste of orange. A fruit ripened on the tree manufactures these substances slightly differently from a fruit that ripens after harvest. For this reason tree-ripened fruit usually tastes best.

All fruits ripen when they are left on the tree or on the vine. And many fruits, including apples, bananas, plums, persimmons, avocados, melons, and apricots—the list is long—can continue to improve after picking. Growers harvest these fruits before they are ripe. They can control the subsequent ripening of the fruit in two ways—they can stop it or start it. Storing fruit at a low temperature or in a low-oxygen atmosphere stops ripening because, in either of those conditions, fruit respires very, very slowly. Fruits always "breathe"—they take in oxygen and give off carbon dioxide. Just before they begin the processes of ripening they start to respire more rapidly and to produce a gas called ethylene. This gas appears to trigger all the chemical functions that result in ripe fruit. Growers start or accelerate ripening by releasing the chemical trigger, ethylene gas, into the air.

You can apply these same principles to the fruit you bring home from the market: refrigerate fully ripe fruit to discourage it from getting over-ripe; or ripen fruit quickly by enclosing it so that the air in the container becomes saturated with natural ethylene gas. More on these techniques later, in "Storing Fresh Fruit," p. 73.

There are a number of fruits that do not continue to ripen after harvest, notably berries and grapes. Because they are picked ripe, they are already soft and vulnerable to bruising when they are crated and shipped. Consequently this type of fruit has a very short storage life and must be speeded to market.

Buying Fresh Fruit

Above all, what you want to know when you buy a piece of fruit is whether or not it was picked *after* it reached maturity. Fruit harvested too early can never ripen, can never be sweet and juicy. Color is the best indicator of maturity and green is often a bad sign, though there are a number of exceptions, some of them obvious: green varieties of apples, melons, grapes, and gooseberries; certain types of nectarines and mangoes; citrus fruits; Kadota figs; greengage plums; and, of course, bananas, which can ripen even if you buy them green as grass. But, with those exceptions, green coloring means immaturity while bright, intense yellows, oranges, and reds are signs of good quality.

Fruit does not have to look picture perfect to be good to eat, though. There are common imperfections that have no effect whatever on the flavor of certain fruits. An example is "scald," the rough, tan patches often seen on the skin of apples. And citrus fruits often bear superficial scars and have dark, mottled patches, neither of which detract from their good quality.

With the exception of those fruits that must be picked and shipped already ripe, most arrive at the market fairly unripe and still quite hard. In a perfect world, this would present no problem at all, because you could buy the fruit, ripen it at home, and eat it at its peak. But no matter how expert you are at identifying mature fruit, and no matter how reliable your grocer, you are bound to end up buying some fruit that has all the appearance of good quality yet will not ripen satisfactorily. The best way to avoid these mistakes is to buy fruit that has already begun to ripen, if only a little. Pick up the fruit—don't pinch—and feel its overall texture. It can be firm but should not be truly hard. And use your nose. A trace of sweet scent is almost a sure sign of beginning ripeness and good quality.

Worse than underripeness is overripeness. An underripe fruit may very well develop perfectly, but overripe fruit is mostly irretrievable. Fruit that is very soft, beginning to darken, and leaking juices may still have some edible parts, but in general it is not a good choice, even sold at reduced prices.

Needless to say, you want to avoid fruit suffering from mold, decay, and mechanical damage. Examine fruit for breaks in the skin, bruised spots, fuzzy patches of mold, and off odors before you buy.

It has become common practice to prepackage fruits for retail sale. The packaging makes it difficult to see or feel and so to judge the fruit you are buying. Whenever possible, shop where fruit is sold in bulk. By choosing fruit piece by piece, you are much more likely to get your money's worth.

Keep in mind that fruits are seasonal. They are harvested in abundance only at certain times of year. You can find strawberries in December, but buying them out of season is much chancier than buying them in May. Out-of-season fruit usually travels over longer distances to get to market and so has been stored and shipped for longer times before it gets to you.

It is not only easier to find high-quality fruits when they are in season, but you can also choose among many varieties of a particular fruit. Some supermarkets stock only three kinds of apples or two kinds of plums, fruits chosen because they survive shipping well, not because they taste the best. If you are just not satisfied with the fruit where you regularly shop, search around for other outlets—greengrocers, farm stands, ethnic markets, and even wholesalers in your area.

The really formidable part of buying fruit is deciding on the right quantity. In midsummer when bushels of bright, sweet fruits at low prices confront you in the supermarket, the temptation is to load the cart with pounds of peaches, melons, and nectarines and worry about what to do with them all later. Unfortunately many of these fruits have a very short storage life, and some will rot before you have the chance to eat them. There are two solutions to the problem. You can either keep your purchases small—only what you can eat within a few days—or you can make plans to preserve the fruits in some way—canning, drying, juicing, freezing, making jelly, or pickling. Instructions on freezing are covered on pp. 74–76 and in the sections on each kind of fruit.

Storing Fresh Fruit

Sort through fruit before you store it. Discard any individual pieces that are decayed or severely damaged. This is a particularly appropriate procedure for fruit bought preboxed or prebagged, like berries and apples.

For convenience, you can wash and dry most fruits before you put them away, to remove dirt and insecticide residues. Citrus fruit and apples frequently have a coating of edible wax that can be scrubbed off with a soft brush, dish detergent, and hot water. Don't, however, wash berries and cherries until you are ready to serve them—too much moisture will hasten decay.

Always handle fruit as gently as possible. Injuries from bruising, cuts, and crushing can make the fruit both unattractive and prone to rot.

Unripe fruits should be left to ripen at room temperature, away from direct sunlight. Keep them stacked loosely in a basket that allows air circulation. Fully ripe fruit needs to be refrigerated.

There is a good method for speeding up ripening that relies on naturally produced ethylene gas to accelerate the process. Fruit stored in a high concentration of the gas will ripen faster than normal. Pack dry, unripe fruit very loosely in a brown paper bag poked full of small holes. Close up the bag and set it on a cool, dry shelf. The bag will hold in the ethylene gas the fruit emits; the holes will let oxygen in and carbon dioxide and moisture out. When the fruit going into the bag is very hard, you can slip in an already ripe apple or banana, which will give off an extra dose of gas. You must check inside the bag each day and remove ripe fruits and ones that show decay or mold.

This provides a very practical way of storing larger quantities of fruit. For example, 5 pounds of unripe bananas left alone on the shelf will ripen at about the same time, and you will have to think of ways of using them up all at once. If you force the ripening of some and leave the rest on their own, you will get batches of ripe fruit in more manageable quantities over a stretch of time.

There are perforated plastic bowls on the market, called fruit ripeners, designed to work the same way as the paper bag, and they look nice while they are doing the job. If you use one, be sure to wash and dry it regularly, so that invisible molds and bacteria are not transferred from fruit to fruit.

Fruit that's past its prime can be poached in a light sugar syrup spiced with white wine, stick cinnamon, or lemon peel.

One hard and fast rule is that fruits and fruit juices should not be stored in galvanized (zinc-coated) containers because acids in the fruit will dissolve the zinc, which is poisonous. Iron, tin, or chipped enamel containers may impart a metallic taste to fruits and fruit juices.

Freezing Fresh Fruit

The fruit you freeze should be ripe but firm. Wash the fruit in cold water, being careful not to bruise it. Fruit such as berries can be put in a colander, then dunked in cold water for the gentlest washing.

Prepare the fruit as you would for serving—pit it, peel it, or slice it if necessary. Cut away the damaged portions of less than perfect fruit, then crush or puree the remainder for the freezer.

Most fruits have very fragile tissues that rupture when frozen. The fruit becomes mushy and the juices leak out. You can prevent this by packing the more delicate fruits, such as melons and pears, in sugar or sugar syrup. The fruit cells will still rupture but they absorb the sweetener and stay plump and solid.

There is evidence that sugar acts on many fruits to preserve the natural flavor components that are otherwise lost during freezing. A sweetened peach, when thawed, tastes much more like a fresh peach than one that was frozen bare. But apples, all berries (except strawberries), figs, pineapple, plums, pomegranate seeds, and rhubarb do just as well in the freezer without sugar.

If you object to sugar, you can choose to freeze any fruit without it. After thawing, the unsweetened fruit will not be as attractive for eating as is but will certainly be suitable for cooking.

Some fruits, like apples and pears, will darken when frozen because of oxidation. They will keep their color if they are treated beforehand with either ascorbic acid or citric acid. Ascorbic and citric acid in crystalline form may be found in drugstores, and grocery stores sometimes carry packaged mixes of ascorbic acid and sugar (follow the directions on the label). Lemon juice is a natural combination of ascorbic and citric acid, but it may mask the flavor of the fruit you are freezing.

Here are the ways different fruits may be treated for the freezer:

• PLAIN PACK. Fruit is prepared and packed without sweeteners. If necessary, sprinkle with a solution of 1 teaspoon ascorbic acid per 1 tablespoon cold water and stir gently to coat each piece before packing.

Certain fruits such as raspberries may be tray-frozen before packing—spread pieces of fruit in a single layer on a baking sheet and freeze them hard. When packed, they won't stick together.

Seal plain-packed fruits in plastic bags or airtight plastic or glass containers. Leave 1/2 inch headroom in rigid containers.

• SUGAR PACK. Sprinkle sugar over prepared fruit. Stir gently with a wooden spoon or rubber spatula until the sugar dissolves in the fruit's own juices, coating each piece. If necessary, sprinkle with a solution of 1 teaspoon ascorbic acid dissolved per 1 tablespoon water just before you add the sugar.

You can tray-freeze certain sugar-packed fruits.

Seal sugar-packed fruits in plastic bags or airtight plastic or glass containers. Leave 1/2 inch headroom in rigid containers.

• SYRUP PACK. Make a syrup by dissolving sugar in very hot water in one of the following concentrations, depending on the fruit and on the sweetness desired:

$$30\% \text{ syrup} = 4 \text{ cups water} + 2 \text{ cups sugar}$$
$$35\% \text{ syrup} = 4 \text{ cups water} + 2^{1/2} \text{ cups sugar}$$
$$40\% \text{ syrup} = 4 \text{ cups water} + 3 \text{ cups sugar}$$
$$50\% \text{ syrup} = 4 \text{ cups water} + 4^{3/4} \text{ cups sugar}$$

Chill the syrup thoroughly for use. You need about 1/2 to 2/3 cups syrup for each 2 cups fruit.

If required, mix ascorbic acid into the syrup just before you pack the fruit. Stir it gently so that no air bubbles get into the liquid.

Seal syrup-packed fruit in rigid, airtight plastic or glass containers, leaving 1/2 inch headroom.

Pieces of fruit tend to float to the top of the syrup, where they can dry out during freezing. To prevent this, place a small piece of crumpled waxed paper in the air space between the syrup and the lid to keep the fruit submerged.

The lids must make an airtight seal. Glass containers need a rubber or plastic ring, either separate or fused to the lid, to be airtight. If you have doubts about the seal on the container, secure it with freezer tape.

• PUREES. Raw or cooked fruit may be pureed in a food mill or food processor before freezing. Or you can simply crush the fruit coarsely. Add sugar or not, depending on how you plan to use the fruit. To prevent darkening, you can sprinkle ascorbic acid directly into the puree and mix. Pack in rigid containers lined with plastic bags. When the fruit is frozen hard, you can pull out the sealed bag and reuse the container.

For the best eating, only partially thaw whole, cubed, or sliced frozen fruit. With a few ice crystals left, the fruit is bracingly cold and springy and ideal for serving.

Your success in freezing fruit will depend a lot on the container you choose. Plastic bags must be vaporproof and moistureproof, the type specifically labeled "freezer bags." The extra-thick bags used with home heat-sealing devices are ideal and of course can be closed even more securely than other bags. Rigid containers should be freezer glass or heavy plastic with wide mouths and very tight-fitting lids. See pp. 349–52 for more on storage containers.

Buying and Storing Canned and Frozen Fruits

Both grades and brand names can be guides to the quality of the fruit inside cans, jars, and freezer cartons. Grade A or Fancy fruits have the best overall appearance and have been packed at the proper stage of ripeness. Grade B fruits, most widely available, are less perfect but still good eating. If the fruit is not graded, a sampling of different brands can guide you to the fruit that suits your own tastes.

Canned fruit may be packed in sugar syrup or in its own unsweetened juice. Choose cans that are in good shape, with no bulging, swelling, rust, or deep dents. Jars should be totally sealed and have no signs of leakage.

Most frozen fruits have sugar added. The packages should feel completely solid when you pick them up; don't buy packages with stains or leaks—the contents have probably thawed and refrozen.

Most canned fruit keeps its quality for one year when stored on a cool (no more than 72° F.), dry shelf. There are fruits that deteriorate more quickly, however: store canned citrus fruits, juices, and mixed fruit salads for no more than six months.

Frozen fruits can thaw very rapidly. Put them in your grocery cart last and take them home as soon as you can. Stored in a 0° F. freezer, they will keep well up to a year.

APPLES

It is difficult somehow to view an apple as a fruit in transition. It seems to have a more fixed identity than a fruit that turns, for instance, from a hard green golf ball into a blushing orange apricot. But though its transformation is less distinct, an apple goes through its rise and decline like every other fruit.

An apple's sturdy flesh does stand up well to storage: virtually all apples grown in the United States are picked between July and November, but suppliers hold back enough of the harvest to keep shipping year round. Usually apples are stored in a cold or in a low-oxygen atmosphere —both conditions slow the ripening considerably.

Nevertheless, apples do continue to ripen in storage and they may become overripe in the short time it takes to get them to market.

Buying Fresh Apples. The best season for buying apples is the fall. Because they store so well, you can buy very good apples almost every day of the year, but the greatest abundance and variety come on the market in the cool-weather months when flavor and texture are best.

Pick the type of apple that suits your purposes. Some are good for baking; some freeze well and some don't; some taste best for just biting into. Here are the most widely distributed varieties of apples and how to use them:

• CORTLAND. Deep red skin and snowy white flesh make this a handsome apple. Its flesh is slow to darken, so it is ideal for salads. A Cortland is sweet, a little bland, but crisp. It works in all kinds of recipes and requires less cooking time than other apples.

• CRAB. Tiny red apples, much too sour to eat out of hand, they make delicious jelly or condiments when cooked.

• EMPIRE. Offspring of the McIntosh, these are very crisp, sweetly scented apples with a pattern of stripes on the red, rather thick skin. They are versatile, but lack the acidic bite desirable in some recipes.

• GOLDEN DELICIOUS. These golden-yellow apples are universally available because of their long storage life and resistance to bruising. Best suited to eating out of hand, they may also be used in cooking. They taste sweet and mild, without any tang.

• GRANNY SMITH. These are large, bright green apples usually imported from the Southern Hemisphere and therefore at their peak in our summer, when it is winter at the bottom of the earth. Good keepers, Granny Smiths are thick-skinned, tart, and very juicy. They are excellent for any purpose and keep well too.

• GRAVENSTEIN. An apple grown mainly to make canned applesauce, the Gravenstein has limited distribution and stores poorly. It is an all-purpose apple that is pleasantly tart when eaten raw.

• IDARED. A large, mild apple, the Idared suits any purpose.

• JONATHAN. The vivid red Jonathan is small, crisp, and tender. It has a rich flavor good for eating out of hand and for all cooking except baking. These apples develop soft spots very readily—check them over one by one even if they are prebagged.

• MACOUN. The purplish-red Macoun (it sensibly rhymes with maroon) does not keep well enough to become a major crop though it is an excellent eating apple, fine-grained and juicy.

• MCINTOSH. The McIntosh has a red and green skin and soft but not mealy flesh. Best eaten raw, it is a nice balance of sweet and tart. It does not store as well as most other varieties and bruises more easily. Be particularly careful to check for soft spots on McIntosh apples.

• NEWTOWN PIPPIN. A yellowish-green apple good for cooked desserts, the Newtown is firm and juicy when eaten as is.

• NORTHERN SPY. The Northern Spy stands up well to all kinds of cooking and baking. It is also a contender for the best-loved eating apple because of its bright flavor and firm texture.

• RED DELICIOUS. This is the most heavily produced apple in the United States. Unfortunately it does not compare well with other apples for lively flavor and crispness. The Red Delicious is for eating fresh, not for cooking or freezing.

• RHODE ISLAND GREENING. A good keeper, the Rhode Island Greening is the apple of choice for making pies and applesauce. This variety develops flavor and holds its texture as it cooks. You also can eat it raw as long as it is fully ripe.

• ROME BEAUTY. Often specified as the best apple for baking, this large round apple is good for all cooking purposes and tastes good raw as well. Its skin has a yellow or green cast and is mottled with red.

• WINESAP. A purplish-red aromatic apple, the Winesap's sharp, rich flavor makes it a good all-purpose apple. It holds up particularly well in storage.

• YORK IMPERIAL. The York Imperial is another great baking apple that keeps its shape (which is naturally lopsided) during long cooking. It is another middle-of-the-road apple, not too sweet and not too tart. The skin—pinkish red on a yellow background—may have pale spots or russeting, neither of which affects the eating quality.

Mature apples have intense coloring for their variety, whether it is yellow, red, green, or all three. Washed-out coloring and shriveled skin mean the apple was picked too soon.

Don't pay too much attention to the luster on the apple—if you are buying it at a store rather than a farm stand, the glow on the apple is probably wax. Producers routinely coat apples with edible wax to make them look inviting. The wax does very little to protect the apple from moisture loss, so it is really just a cosmetic. You can eat the wax or scrub it off, using a soft brush, warm water, and detergent.

Avoid apples with punctures or cuts in the skin. Other defects on the skin, pale or tan flecks, do not detract from the apple's eating quality.

An overripe apple has soft, mealy flesh. The apples you buy should feel very firm and not yield to finger pressure. Both Red and Golden Delicious apples get overripe in only a few days if stored at room temperature.

Check every apple you buy for bruises—small depressions in the skin, usually, but not always, soft and discolored. All apples are susceptible to this kind of injury. Bruising affects the apple inside and out, creating darkened patches of dry, grainy flesh.

Storing Fresh Apples. All apples ripen very quickly at room temperature. If the apples you buy are too firm or too sour, you may want to leave them out for one or two days. Once an apple is perfectly ripe, it should be stored in the refrigerator, preferably in a plastic bag. If you wash the apples beforehand, dry them well before storing.

An apple stays in peak condition in the refrigerator two to four weeks. When it begins to soften, it is better for cooking than for eating raw.

If you have a place in your home that stays between 32° F. and 45° F. during the winter, you can keep apples for several months. This only works if you start with perfect apples and keep them above freezing at all times. Wrap each apple in crumpled tissue or newspaper and pack it in a large container, such as a plastic garbage can, with a tight-fitting lid. Don't open the container too often, but when you do, pull out any apples that have begun to soften or rot. Apples can keep anywhere from two to six months this way, depending on their variety, the temperature, and the humidity. The closer to 32° F. the better they will do.

Freezing Fresh Apples. Wash, peel, core, and slice apples for freezing. To keep them from darkening while you prepare a whole batch, coat slices with a solution of 1/2 teaspoon ascorbic acid per 3 tablespoons cold water. You can also soak the slices in a brine of 1 gallon cold water and 2 tablespoons salt. A third antidarkening method is to steam-blanch the apples for 1 1/2 minutes before freezing.

The simplest way to freeze apple slices is to use the Plain Pack. Simply tamp the apples firmly into a rigid container, leaving 1/2 inch head space, and freeze. These apples can be seasoned when they are thawed and cooked. A shortcut for piemaking is to line a pie pan with plastic or foil, pack it with apple slices, wrap the apples, and freeze them. The pie-shaped package of apples can be removed from the pan and stacked in the freezer until needed. To make the pie, take off the wrap, put the disk of frozen apples directly into the pie dough in the pan, add the sugar and spices, and bake.

When you plan to serve the apples uncooked, freeze them in 40 percent syrup (see p. 75).

Apples keep ten to twelve months at 0° F.

Canned Apples. Canned apples may be found plain, sweetened, or prepared as pie filling. On the whole, these products taste all right but, since fresh apples are so easy to come by, there is not much point in foregoing the vitamins and minerals that are lost in processing.

Canned applesauce is a convenience food often of very high quality. It is available sweetened and unsweetened. You can judge applesauce packed in jars by the texture, which should be fine, and the amount of separation, which should be minimal.

Canned apple products can be stored on a cool, dry shelf up to twelve months. Once opened, they keep as long as two weeks if stored well covered in the refrigerator.

APRICOTS

Buying fresh apricots is a shot in the dark. More than most fruits, the full-flavored apricot depends on tree ripening. Unfortunately, tree-ripened apricots are so soft that they can't be successfully shipped even short distances. So, unless you live in Santa Clara County, California, or thereabouts (where 96 percent of all U.S. apricots grow), the best you can hope for is a fresh apricot harvested at the right degree of maturity. Ripened off the tree, these fruits will be soft and sweet but very bland. If you crave the incomparable velvety tang of a perfect apricot and don't live in the shade of an apricot orchard, perhaps the closest you can come is to buy the fruit dried. Dried apricots (discussed on p. 116) are richly flavored both raw and cooked in desserts and sauces.

Buying Fresh Apricots. June and July are the peak months for fresh apricots. Look for golden-orange fruit with no greenish tint, and don't settle for pale yellow. Be sure the flesh has a little give—this first sign of ripening is your only assurance that the apricot will ripen at all.

Storing Fresh Apricots. Keep apricots at room temperature until they are fully ripe—soft but not mushy. When they are ripe, keep them in the refrigerator in a plastic bag for no more than two or three days.

Freezing Fresh Apricots. Wash ripe apricots and remove the pit. You can peel them; drop them into boiling water for 30 seconds to soften the skin, if necessary.

For best results, pack apricots in 40 percent syrup (see p. 75) with ¾ teaspoon ascorbic acid added per quart. They keep ten to twelve months at 0° F.

Canned Apricots. Apricots picked for canning are usually left to ripen on the tree a little longer than those to be sold fresh. Consequently, in many cases canned apricots are an improvement over fresh. They are also very close to equivalent in nutritional value.

Canned apricots keep up to twelve months on a cool, dry shelf.

ASIAN PEAR

You may encounter fruits of different shapes and colors—round or pearlike, green or russet—all called Asian pears or for that matter called apple pears or Chinese pears. What they have in common is mild flavor, crisp, nicely grained texture, and a lot of juice. Treat them as you would delicate apples.

AVOCADOS

As the avocado tree spread from its birthplace in Latin America to other parts of the world its eccentric fruit acquired a cargo of common names: those that were complimentary—custard apple, butter pear; a couple that were merely descriptive—alligator pear and shell pear; and one that altogether dodges responsibility—"pears which are unlike pears."

All this confusion is understandable. The leather-skinned avocado is a fruit that tastes like a vegetable. Its smooth yellow-green flesh ripens perfectly off the tree, so almost any avocado you buy can be creamy and delicious if handled properly. There are two types of avocado you are most likely to find: the Hass, which is plump and has pebbly skin that turns black as the fruit ripens; and the Fuerte, which has thin green skin and a pronounced pear shape. Hass avocados are by far the best-tasting. They are available all year, but the greatest supply comes in late winter and early spring.

An avocado is ripe when it is soft enough to yield to gentle pressure. If a light prod of your finger leaves a dent in the fruit or if the skin seems too big for the flesh the avocado is overripe.

Buying Avocados. Most avocados come to market quite hard, and so resistant to bruises. As long as you don't need to eat it immediately, a firm but not rock-hard avocado is a good choice. Look for fruit that feels heavy for its size. Examine the avocado for the soft, dark spots that are signs of bruising and damaged flesh. Tan patches on the skin, on the other hand, have no effect on the quality of the fruit.

To pick a ripe or nearly ripe fruit, test for softness with a gentle poke at the top of the stem end—that way you are less apt to leave a bruise.

Storing Avocados. Leave avocados at room temperature to ripen. Uncovered, very firm fruit should soften in 3 to 6 days. You can speed up ripening by wrapping an avocado in foil or brown paper. A ripening bowl works too. Never store unripe avocados in the refrigerator—they will never soften properly if they are chilled.

As soon as they are fully ripe, store avocados in the vegetable drawer in the refrigerator. They should last ten days to two weeks.

When you are using only a part of an avocado, you can store the rest for several days. Leave the skin on and the pit in the fruit and brush the cut side with lemon or lime juice to keep the flesh from darkening. Wrap the half avocado in plastic film and refrigerate.

Freezing Avocados. Puree avocados for freezing. Mix in 1 tablespoon lemon juice for every 2 avocados. Pack in rigid containers with 1/2 inch head space, seal, label, and freeze. Avocados keep three to six months at 0° F.

BANANAS

Bananas are the most popular fresh fruit in the United States. Air-conditioned ships, some bearing as many as 5 million pounds of bananas, bring the fruit up from Latin America every month of the year.

Perhaps bananas are so well loved because they are the only fruit that tastes better ripened *off* the tree than on. However green they are when picked, they will eventually become yellow and sweet to eat. It is nearly unheard of to bite into a bad banana.

The beloved banana takes on new charm when you learn its growers' vocabulary: 8 or more bananas attached at the stem are a "hand"; 3 or more is a "cluster"; each piece of fruit is a "finger."

Buying Bananas. You can choose bananas in any stage of ripeness—they can be totally green, green just at the tips, or yellow flecked with brown—without going wrong. Even overripe, brown bananas can be put to good use (see "Storing Bananas"). There is an advantage to buying bananas that are either green or yellow without any brown because these are firmer fruits, less likely to bruise as you transport them home.

Look for plump rather than skinny bananas that still have some stem attached. (When the stem is removed the banana may decay before it is right for eating.) Don't buy fruit with bruises—dark, soft depressions in the skin. Don't buy bananas that look dull and gray—they have probably been chilled or overheated during storage.

Red bananas from Honduras and Haiti occasionally turn up in urban markets. They are often larger than yellow bananas, and they tend to ripen more slowly, but they are virtually identical in flavor and texture. As red bananas ripen, the skin color deepens and the fruit yields to gentle finger pressure.

Storing Bananas. Bananas do best simply left in the open at room temperature. Keep them away from high heat and direct sunlight, but otherwise they require no care at all.

Once a banana has reached the perfect stage of ripeness, you can put it in the refrigerator. The skin will get dark but the flesh will be tasty several more days.

When not using the whole banana, slice off what you need and leave the rest in its skin; wrap the remainder and refrigerate. The cut end will darken but otherwise the fruit should keep four or five days.

If your store has only green bananas and you need ripe ones, you can speed up ripening by putting them in a paper bag or a fruit-ripening bowl. You may find that the banana's skin color is no longer a perfect measure of ripeness, though. Quick ripening tends to leave the peel a bit green even when the flesh is completely soft and sweet.

Everyone has his own opinion about when a banana is just right for eating out of hand. When it is yellow all over and the skin is freckled with brown spots, the fruit has gotten as sweet as it ever will. But some people like the firmer flesh of a slightly underripe banana, one that still shows green at the tips. A very ripe banana usually has a lot of brown on the skin and its flesh has begun to darken. At this stage the banana is ideal for most cooking purposes.

Slightly underripe, firmer bananas are best if you plan to bake, sauté, or broil the fruit whole. Use ripe and very ripe bananas for baking quick breads and cake.

Freezing Bananas. Bananas do just fine in the freezer, as long as you plan to use them in cooking when they are thawed. Puree the fruit and pack them with sugar and ascorbic acid. Use 1/3 cup sugar and 1/4 teaspoon ascorbic acid for each cup. Label the bananas with the amount of sugar used so you can adjust your recipes accordingly. Mashed bananas freeze well without sugar too—just add 1 tablespoon lemon juice or 1/8 teaspoon ascorbic acid per cup. Seal the puree in a rigid container, leaving 1/2 inch head space.

Bananas keep up to three months at 0° F.

BLACKBERRIES AND RASPBERRIES

Raspberries, blackberries, and their kin—boysenberries, loganberries, dewberries, youngberries—are so fragile you could almost crush them with a look.

They must be ripened on the bush and picked when they are already soft, then hurried to market before they spoil.

Buying Fresh Raspberries and Blackberries. First assure yourself that the berries were mature and ripe when harvested. The tiny seed-bearing spheres that make up each berry are called drupelets. Green, pale, or off-color drupelets are a sign of immaturity. If the berries still have green "caps" of leaves, they were picked too soon.

You also want to avoid overripe berries. They look dull and mushy. Mold on the fruit is evidence that it has been stored too long, as well.

Leaky berries are either overripe or damaged. Staining on the carton means the berries have been leaking juices.

What you want are bright, clean, dry berries that are plump and deeply colored. Berries really should look picture perfect when you buy them. Their peak season is in June and July but they are readily available in May and August, and there is a fall crop as well.

A pint of berries serves 2 to 4 people.

Storing Fresh Raspberries and Blackberries. Sort through the berries and discard any crushed or moldy ones. Because they crush so easily, raspberries and blackberries should be stored in a single layer rather than dumped together in a carton. Keep the berries uncovered on the refrigerator shelf—not in the vegetable drawer—and they should last one or two days. Do not wash them until you are ready to eat them.

Freezing Fresh Raspberries and Blackberries. It is easy to freeze these berries and the thawed fruit is still appetizing if not quite as delectable as fresh.

If you plan to use them for cooking, simply spread washed berries on a tray and freeze them individually. Pack the frozen fruit in sealed plastic bags. You can then take out berries as you need them and reseal the bag.

Soft berries also freeze very well covered in a 40 percent sugar syrup (see p. 75). Pack them in rigid, airtight containers and leave 1/2 inch head space. This is the best method if you want to serve the thawed berries without cooking.

At 0° F., berries keep ten to twelve months.

BLUEBERRIES

Blueberries are fairly sturdy but, like softer berries, must ripen completely before they are picked. They are just firm enough to make shipping and storing them a bit easier. They are still very perishable, though, and need to be handled with a light touch to keep them fresh as long as possible.

Blueberries have a place in folk medicine, especially in Russia, as a preventive and cure for flux and other abdominal misfortunes. In summer, when they are in season, feel free to put aside the cod liver oil and set your stomach to rights with fresh blueberries.

Buying Blueberries. Blueberries may be light or dark blue, but they will be completely blue, with no reddish tinge, if they are fully mature. The silvery sheen on blueberries is a natural protective coating you want to see. Overripe berries are soft, watery, and often moldy. The season for blueberries is June, July, and August.

Blueberries should be clean and very dry—too much moisture causes early decay. Avoid berries that look dull, shriveled, or sticky.

Storing Blueberries. Keep blueberries refrigerated all the time. Store them, unwashed, in a rigid container covered with plastic wrap and they can last up to two weeks.

Freezing Blueberries. Blueberries seem designed to be frozen. They come out of the freezer only a little less bright and juicy than when they went in.

If you plan to use them in cooking, pack the washed berries in rigid containers, leaving 1/2 inch head space and freeze them as is. For serving uncooked, pack blueberries in 40 percent syrup (see p. 75),and seal in rigid containers.

Blueberries keep well up to one year at 0° F.

NOTE: *The storage and freezing instructions for blueberries also apply to huckleberries and elderberries.*

CANTALOUPES

Pyramids of aromatic melons piled high in grocery stores and fruit stands mean that summer has truly arrived. The khaki-colored cantaloupes, in great abundance in June, July, and August, are inexpensive and delicious, especially those grown in California's hot, dry Imperial Valley, so like the cantaloupe's homeland in the deserts of Persia.

When they ripen off the vine, cantaloupes don't get any sweeter than they were at harvest, but they do soften and develop flavor. A mature melon, given half a chance, can almost always become a luscious fruit under your care.

Buying Cantaloupes. Cantaloupes, more than most summer fruits, give off many signals that they are good and ready to eat.

As a cantaloupe matures, a layer of specialized cells grows between the stem and the fruit. This layer makes it possible to gently pluck the melon from the vine, leaving a clean break. You can easily identify a mature cantaloupe in the market by looking at its stem end—there will be a completely smooth, circular indentation with no ragged edges.

Buy cantaloupes that are at least 5 inches in diameter and well shaped, without dents or flattened sides. They should not be bruised, cracked, shriveled, or show any mold.

The skin of a mature cantaloupe also has a very distinct texture that looks like thick netting. Any break in the skin is an opening for decay organisms. Avoid melons with leathery, off-color patches of skin because the defect probably penetrates into the flesh as well.

You probably won't find a perfectly ripe melon at the store, but you can look for one that is well on its way. The skin beneath the coarse netting should be more yellow than green and the melon should have a sweet scent. Overripe melons feel flabby and smell too sweet.

Average cantaloupes yield 2 to 3 servings per melon.

Storing Cantaloupes. Most cantaloupes benefit from two to four days of ripening at home. Leave them at room temperature, away from heat and direct sunlight. And treat them gently—they are not as tough as they look.

A cantaloupe is ripe when it is yellow, fragrant, and feels springy if you press it between your palms. Pressing the end of the melon with your thumb is not a reliable test of texture because the end is more likely to be spongy from decay or injury. Seal a ripe cantaloupe in a plastic bag and store it in the warmest part of the refrigerator. (The bag keeps the melon from exchanging odors with other foods.)

Store cut melons in the refrigerator, seeds in, skin on, also very tightly wrapped.

A whole ripe cantaloupe keeps a week to ten days in the refrigerator. Cut cantaloupe should be eaten in one or two days.

Freezing Cantaloupes. Seed the cantaloupe, then either cut it into cubes or scoop it out with a melon baller. Pack the pieces in a rigid container and cover them with a 30 percent syrup (see p. 75), leaving 1/2 inch head space. Store frozen cantaloupe ten to twelve months at 0° F.

The thawed melon is very soft. You should serve it only partially thawed for better texture. This is also true of commercially frozen melon balls.

CASABAS

The casaba is a large, round melon that comes to a sharp point at the blossom end. Its golden rind is deeply furrowed. The best casabas are on the market in September and October.

Buying Casabas. The color of the melon changes from green to yellow as it ripens. (Avoid casabas that are neon yellow because they have probably been dyed, making it difficult to judge ripeness.) Another sign of ripening is a slight softening at the blossom end, opposite the stem end. Don't buy a spongy melon, though. It should be firm and have no cuts or bruises on the rind. Casabas, by the way, do not give off a scent even when they are perfectly ripe, so you must rely solely on the feel of the melon.

Storing Casabas. Leave underripe melons out at room temperature for two to four days. When they are golden and yield slightly to the pressure of your palms, refrigerate them seven to ten days. Wrap and store cut melon no more than two days.

Freezing Casabas. Follow directions for freezing cantaloupe.

CHERIMOYA

The cherimoya is vaguely heart-shaped with a broad fish-scale pattern on its lime-green skin. Chilled, the white pulp tastes like strawberry-banana-pineapple sherbet. Ripe cherimoya are uniformly green and yield when pressed between your palms. Though a few brown patches are acceptable, overall browning means the fruit is overripe. The unripe fruit is durable enough to ship well and is marketed in California and a few major urban centers in winter and early spring. Once ripe, refrigerate no more than one or two days.

CHERRIES

Mahogany-red cherries dangling like small lustrous bells from their stems are as tantalizing as a fruit can be. Their deep gloss and sweet meat are ready right from the tree—most cherries you find have been picked at full ripeness and maturity.

Buying Fresh Cherries. Sweet cherries for eating out of hand and sour cherries for cooking are in season from May through July and sometimes into August.

Color is the best guide to eating quality. Very dark cherries are the sweetest: they may be dark red, purple, or even black. Varieties of sour cherries are lighter red but should have bright, uniform coloring.

Cherries should be clean, plump, firm, and shiny. Buy cherries with stems attached because once they are removed the fruit is open to decay organisms. Avoid fruit with cuts, bruises, or stale, dry stems. Pick up and examine cherries a handful at a time. Because the fruit is so dark, discoloration and leaking are otherwise difficult to spot.

Storing Fresh Cherries. Cherries need cold temperatures and high humidity but should not be wet. Pour them unwashed into plastic bags and keep them in the refrigerator. Cherries taste best eaten at room temperature, though, and you can leave them out for several hours without drastically reducing their storage life.

Cherries should keep well two to four days at home.

If you come into a big quantity of cherries, too much to eat up in a few days, the least complicated thing to do is to poach them.

Canned Cherries. Pitted sour cherries come packed in water or in thick sugar syrup for ready-to-use pie filling. Canned sweet cherries may be either light red, which are usually packed with pits, or dark red, which are most often pitted.

Canned cherries are quite unlike the fresh, but they do have bright flavor and good, meaty texture, appropriate for pies and sauces. Grade A sweet cherries are tender, thick-fleshed, and deeply colored. Lower grades tend to be soft and dull red.

Maraschino cherries are hardly cherries at all, pitted, artificially colored, and packed in flavored syrup as they are.

Unopened, canned cherries keep up to one year on the shelf without noticeable loss of quality. After opening, keep cherries refrigerated in covered containers. Sweet and sour cherries keep for a week after opening; maraschinos last six to twelve months in the refrigerator.

Freezing Fresh Cherries. Unless you take the pits out of cherries, they will take on an almond flavor in the freezer. There are utensils designed for cherry pitting or you can use some makeshift device like a paper clip (use the rounded end to scoop the pit out of the destemmed fruit).

To freeze sour cherries for pie filling, stir ¾ cup sugar into every quart of pitted, whole cherries. Seal in plastic bags or pack in rigid airtight containers, leaving ½ inch head space. Freeze whole, pitted sweet cherries in 40 percent syrup (see p. 75) with ½ teaspoon ascorbic acid added per quart of liquid.

Cherries keep ten to twelve months at 0° F.

CRANBERRIES

It should come as no surprise that cranberries peak in November, right before Thanksgiving. These firm scarlet berries (yes, they are native to America and were used widely by the same Indians who catered the first Thanksgiving) are on the market from October through December.

Cranberry bogs are kept dry while the berries grow, then flooded knee-deep at harvest time. A special machine shakes the underwater vines until the loosened berries float to the surface. Workmen wade into the brilliant red mass and scoop up the crop for packing.

Buying Fresh Cranberries. Cranberries should look shiny and plump. They may be light or dark red, depending on variety, but their color should be bright. Don't buy soft or shriveled berries or any with brown spots. Cranberries are best when they are quite firm—resilient enough to bounce off a hard surface.

Storing Fresh Cranberries. Cranberries keep up to one month in a plastic bag in the refrigerator. Sort through them regularly to weed out soft or browning berries. Cooked cranberries keep another month in a covered container in the refrigerator.

Freezing Fresh Cranberries. You can freeze berries whole without sugar or syrup. Pack them in a plastic bag or in a rigid container, leaving 1/2 inch head space, and freeze. They keep ten to twelve months at 0° F.

CRENSHAWS

Crenshaws are large melons, 7 to 9 pounds. They are on the market from July through September. The rind turns from green to gold as it ripens, although, late in the season, melons show a considerable amount of green even when ripe. The rind is lightly ribbed and comes to a blunt point. A fine crenshaw is smooth and lush and the best of all melons.

Buying and Storing Crenshaws. Pick a melon that is heavy for its size and has smooth, unbroken skin. Avoid crenshaws with bruises or mold on the surface. Try to find one that shows signs of ripening: the melon should be firm, but not hard, and have a sweet scent.

Ripen crenshaws at room temperature for two to four days until the blossom end is slightly softened and the aroma is strong and fruity. Store whole ripe melons in a plastic bag in the refrigerator one to two weeks. Wrap cut melons in plastic and store two to three days.

Freezing Crenshaws. Freeze according to the directions for cantaloupe (p. 86).

CURRANTS

Fresh currants are dainty red berries on the market in the midsummer months. Not to be confused with dried currants, which are a variation on raisins, fresh currants are plump, translucent, and so distinctly tart that they are used primarily for jelly.

Black currants are another summer fruit, seen only rarely in the United States.

Buying and Storing Currants. Currants should be firm, whole, dry, and free of hulls and stems. Small berries are best.

Store unwashed currants in an uncovered bowl or basket on the refrigerator shelf for one to two days.

Freezing Currants. Freeze currants according to the directions for cranberries (p. 89). They will not retain their texture as well as cranberries do in the freezer but should be fine for cooking.

DATES

Dates are small, sugary sweet fruits that are always, to some degree, processed before they come to market. They may be dried if they are too moist, or hydrated if they are too dry. In either case, natural ripening comes to an end.

Date palm groves are a familiar sight in the deserts of Southern California. These towering palms bear fruit from childhood (age eight) to well over a hundred.

Buying Dates. Dates are sold soft, semisoft, and dried. Whichever kind you buy, look for plump, springy fruit with clean glossy skins. Even dried dates, though firm, should not feel hard.

Avoid dates that are sticky or have crystallized sugar on the surface. The fruit should smell fresh, not sour.

Storing Dates. Soft and semisoft dates are perishable. The best way to handle them is to seal them very tightly—they quickly absorb odors from other foods—in plastic bags and keep them in the refrigerator or freezer. This is a good idea even for dried dates.

Hard dates keep well up to one year in the refrigerator and, if carefully packaged, indefinitely in the freezer at 0° F. Softer dates may last no longer than a few weeks in the refrigerator.

FEIJOA

An import from New Zealand though it is native to southern Brazil, this small, oval fruit has dark green, waxy skin and a pineapple scent. Ripe fruit, which yields to light finger pressure, should be refrigerated no more than one or two days.

FIGS

Fresh figs have perhaps the briefest life of any fruit on the market. They must be picked mature and ripe, for they do not ripen as well off the tree. Once harvested in California, they are sent by air to other parts of the country. Even when they are handled with kid gloves and held at constant low temperatures, figs spoil within seven to ten days of harvest.

The season for figs stretches from mid-June to mid-October.

Buying Fresh Figs. There are many varieties of figs and they all look different. Calimyrna figs are large and have yellow skin; Mission figs are purple-black; Kadota figs are yellowish green.

Sniff fresh figs before buying. A sour odor means the fruit has begun to ferment and spoil. The fruit should be clean and dry. The peel is extremely thin and tender, but of course should be unbroken. A fig that meets all other requirements but has a slight, still moist tear in the skin is acceptable, as long as you eat it the same day you buy it.

A ripe fig feels soft and yielding but not mushy. What you are going to find most often are overripe figs that, instead of being rounded, have begun to collapse inward. Avoid these, but on the other hand, don't buy a very firm fig—it won't ripen properly.

Fresh figs really must be kept cold. If they are displayed at room temperature in the store, you should buy them only if you plan to eat them immediately.

Canned Figs. Kadota figs packed in light syrup are most widely available. Drained and served with cream, they are a naturally sweet breakfast fruit. Grade A figs are whole rather than split or broken, a sign that they were packed at the right moment of ripeness. Unopened canned figs keep up to one year on the shelf; opened cans keep one week in covered containers in the refrigerator.

Dried Figs. See "Dried Fruits," p. 115.

Storing Fresh Figs. Keep figs in a plastic bag in the coldest part of the refrigerator. Plan to use them within one or two days.

Freezing Fresh Figs. You can peel figs or not according to taste. Slicing the figs gives you the chance to sample the flesh to make sure the fruit is sweet all the way through, but you can freeze them whole, too.

You can use a dry pack, with or without sugar. Seal in rigid containers, leaving ½ inch head space. At 0° F. figs keep ten to twelve months.

GOOSEBERRIES

A gooseberry looks like a small grape with white stripes; it is usually green, but there are red, white, and yellow varieties, and some with a pelt of light fuzz. The gooseberry is very sour and really must be sweetened to be palatable. They are in peak supply in June and July.

Buying and Storing Gooseberries. Larger berries are preferable and should be firm. Don't buy soft, leaking berries.

Refrigerate gooseberries in a rigid container, covered with plastic wrap. They keep two to three days.

Freezing Gooseberries. Remove the berries' stems and tails. If you intend to use them in cooking, remove stems and tails and pack whole, with no sugar, in an airtight rigid container, leaving ½ inch head space. Otherwise, pack the whole berries in 50 percent syrup (see p. 75) in a rigid container, leaving head space.

Store gooseberries ten to twelve months at 0° F.

GRANADILLA

The granadilla is also called passion fruit, because its blossom suggests symbols of the Crucifixion. The fruit itself, which is harvested in the fall, is egg-sized and has thick, purple, deeply wrinkled skin. Ripe fruit has a strong, sugary scent and should be firm but not hard. Refrigerate ripe granadilla no more than one day.

GRAPEFRUIT

Grapefruit's brisk flavor is a real eye-opener in the morning, which may account for its great popularity. Furthermore, the grapefruit you find in the market are usually very good. Far from needing to be rushed to market in breathtaking balance between undermaturity and overripeness, grapefruit ripen fully on the tree and can even stay there hanging on the branch, sweet and stoic, for as long as a year before harvest.

After picking, life is a little tougher. Grapefruit are washed, waxed, and treated with fungicide (they are prone to fungus diseases). Although the skin color of grapefruit has no relation at all to ripeness—it may be a vivid green on the outside but delicious on the inside—packers often use a little ethylene gas to rid the fruit of green pigment and improve its appearance for market. Unfortunately the gas treatment detracts from grapefruit's storage life, but not enough to affect home use.

The best grapefruit come from Texas and from Florida, particularly the Indian River area on the state's eastern coast. These sweet, thin-skinned Southern fruits are on the market September through June.

Buying Grapefruit. You will find grapefruit with either white or pink flesh—there is virtually no flavor difference between the two. Grapefruit also come seeded and seedless. The seeded varieties have a little more flavor; seedless are easier to eat.

Grapefruit are always ripe and ready to serve when you buy them, so what you are looking for are signs of flavor and juiciness. Most important, a grapefruit should feel heavy for its size, which means it will be full of juice. It should be firm and resilient, not soft. Usually a thin-skinned grapefruit is better than one with coarse, spongy skin. If it comes to a point at the stem end, the grapefruit probably has thick skin. Green, spotty, russeted, scratched, or whatever, the appearance of the skin is totally unimportant. Soft patches at the stem end and a soft breakable peel are signs of decay.

Storing Grapefruit. Grapefruit last longest when they are chilled. You can store the fruit in a perforated plastic bag in the warmest part of the refrigerator for ten days to two weeks.

You can also store grapefruit at room temperature. They last about a week on the shelf.

A cut grapefruit should be tightly wrapped in plastic and refrigerated for no more than two or three days.

Freezing Grapefruit. Since fresh grapefruit are available all the time and preparation for freezing them is time-consuming, you should think twice about doing it. But if you have a good reason, here is how. Peel the fruit and cut or pull it gently into sections. Remove the membrane around each section with a sharp knife and remove all seeds, working over a bowl to catch juices. Pack the segments in a rigid container and cover with a 40 percent syrup (see p. 75) made from the fruit's juices and water. Add 1/2 teaspoon crystalline ascorbic acid to each 1 quart syrup for best results.

Seal, label, and freeze at 0° F. Grapefruit keep ten to twelve months in the freezer.

Another approach to preserving grapefruit is to extract the juice and freeze that. Squeeze the juice in such a way that no oil from the rind is released and strain out any seeds. Add 1/4 teaspoon crystalline ascorbic acid per quart of juice and sugar as desired. Pour the juice into glass containers for freezing, leaving some 1/2 inch head space.

GRAPES

Grape growing is the biggest food industry in the world. Grapes grow in almost any climate and can be put to a number of noteworthy uses. There are grapes for the table, grapes to make raisins, grapes grown for juice, and grapes destined to become wine.

Table grapes come in many different varieties, each with a fairly distinct character. Thompson seedless grapes are the mild green grapes you find most often. They are in season June to November. Ribier grapes, available July through February, are large and dark purple; they have seeds and taste neither very sweet nor sour. Tokays are big firm grapes with seeds; dark red and mild; they are on the market September through December. The pale reddish-purple Emperor grapes are in season through winter and spring; they are also quite mild. Many other types may be found at different times and in different local markets, but their distribution is limited.

Buying Grapes. Grapes do not ripen off the vine, nor do they grow sweeter or juicier after harvest. What you are looking for are the freshest, brightest fruit you can find.

The best test is to pick up a bunch of grapes by its stem and shake it very gently. If the grapes are firmly attached, they are fresh. If grapes drop from their stems, the bunch has been in storage too long. Grapes should be plump, not shriveled; the skin should be dry and unbroken.

Choose grapes well colored for their variety, with no brown or gray undertone. Green grapes, however, are sweetest when they have a yellowish cast, rather than being icy green.

Storing Grapes. Grapes need cold temperatures and high humidity. Remove any shriveled or damaged grapes from the bunch and keep the rest in a plastic bag in the coldest part of the refrigerator. They should keep three to five days.

Don't wash them for storage but wash them very meticulously before serving because grapes have been doused with many different chemicals before they get to you.

Freezing Grapes. Grapes lose a lot of spunk in the freezer but still may be suitable for gelatin salads, jam, or chutney.

Wash and stem the grapes. Leave seedless ones whole. Seeded grapes should be halved and pitted. Pack in 40 percent syrup (see p. 75) in rigid containers, leaving 1/2 inch head space. Use a plain pack if you want to use the grapes for jelly or juice. Grapes keep ten to twelve months at 0° F.

Grapes frozen only briefly make a very special sweet snack. Wash the grapes and, while they are still wet, dunk them in egg white, then roll them in sugar. Tray-freeze them for 30 to 40 minutes and serve.

GUAVAS

A few of these small tropical fruits are grown domestically and sold fresh. They come in many colors—yellow, red, purple, or even black. They can be eaten raw but are more often available in this country as jelly or paste.

Fresh guavas should be chosen firm for cooking. Let them ripen at room temperature until they yield to gentle pressure for eating out of hand. Once ripe, guavas smell irresistible; they keep up to two weeks in the vegetable drawer in your refrigerator.

HONEYDEWS

The honeydew melon is a pale, smooth globe with dense green flesh which, under the right circumstances, is as succulent as any fruit can be. Honeydews have a nice long season, from June through October, and it is not impossible to find good ones in most other months of the year.

Like other melons, honeydews must build up all their natural sugars while still on the vine. They ripen after harvest, becoming softer and juicier, but will never be sweeter than on the day they are picked.

Buying Honeydews. Pick up the honeydew (it weighs between 4 and 8 pounds) and run your hands over its surface. A very slight texture, a feeling of light velvet, means a mature melon. A slick, bald skin is a sign the melon was harvested too soon. Take a long, appraising look at the rind: you want a yellowish-white or cream-colored melon—dead white or greenish skin means immaturity. If the stem end is just a little springy to the touch, the melon has already begun to soften and, when treated right, will ripen all the way.

There will be times when you can't find a single honeydew in the bin that looks just right. Don't buy what is there and hope for the best—an immature honeydew stays hard, dry, and tasteless until it spoils.

Some retailers slice honeydews into sections for sale. You can then look the flesh over and decide whether it looks soft and juicy. You should be able to smell the fragrance of the melon even through its wrapping.

Storing Honeydews. Leave a honeydew at room temperature for two to four days until it shows all the signs of being fully ripe—creamy color, soft blossom end, and sweet scent. Store ripe melons in the refrigerator in a plastic bag. They should last up to a week.

Cut melons must be snugly wrapped in plastic and refrigerated. They will keep three to five days.

Freezing Honeydews. Follow directions for freezing cantaloupes (p. 86).

KIWIFRUIT

Kiwifruit look like brown, furry eggs. Inside is a wonderful surprise—bright green jewellike flesh with the flavor of sweet gooseberries. They store and ship particularly well, so growers can afford to leave the fruit plenty of time to mature on the vine.

Kiwifruit are in season nearly year round. When they are not coming from California, they arrive from New Zealand, their other main growing area.

Buying Kiwifruit. Look for plump fruit with a firm but not hard texture. There should be no shriveling, soft spots, or mold on the fruit. The color of the skin is quite uniform and not much of a clue as to the state of the fruit.

Storing Kiwifruit. Leave kiwifruit at room temperature for a couple of days until they are completely ripe. Their texture is the key to ripeness. Soft but not spongy, a ripe kiwifruit feels about the same as a ripe peach.

Store ripe kiwifruit in the refrigerator up to one week.

You can use kiwifruit exactly the same way you would use any other fresh fruit. Peel it, slice it, and serve it as is, or bake it into muffins, add it to fruit salads and desserts, or poach it for a little extra storage life. Sliced in cross section, kiwifruit has a purple starburst at its center that makes it very decorative in fruit tarts or as a garnish.

Freezing Kiwifruit. Peel kiwifruit and pack in a rigid container, whole or sliced, in 40 percent syrup (see p. 75) to which has been added 1/2 teaspoon ascorbic acid per quart of liquid. Leave 1/2 inch head space, seal, label, and freeze. Kiwifruit keep ten to twelve months at 0° F.

KUMQUATS

Kumquats resemble tiny football-shaped oranges. They are eaten whole, skin and all, except for the seeds. Choose plump kumquats, with perfect skin, that are firm but not hard. You can store kumquats at room temperature one week or in the refrigerator up to three weeks.

LEMONS

Lemons are the most versatile of fruits. Lemon juice can animate the flavor of zucchini, bake into a tangy cake, perk up a glass of iced tea, and marinate a leg of lamb.

Lemons are in continuous supply because they grow during all seasons of the year and because they store well up to 6 months in controlled-temperature warehouses.

What makes lemons desirable is their acidity, so ripeness is beside the point. Growers pick lemons that are the right size to have developed plenty of juice.

Buying Lemons. Almost all lemons that come to market are good quality. You can judge them for juiciness, though. Look for glossy, fine-grained skin. Also, the fruit should feel firm and heavy for its size. A green tinge to the peel means the lemon is particularly fresh and will store a bit longer than a thoroughly yellow fruit. Check for soft spots at the stem, which indicate internal decay. Avoid lemons with thick skin, hard skin, or a spongy texture.

Storing Lemons. Store lemons in a plastic bag in the refrigerator. They last two to three weeks when chilled.

A cut lemon can keep almost as long if you store it in a small tightly lidded jar in the refrigerator.

Lemons come to market with an edible but artificial coating of wax on the skin. You can wash it off in detergent and warm water before using the peel in your recipes.

You can save even a dry, shriveled lemon from the garbage can by simply extracting the juice. (Juice squeezes out more easily if you press the whole lemon on the kitchen counter and roll it around under your palm first.)

Freezing Lemons. There are two ways to freeze lemons. You can extract the juice and freeze that. One lemon yields 2 to 4 tablespoons of juice, depending on size and juiciness. Pour the juice of each lemon into an ice cube tray compartment and freeze at 0° F. Take the frozen cubes out and seal them in an airtight plastic bag. You can remove what you need from the bag and reseal it.

You can also freeze a whole lemon. When it is thawed it will be very soft and you can easily extract all its juices. It is also possible to slice it. The texture will be mushy, but you can use the slices in beverages.

Lemons and lemon juice keep three to four months at 0° F.

LIMES

Persian limes are the tart green citrus fruits that have all the versatility, if not the popularity, of lemons. They are conveniently seedless. Green limes are available every month of the year but are particularly abundant (and usually cheaper) in the summer months.

Key limes are small, oval, and yellower than Persians. They are the basis for the deservedly famous Key lime pie. It is very rare to find these fruits outside their growing area in Florida.

Buying Limes. Persian limes should have shiny green skin and feel heavy for their size. Yellowing skin means the juice is losing its bite and the fruit is aging.

Key limes' light yellow skin should be very thin and fine-grained.

Neither variety should have purple or brown scabs on the skin or any soft, watery spots. These are symptoms of disease and injury.

Storing Limes. Keep limes in a plastic bag in the refrigerator or loose in the vegetable crisper. They should keep three to four weeks but will have lost a lot of flavor at the end of that time. When the end of a lime's storage life is near, the skin becomes pitted.

Cut limes stay juicy stored in a tightly lidded jar in the refrigerator.

Freezing Limes. Freeze only the juice of limes, using the technique described for lemon juice (p. 97). Lime juice keeps three to four months at 0° F.

LITCHIS

Just picked, the litchi looks like an outsized strawberry, but soon after, the rough, hard skin browns. Fresh litchis have white, translucent flesh with a slightly acidic cherry flavor. The unpeeled fresh fruit keeps a week in the refrigerator. Dried, it's called a litchi nut and can be refrigerated for several months. But litchis are most often available canned in syrup.

MANGOES

A mango is an oval or kidney-shaped fruit with a warm rosy color. In the tropics mangoes are as common as apples in cooler climates, but in the United States the fruit is not sold everywhere. Fortunately, at the height of its season in June and July, the mango is not discouragingly expensive.

Its delicate flesh is both sweet and tart. The mango does present a challenge to the newcomer—its large pit grips the edible pulp with maddening tenacity. To slice open the fruit, cut around the pit, leaving some of the flesh on. You can then, happily but messily, take the pit in hand and nibble inward.

Buying Mangoes. The smooth outer skin of the mango starts out green and turns yellow and red as it ripens. Never buy a completely green mango because it could easily be an immature fruit that will never ripen. (There is one winter variety of mango—varieties abound, as with apples —that stays mostly green even fully ripe, but does show some red and yellow coloring.)

Most mangoes have black spots on the skin that increase in number as the fruit loses its green cast. The defect is only skin deep until the very final stages of ripeness, when it may penetrate the flesh. It is not necessary to reject all fruit with black spots, but the fewer the better.

Storing Mangoes. For eating out of hand, ripen mangoes at room temperature until they feel fairly soft. An underripe mango can have an odd, chemical taste eaten raw. If your kitchen stays as warm as 80° F., ripening mangoes can sometimes develop a strong, less desirable flavor.

Once ripe—yellow and orange-red; nicely softened and sweetly scented—refrigerate a mango in a plastic bag for no more than two or three days.

Green and half-ripe mangoes are specified in recipes for chutney and a few exotic cooked dishes.

Freezing Mangoes. Peel, pit, and slice mangoes. For fruit you intend to eat raw, use the sugar pack (see p. 75), sealing sugar-coated slices in a rigid container with 1/2 inch head space.

Pack mango slices in 30 percent syrup (see p. 75) if you want to use the thawed fruit as a sauce for ice cream, yogurt, and puddings.

Mangoes keep ten to twelve months at 0° F.

NECTARINES

No matter what you have heard to the contrary, a nectarine is not a cross between a peach and a plum. It is very much itself and a princess of a fruit at that.

From June through August you will find lots of nectarines on the market. They are not only delicious; they are also loaded with vitamin A. Children who have balked at green vegetables all winter long can hardly be restrained from eating up any nectarines left lying around.

Buying Fresh Nectarines. A nectarine often has a wash of autumn red over a large part of its very smooth skin but may be almost all yellow and still be sweet and juicy. The background color can be yellow or yellowish green depending on the fruit's variety. Though this hint of green suggests immaturity, it may mean nothing of the kind. Even more confounding is the fact that some nectarine trees produce fruit that are blushing red even before they are mature. It is true that the coloring on a mature fruit is intense—much richer-looking than on one picked too soon —but this is not an entirely reliable guide. Nectarines are routinely coated with edible wax to make them more attractive and the waxing literally glosses over defects of color. In judging a nectarine, it is wise to ignore most of the evidence of your eyes and rely on other senses: an immature nectarine looks dull and feels very hard. Shriveling and wrinkled skin are a sure sign of a pithy, tasteless fruit. Avoid nectarines with soft spots or broken skin. "Russeting," flecks of tan on the skin, has no effect on the quality of the fruit.

Your only reasonable choice is to buy fruit already in the first stages of ripening. Very gently probe along the seam that runs the length of the nectarine. If the flesh there is yielding, the fruit has begun to soften. And a sweet scent is practically a guarantee of good fruit.

Storing Fresh Nectarines. Nectarines ripen completely in two or three days at room temperature.

Store ripe nectarines in a plastic bag in the coldest part of your refrigerator. They will keep three to five days.

A cut nectarine should be wrapped tightly in plastic and stored in the refrigerator no more than one or two days. Cover the cut edge with lemon juice to keep the flesh from turning brown.

A nectarine can do anything a peach can do, including fill a pie. Don't wait until extra nectarines are very soft before you decide what to do with them. Overripe nectarines can have a bitter taste. Poach them or puree them while they are still slightly firm.

Freezing Fresh Nectarines. Nectarines do not hold up in the freezer very well, but if you have really lovely, firm fruit you want to preserve, you can try it.

Wash, pit, and slice the nectarines. Pack them in rigid containers covered with a 40 percent syrup (see p. 75) with 1/2 teaspoon ascorbic acid added per quart of liquid. Leave 1/2 inch head space.

A less risky method is to freeze nectarines as a puree. They must be peeled first. To do this, plunge the whole fruit into boiling water for 30 seconds. Transfer them immediately to an ice-water bath. The skins should slip off easily. Remove the pit and crush the pulp with a fork. Stir 1/2 cup sugar into every 2 cups pureed fruit. Pack in rigid containers, leaving 1/2 inch head space.

Nectarines keep ten to twelve months at 0° F.

ORANGES

Driving along the highways of central Florida is a lesson in astronomical numbers. As you pass through miles and miles of orange groves you can see what appear to be hundreds of oranges on each of tens of thousands of trees. Fewer, but also awesome, are the groves of California. Each year's crop is tens of billions of oranges.

There are so many trees because Americans drink oceans of orange juice. There are so many oranges on each tree because the fruit grows and matures at an unhurried pace.

An orange does not ripen in the same sense that most other fruits do. It must be mature on the tree to be good, for it does not get softer or sweeter after picking. In fact there are statutes making it illegal to ship an immature orange, but the citrus grower has little reason to harvest oranges too soon. An orange grows at an extremely slow pace and once it matures remains stable for months. A grower can virtually store mature fruit on the trees and pick it early or late with no loss of quality.

The natural color of an orange hinges on only one thing—temperature. The fruit grows orange as the air grows cool. It can even switch back and forth. A single fruit, still on the tree, can be neon orange in January and turn green again in warm March weather.

The citrus industry believes, rightly or wrongly, that the American shopper cannot be made to understand how oranges work and will always reject green ones. Therefore it is common practice to dye green oranges orange for the fresh market. Each fruit so treated must be stamped with the words "Color Added." California's climate produces orange oranges that are not dyed.

Nearly all oranges you find in the store are waxed.

Buying Fresh Oranges. Fresh oranges come in several varieties that have different peak seasons and distinct individual characteristics.

• VALENCIA. Valencia comes nearest to being a dual-purpose orange. It is good for juicing and good for eating out of hand. The rind is thin, the seeds few, and the flavor bright. Valencias are on the market every month of the year, but most abundantly from April through October.

• NAVEL. Navel oranges are the most approachable. The thick skin zips off with an almost effortless tug; each segment separates neatly from the rest of the fruit; and it contains no seeds. The flavor is both tart and sweet. You can recognize a navel orange by its blossom end, which looks precisely like a belly button. Their season starts in the fall and ends in the spring.

• HAMLIN. Hamlins are thin-skinned and good for juice. They are usually green at harvest—October to January.

• PARSON BROWN. Sharing the winter months with Hamlins, Parson Browns are another good juice variety.

• PINEAPPLE. Pineapple oranges have lots of flavor and lots of seeds. Use them for fresh juice and for eating out of hand. January and February are their peak months.

• TANGELO. A tangelo is a hybrid of grapefruit and tangerine stock. It has its own piquant flavor, but in every other respect resembles an orange. Tangelos are on the market from October through January.

• TANGERINE. A small citrus with loose-fitting skin, the tangerine has a flavor with a little more zip than its larger relatives. It is good eating and makes wonderful juice, too. Peak months for tangerines are November, December, and January.

• TEMPLE. A temple is not exactly an orange, but close enough: it is a hybrid of tangerine and orange. Distinctly sweet, it is easy to eat out of hand because of its loose-fitting skin and readily parted segments. It is in season December through April.

• BLOOD. The blood orange gets its name from the deep red color of its flesh: a cut blood orange looks as though it might actually bleed to death. Available March, April, and into May, it is an outstanding eating and juice orange with very soft pulp and no bitterness, even in the peel; most blood oranges are imported from Sicily.

• JAFFA. Very juicy oranges, prized for their flavor, Jaffas are imported from Israel. They are most available in early spring. Jaffas keep longer than other varieties.

• CLEMENTINE. Another cross between tangerine and orange breeds, the tiny seedless clementine is in season in midwinter. The best come from North Africa, especially Morocco.

• MINEOLA. From California in the spring comes this large tangerine hybrid, heavy with sweet, winy juice.

Choose oranges not for color but for feel. They should be firm and heavy for their size. Avoid oranges with soft, soggy spots or mold on the surface. The skin should be smooth and fine-grained. Brown mottling on the peel has no effect on the quality of the fruit. The peel itself is a good source of fresh orange flavor. Scrape off the deeply colored skin with a vegetable peeler or a zester, being careful to avoid the bitter white underlayer, and use in baking, sauces, salads, and drinks.

Storing Fresh Oranges. Oranges keep well several days at room temperature. Stored in a plastic bag in the refrigerator, they should keep ten days to two weeks.

A sliced orange should be wrapped tightly in plastic and refrigerated. It will keep two or three days.

Canned Oranges. Most canned oranges are of a variety not usually found fresh in the United States—mandarin oranges. The small, plump segments are very sweet. These and other canned oranges keep for up to six months on the shelf. Opened, they should be stored, well covered, in the refrigerator for no more than a week.

Fresh Orange Juice. The original, hand-operated, no-moving-parts juicer with a conical hump in the center will get juice out of oranges but takes a lot of muscle. There are a number of mechanical juicers you can use to squeeze fresh oranges more quickly, but take care to choose one that doesn't also press the bitter-tasting oil from the rind.

Pour fresh juice (seeds removed) into a glass container, cover, and refrigerate. The complex flavor and virtually all the vitamin C will last at least two to three days.

Freezing Fresh Oranges. Fresh orange sections can be frozen, using the same method described for freezing grapefruit (p. 93).

To freeze juice, seal juice in a tightly lidded glass container, leaving ½ inch head space per pint. Add a pinch of ascorbic acid crystals if you are going to store the juice more than a few weeks.

Orange juice keeps three to four months at 0° F.

PAPAYAS

Papayas look more like squash than fruit. The main type found in U.S. markets are pear-shaped and weigh in around 1 pound. They are distinctly yellow when ripe.

The flavor is sweet and bland and musky.

Buying Fresh Papayas. Select fruit that is at least half yellow. The skin should be smooth with no bruises, pockmarks, cuts, or other damage. Choose papayas that are still firm.

Storing Fresh Papayas. A firm papaya can be left to ripen completely at room temperature for three to five days. A fairly soft, fragrant, fully ripe fruit keeps up to two weeks in a plastic bag in the refrigerator.

Scoop out the sticky black seeds and a papaya becomes a natural cup for serving fruit, chicken, or seafood salads. You can bake a papaya in the skin to serve as a side dish—season it with honey, cinnamon, and butter.

Freezing Fresh Papayas. Follow directions for freezing cantaloupes (p.86).

PEACHES

There are dozens of varieties of peaches, but the differences among them are not so great that you need to distinguish one from the other. Some have yellow flesh and some have white, but they can be equally succulent. Varieties grown in warm climates are usually tastiest—Georgia's reputation for wonderful peaches is well earned.

All the peaches sold fresh are "freestones," meaning their pits slip easily away from their flesh. "Clingstones" or "cling" peaches have a firmer hold on their pits and are most often canned.

Fresh peaches begin to appear in stores in the middle of May. July and August are the peak months, but some peaches roll in as late as September.

Buying Fresh Peaches. A peach is one of those fruits that does not turn sweeter after harvest, though it will become softer and juicier. It must be picked fully mature to be any good at all. You can spot a mature peach by its color: not the red blush you often see, which only develops on certain varieties, but the background color, which should be yellow or cream. Avoid peaches with any hint of green. You should be warned that it is not always easy to find tree-ripened fruit. Many, many peaches come to market before their proper time and these will always be pithy, dry, and unpalatable.

Skip very hard peaches and look for ones that have begun to soften and to give off a sweet aroma. Don't buy peaches with shriveled, wrinkly skin. These are either immature or have been mishandled. Especially beware of a circle of dead, tan-colored skin on the peach, a symptom of internal decay.

Pick peaches that are clean, well shaped, and free of bruises and other damage. The fruit should also have a distinct seam on one side—another sign of maturity.

Peaches are often supercooled to around 32° F. right after harvest, a practice that keeps the fruit firm and disease-resistant during storage and shipping. Occasionally the peaches suffer from cold injury if the temperature was too low for too long. When this happens, the peaches turn dry and mealy on the inside while looking perfectly healthy on the outside. If you buy one of these injured peaches when it is still cold, the fruit may look good and feel springy and properly heavy, but it will deteriorate very quickly. Therefore, don't trust peaches that are still cold.

Storing Fresh Peaches. Peaches ripen fairly quickly. A day or two at room temperature should bring a mature peach to its peak. When ripe, it will be firm but yield to gentle pressure and it will have a strong, sweet scent.

Ripe peaches need cold temperatures and high humidity. Store them in plastic bags in the refrigerator three to five days.

A squeeze of lemon juice on sliced peaches prevents darkening. Coat cut peaches with juice, then wrap them very securely in plastic. Keep them in the refrigerator no more than two days.

Leftover fresh peaches stand up to every kind of treatment. You can poach, broil, pan-fry, or puree a peach.

Canned and Frozen Peaches. Processed peaches are no match for fresh. On the other hand, really good fresh ones are not always easy to find.

Canned cling peaches are smooth; freestones are softer and have raggedy edges. They are packed in sugar syrups ranging from heavy to light, and some come in just their own juices. A can of peaches can be stored up to a year on a cool, dry shelf. Out of the can, store the peaches, covered, up to one week in the refrigerator.

Frozen peach slices are usually sweetened and treated with an antioxidant to preserve their color. They taste closer to fresh than canned peaches do. Store them at 0° F. as long as one year. Thaw only what you need and keep the rest frozen.

Freezing Fresh Peaches. Wash, peel, and pit firm, ripe peaches. To peel them quickly, first plunge them in boiling water for 30 seconds, then transfer them to ice water. Peeling them without this treatment leaves a smoother surface but takes more time.

Peaches keep their flavor better when frozen in syrup. Use a 40 percent syrup (see p. 75) with 1/2 teaspoon ascorbic acid crystals added per quart. Pour a couple of inches of syrup into a rigid freezer container and slice peaches right into the liquid. When the container is nearly full, add more syrup to within 1/2 inch of the top. Put a crumpled piece of waxed paper between the lid and the fruit to keep the peaches submerged.

Frozen peaches keep ten to twelve months at 0° F.

For the most pleasing texture, serve frozen peaches when they are only partially thawed.

PEARS

The pale, sweet, slightly grainy flesh of the pear actually ripens nicely off the tree. Pears are picked when mature but still very firm, and so proof against a few bruises and a few months of commercial storage. The pear's hardiness helps keep it in the stores nearly year round.

Buying Fresh Pears. The four most widely available pear varieties are fairly easy to recognize:

• BARTLETT. This is the main summer pear. It is bell-shaped, green turning to yellow when ripe, and sometimes a bit lumpy-looking. The fine-grained flesh has a sweet, winy flavor. Spectacularly good Bartletts are imported from Chile in early spring.

• ANJOU. The least pear-shaped pear, the Anjou is more of a stubby oval. Its greenish-yellow skin is no clue to its ripeness. You can find it all winter, from October to May.

• BOSC. The Bosc is unmistakable. It is long and tapered, with skin the color of cinnamon. The flesh is both sweet and very juicy. You will find Bosc pears from October through early spring.

• COMICE. The most easily bruised pear, the Comice is perhaps the one most worth protecting. Its rich, sugary flavor is considered the best. The Comice is plump, greenish yellow, and often has a blush of red on the skin. It is available October through January.

• SECKEL. These small pears are a bit gritty and unpredictable, but when they are good their spicy flavor is a particular treat. Cold storage plays havoc with Seckel pears, so buy them only at the height of their season, September and October.

• FIRELLI. These have a freckled blush, come in the fall, and are great eating.

Choose pears that are firm, well colored, and clear skinned. You don't have to be too cautious because most pears you buy will, when ripe, be of very high eating quality. However, avoid pears with wrinkled skin or overly soft flesh.

Storing Fresh Pears. For eating out of hand, ripen pears at room temperature until the sides yield to a little pressure. This takes from one to four days. Pears meant for cooking should be a bit firmer.

Store ripe pears three to five days in a plastic bag in the refrigerator. Wrap sliced pears in plastic and refrigerate no more than one or two days. To prevent discoloration, coat the cut surface of the pear with lemon juice.

Freezing Pears. Freezing pears takes a little more effort than freezing most other fruits. Wash, core, and slice pears that are firm but not hard. Submerge the slices in boiling 40 percent syrup (see p. 75) for 1 minute. Drain and cool the fruit, then pack it in rigid containers. Cover the slices with cold 40 percent syrup to which 3/4 teaspoon ascorbic acid crystals per quart has been added.

Sealed airtight, frozen pears keep ten to twelve months at 0° F.

PERSIAN MELONS

Persian melons are like very large cantaloupes, but they have a much finer texture to their rind. The flesh is orange and delicately flavored.

You can buy Persians using the same guidelines as for cantaloupes (see p. 85–86). There is one important difference, however. The background color of the Persian melon stays light green even when it is mature and ripe.

To ripen, store, and freeze Persians, follow the instructions for cantaloupes.

PERSIMMONS

Persimmons are slick-skinned, with a luxurious orange-red color. The flavor is sweet and rich when the fruit is ripe, but biting into an underripe persimmon is a sorry experience. Until it is ripe, a persimmon contains an astringent chemical compound that causes a puckery feeling in your mouth.

Buying Persimmons. Look for plump, smooth fruit with the stem cap still attached. If it is too soft it will bruise too easily, so pick a persimmon that is at least a little firm.

Storing Persimmons. There are two methods for ensuring a ripe persimmon. You can place the fruit in a perforated paper bag along with a ripe apple or banana. Another way is to wrap the persimmon in aluminum foil and place it in the freezer overnight. After a few hours of thawing it will be ready to eat.

The flesh of a ripe persimmon is very soft, almost like jelly. Ripened the first way, the fruit can be stored another day or two in the refrigerator. If you use the freezer technique, eat the persimmon as soon as it is thawed.

Freezing Persimmons. Wash, peel, and puree persimmons. Stir in 1/8 teaspoon ascorbic acid crystals per quart of fruit. Homegrown persimmons don't need sugar, but store-bought varieties do better with 1 cup sugar added per quart. Pack in airtight, rigid containers.

Persimmon puree keeps ten to twelve months at 0° F.

PINEAPPLES

A pineapple is like an elephant—its appearance is so familiar, you forget how odd it really looks. The skin of the pineapple is like tree bark and its fan of gray-green leaves is thick and tough. It must have been the alluring scent that persuaded the first person to open up a pineapple and discover the juicy yellow flesh inside.

Available year round, pineapples are most abundant in spring. Hawaiian pineapples are generally considered superior to those from Puerto Rico and other tropical ports, but most of the Hawaiian crop ends up in cans.

Buying Fresh Pineapples. Look for a plump, fully ripe pineapple that yields slightly to finger pressure and has flat, shiny eyes on its rind. The color may be green or yellow—it does not relate to the ripeness of the fruit—but it should look bright and not dull. Don't buy fruit with soft spots or mold on the surface.

The fruit should smell sweet and strong. A sour odor means the pineapple has begun to spoil.

Storing Fresh Pineapples. Store a whole ripe pineapple in a plastic bag in the refrigerator. It will keep three to five days. Treat the fruit gently—the skin is much more vulnerable to bruising than it looks.

Cut chunks of pineapple, amazingly, keep better than the whole fruit. Store them in an airtight container in the refrigerator and they should last up to a week.

If the pineapple you bring home needs ripening, leave it at room temperature for one or two days. When it is ready to eat, the leaves pull out easily.

The best way to add a couple of days to a pineapple's storage life is to sauté the fruit in butter and sprinkle it with a little brown sugar. The cooked fruit keeps well for three or four days.

Canned Pineapple. Canned pineapple, especially the kind packed in its own juice, is a pretty tasty dish. Fancy-grade fruit comes from the bottom of the pineapple, which has deeper color, softer fibers, and sweeter flavor than flesh at the top of the plant.

Unopened cans of pineapple keep up to one year on a cool, dry shelf. Opened fruit should be stored in a covered container in the refrigerator for no more than one week.

Freezing Pineapple. Peel and core the fruit and cut it into bite-size pieces. Pack the pieces tightly into a plastic bag or rigid container—enough natural juice will accumulate around the fruit to almost cover it. Sugar is optional.

Pineapple served before it is completely thawed has a nice frosty taste.

PLANTAINS

A plantain is a kind of banana used more like a vegetable than like a fruit. The thick, coarse skin is greenish and mottled. Plantains are never eaten raw.

Buying Plantains. Look for plump, undamaged fruit, but don't be too concerned by the appearance of bruising—the skin is tough enough to protect the flesh from most bumps. The skin turns yellowish brown and finally black as the fruit ripens. Recipes may call for plantains in any stage of ripeness—*tostones,* the traditional, twice-fried plantains, start with green fruit.

Storing Plantains. You can leave plantains at room temperature until they are ripe. It takes about one week to get from totally green to yellow-brown and another week before the fruit is black. Store very ripe fruit in the refrigerator three to five days.

Plantains can be fried, poached, or baked in their skins (cut a slit in the peel to allow steam to escape). Serve them like potatoes as a side dish with meat.

PLUMS AND FRESH PRUNES

Tragedy, Elephant Heart, Sharkey, Satsuma, El Dorado—you can find these at the grocery store all summer long, but they are not on the paperback best-seller rack. They are types of plums and fresh prunes, fruits with taut, gleaming skins and soft, juicy flesh.

Hundreds of varieties of plums and fresh prunes are grown. The Santa Rosa, most often seen, has deep red skin and yellow or red flesh, but this is by no means standard. The Nubiana is purple-black with amber flesh; the Wickson is distinctly yellow inside and out; Italian plums have navy-blue skin clouded with a natural white bloom and the flesh is green.

Plums and fresh prunes are essentially the same fruit. The only important differences are that fresh prunes are freestone, meaning the pit separates easily from the pulp, and that fresh prunes don't ferment when they dry.

Buying Plums and Fresh Prunes. The varieties of plums and fresh prunes can taste as different as they look. Some, like Queen Anne, are very sweet, and some, like Damson, are quite tart. Most varieties are somewhere in the middle. You should sample a number of different kinds and choose your own favorite.

Plums and fresh prunes only soften after harvest—they don't change color or become any sweeter. Look for deep color and firm but resilient texture. They should seem ready to eat when you buy them. Avoid plums that are very hard or shriveled. They are probably immature. Also avoid plums that are too soft, leaky, brownish, or punctured.

Storing Plums and Fresh Prunes. Put plums and fresh prunes in a plastic bag and store them in the refrigerator. They keep well three to five days. If they are very firm, you can leave them at room temperature to soften, but for no more than one day.

Extra plums and prunes can be poached for longer storage.

Freezing Plums and Fresh Prunes. You can freeze plums whole. Wash them and seal them in a plastic bag.

To serve them, dip each plum in cold water for 10 seconds, then slip off the peel. You can then thaw the fruit as is or in a 30 percent syrup (see p. 75).

To freeze plums and prunes for future use in pies, wash, pit, and slice the fruit. Pack in 40 percent syrup, with 1/2 teaspoon ascorbic acid crystals added per quart of liquid, and seal in rigid containers, leaving 1/2 inch head space.

Whole, unsweetened plums and prunes keep three months at 0° F.; packed in syrup, they keep ten to twelve months.

POMEGRANATES

The pomegranate is a hard-skinned fruit about the size of a large orange. Its edible parts are the seeds and the tangy, jelly-like pulp that surrounds each seed. The seeds and pulp together look like brilliant red corn kernels and are gloriously messy to eat.

Pomegranates are on the market only briefly—from late September into November.

Buying Pomegranates. The skins of pomegranates tend to look a bit scuffed, but they should have deep color, either yellow-orange or reddish purple. Look for large fruits that feel heavy for their size and have clean, unbroken rinds.

Storing and Freezing Pomegranates. You can store the whole fruit in the refrigerator two to three weeks, but it is more convenient to extract the seeds and store them alone.

Cut the fruit in half and scoop out the pulp-coated seeds, leaving behind all the bitter, yellowish membranes that hold them in place. The juice squirts and stains very readily, so be cautious.

Put the seeds in a tightly lidded jar and store them in the refrigerator for two weeks or in the freezer up to one year.

Eat the seeds out of hand or sprinkle them in salads.

PRICKLY PEARS

Being the fruit of the cactus, a prickly pear is born with sharp spines, but someone snips them off before the fruit reaches market. The shorn skin is rather hard and greenish, the fruit itself small and oval. The flavor suggests watermelon and apples.

Choose prickly pears without cracks or soft spots. They turn from green to yellow or orange-red as they ripen, but the rind stays firm. Ripen at room temperature, then refrigerate up to three days.

QUINCES

A quince looks like a green apple with a slightly misshapen stem end. A bit too tart to eat raw, cooked and sweetened quinces have a deep, spicy flavor more complicated than that of an apple or pear.

They are in season in the late fall and early winter.

Buying Quinces. They yellow as they ripen, so choose quinces that are most nearly golden. Like good apples, they should be firm but not hard. They bruise easily, which makes them spoil too soon, so buy fruit with little or no signs of damage.

Storing Quinces. An underripe quince should be stored at room temperature. You can keep ripe quinces in a plastic bag in the refrigerator for two to three weeks.

Quinces bake beautifully, keeping their shape and turning a pretty pink inside. You can poach them, too, or use them for jelly.

Freezing Quinces. Wash, peel, core, and slice quinces for freezing. Mix 1/2 cup sugar into each quart of fruit, gently coating each slice. Pack in a plastic bag or rigid container, leaving 1/2 inch head space.

Quinces keep ten to twelve months at 0° F.

RHUBARB

Rhubarb has the nickname "pieplant," but when you see what looks like red celery in the grocer's case it may be hard to believe these long crisp stalks can make a luscious dessert.

Rhubarb is in good supply January through August. There are two kinds on the market: field-grown, which is deep red; and hothouse, which is pink and milder in flavor.

Buying Rhubarb. Stalks of rhubarb should be firm, crisp, thick, and well colored. Avoid rhubarb that is wilted, pithy, stringy, or rough-textured.

Storing Rhubarb. Keep rhubarb cold. Put the stalks in a plastic bag and store them in the refrigerator three to five days.

Rhubarb, cut into 1-inch pieces and stewed in a little sugar and water, makes a delicious dessert or sauce that keeps up to one week, well covered, in the refrigerator. (If you sweeten stewed rhubarb to taste after it is cooked, you'll need less sugar.) Also use rhubarb to make jam or chutney. Do not eat the leaves, which are toxic.

Freezing Rhubarb. Wash stalks and remove all leaves. Cut the rhubarb into 1-inch pieces and seal in a plastic bag.

Rhubarb keeps ten to twelve months at 0° F.

SAPOTE

Another name for sapote is Mexican custard apple, with reference to its origins and to its sweet, milky pulp. It looks like a round green apple with no dimple at the blossom end. Ripen sapotes at room temperature until the flesh yields to slight pressure, then refrigerate three to five more days.

STAR FRUIT

In cross section, the waxy, yellow star fruit forms a distinct five- or six-pointed star. The star fruit, also called carambola, should ripen at room temperature until it is intensely yellow. Strip the brownish fibers off the tops of the fruit's ridges and eat the peel and the slightly sour pulp, but not the core or seeds. Ripe star fruit can be refrigerated up to a week. They are good for jams and jellies.

STRAWBERRIES

Strawberries are unique among berries. They can be cultivated in a broad range of soils and climates so they are harvested fresh, somewhere, every month of the year. When they are picked, also unlike other berries, their green caps stay on.

Buying Fresh Strawberries. Strawberries should be entirely deep red (not greenish), plump, and moist-looking. Don't buy shriveled strawberries or ones with dry, browning leaf caps. They should be firm, not soft and bruised or leaking juices. Pass up berries in stained or sticky containers. Ignore very large containers of berries; they can be crushed by their own weight. The best strawberries have a full, sweet scent.

Storing Fresh Strawberries. Sort through the strawberries and remove any damaged or moldy fruit. Refrigerate unwashed strawberries, green caps still intact, covered with plastic wrap, three to five days.

Freezing Fresh Strawberries. Carefully wash and hull fully ripe strawberries. Gently stir 2 cups small whole or sliced berries in a bowl with 1/3 cup sugar and mix thoroughly. Pack in a rigid container, leaving 1/2 inch head space.

TAMARINDS

The tamarind is a brittle brown pod containing shiny seeds and gristly fibers suspended in a soft, reddish pulp. Only the pulp is edible and it is used somewhat like lemon juice in Indian and Southeast Asian cooking. The whole pod may be sold in ethnic markets, but more common are compressed bricks of pulp, which should be somewhat pliable. Tamarind paste can be refrigerated for two to three weeks.

UGLIFRUIT

Uglifruit actually is ugly. A citrus fruit that looks like a bruised, misshapen grapefruit, it is on the market in the winter months.

An uglifruit is very juicy, so it should be heavy for its size. Don't be put off by its loose skin and spongy texture. That is the way it is supposed to feel. Treat uglifruit like other citrus: eat the sections raw or squeeze it for juice.

Store and freeze uglifruit according to the directions for grapefruit, p. 93.

WATERMELONS

A watermelon has two incomparable assets as a fruit: it is large enough to feed a lot of people and juicy enough to quench a lot of thirst. This gigantic, elongated melon can weigh from 10 to 50 pounds. Its peak season is during the hot weather months.

Charleston Gray is the most widely distributed variety of watermelon—the rind is light green with darker green veins, the flesh crisp and red. The Cannonball watermelon has a solid dark green rind. Klondike watermelons look like Charleston Grays but are even sweeter. The Sugar Baby is a smaller melon, usually 8 to 10 pounds. There are other types with startling orange-yellow flesh, but they taste just like more ordinary watermelons.

Buying Watermelons. When you pick out whole melons, look for deep-colored rinds with a dull, waxy bloom. The bottom side of the watermelon, where it touched the ground as it grew, should look yellow rather than white or pale green. The watermelon should be well rounded and symmetrical. Neither end should be flattened. It should feel firm but not rock-hard.

Cut watermelon is easier to judge. The flesh should have intense, uniform coloring with no white streaks. The seeds should be dark and hard. Too many pale seeds mean a bland, immature melon. The flesh should look crisp, neither too mushy nor too dry, and should give off a light, sweet fragrance.

Storing Watermelons. Watermelons won't ripen or sweeten after harvest but you can store them at room temperature for two to three days.

In the refrigerator, a whole melon keeps up to a week.

Wrap cut melon tightly in plastic wrap and refrigerate no more than three or four days.

Freezing Watermelon. Remove all the rind and seeds from the flesh. Cut the fruit into cubes or balls. Pack in rigid containers and cover with 30 percent syrup (see p. 75).

Frozen watermelon has a better texture when it is served just partially thawed.

Watermelon keeps ten to twelve months at 0° F.

DRIED FRUITS

Dried fruits are wonderful foods—intensely sweet, chewy, and loaded with vitamins. They last a good long time on the shelf because, of course, they have been preserved by drying, either sun-dried or dehydrated in heating ovens. Drying reduces the moisture in the fruit to such a low level that the microbes that would normally spoil the food can no longer function.

The enzymes in the fruit to be dried are usually, but not always, deactivated by the application of heat or sulfur dioxide. Many commercially dried fruits have some preservative added as well. The label lists all additives. Fruits labeled sun-dried probably have not been pretreated, but the label will tell you for sure. Sun-dried fruits may not always look as glossy as oven-dried but usually taste much better.

Buying Dried Fruit. The most commonly available dried fruits are:

• APRICOTS. You can find both dehydrated and sun-dried apricots, pitted and halved. The sun-dried kind are tougher to chew.

• CURRANTS. Dried currants are not the berries of the same name, they are a variety of tiny grape.

• FIGS. Both amber-colored Calimyrna figs and dark purple Mission figs are sold dried.

• PRUNES. Prunes come sweetened or unsweetened in four different sizes and are either pitted or whole. They are moister and softer than other dried fruits.

• RAISINS. Raisins may be dark or golden. Dark raisins are made from Thompson seedless grapes. Golden raisins may be made from Muscat variety or bleached Thompson seedless grapes. Sultanas, made from a seedless yellow grape, are sweeter and softer than Thompson raisins. You can use dark, golden sultanas, or dried currants in place of raisins in any recipe.

• OTHER FRUITS. Dried peaches, pears, pineapples, apples, and bananas are often sold prepackaged in supermarkets and in bulk in specialty shops.

All dried fruits should be somewhat soft and springy; if you can see the fruit, look for bright, appetizing colors. You can squeeze even boxed fruit and feel if it is too brittle. Don't buy a box that has a dry rattle when you shake it.

Storing Dried Fruit. Dried fruit can last a very long time, but only under the right conditions.

Moderately high temperatures, even 75° F., shorten shelf life—the fruit becomes hard, it darkens and loses vitamins. Exposure to too much moisture has the same effect.

Store unopened packages of dried fruit no more than a month at room temperature. For longer storage, keep packages in the refrigerator.

Transfer fruit from opened packages into plastic bags or tightly lidded jars. Seal the fruit airtight and refrigerate.

In the refrigerator, dried fruit keeps six months to one year.

Dried fruit that has begun to harden is not a lost cause. You can soak, steam, or poach it and bring it back to life.

Soak raisins or other dried fruits in sherry or brandy and add them to cakes, breads, fruit compotes, caramel ice cream topping, rice pudding, or serve as a side dish with ham or chicken.

Prunes poached in 30 percent syrup (see p. 75), with cinnamon and lemon slices added, are sweet and spicy.

To use up tidbits of dried fruit, cut them into little pieces with scissors dipped in warm water and add them to hot cereal, cookie batter, pancake batter, poultry stuffing, ham or chicken salad, cheese spreads, and even to rice as it cooks.

You can also make juice from dried fruit. Simmer 1/4 pound fruit in 1 cup water over low heat for about 45 minutes. Store the fruit and liquid together in the refrigerator in a tightly lidded jar up to two weeks. Strain out the chilled liquid as a beverage and use the stewed fruit as described above.

JELLIES, JAMS, AND PRESERVES

The essence of any fruit can be preserved as jelly or one of jelly's relatives.

Most fruit contains pectin, a natural substance which, when heated, solidifies juices into the soft, translucent gel that spreads so sweetly on a slab of bread.

Buying Jelly, Jam, and Preserves. Here are the types of fruit jellies you see in the store and, for that matter, on the home canner's shelves:

• JELLY. Jelly is clear and smooth. It is made from strained fruit juices.

• JAM. Jam is jelly that contains cooked fruit pulp.

• MARMALADE. Usually made from citrus fruits, marmalade is a jelly that contains the fruit's rind. The rind introduces an edge of bitterness to offset the sweetness.

• PRESERVES. The cooked fruit in preserves is heated only slightly, so it keeps its shape in the jelly.

• FRUIT BUTTER. Fruit butter is a spicy puree of fruit, usually apples.

• CONSERVE. Jam is called conserve when it is made from more than one type of fruit.

Most commercial and homemade jellies have extra pectin and lots of sugar added for flavor and consistency. Some commercial brands contain other additives—all are listed on the label.

Storing Jelly, Jam, and Preserves. Unopened jars of jelly, on a cool, dry, dark shelf, should keep up to twelve months.

Opened jars must be refrigerated, well covered. At room temperature they are prone to lose a lot of flavor and to grow mold.

You can refrigerate an opened jar of jelly at least six months.

You can use up the few teaspoons of jelly left clinging in the jar in a couple of ways. Scrape it out and spread it on cooked carrots, parsnips, or winter squash as a glaze. Glaze ham, pork, lamb, or chicken with jelly. Mix a little jelly into barbecue sauce, tomato sauce, or baked beans for a subtle change of flavor. Whip jelly into butter to serve over pancakes or waffles.

Make tiny jam tarts with homemade pie dough or the little store-bought pastry cups.

STORAGE TIMETABLE FOR FRESH FRUIT

Type of Fruit	Ripen at Room Temperature	Time in Refrigerator	Months in Freezer
APPLES	YES	2–4 weeks	10–12
APRICOTS	YES	2–3 days	10–12
ASIAN PEAR	YES	2–4 weeks	10–12
AVOCADOS	YES	10–14 days	3–6
BANANAS	YES	1 week	3
BLACKBERRIES	NO	1–2 days	10–12
BLUEBERRIES	NO	1–2 weeks	10–12
CANTALOUPES	YES	7–10 days	10–12
CASABAS	YES	7–10 days	10–12
CHERIMOYA	YES	1–2 days	10–12
CHERRIES	NO	2–4 days	10–12
CRANBERRIES	NO	1 month	10–12
CRENSHAWS	YES	7–14 days	10–12
CURRANTS	NO	1–2 days	10–12
DATES	NO	variable	10–12
ELDERBERRIES	NO	2–3 days	10–12
FEIJOA	YES	1–2 days	10–12
FIGS	NO	1–2 days	10–12
GOOSEBERRIES	NO	2–3 days	10–12
GRANADILLA	YES	1 day	10–12
GRAPEFRUIT	NO	10–14 days	10–12
GRAPES	NO	3–5 days	10–12
GUAVA	YES	2 weeks	10–12
HONEYDEWS	YES	1 week	10–12
HUCKLEBERRIES	NO	1–2 weeks	10–12
KIWIFRUIT	YES	1 week	10–12
KUMQUATS	NO	3 weeks	10–12
LEMONS	NO	2–3 weeks	N/R[1]
LIMES	NO	3–4 weeks	N/R[1]
LITCHIS	NO	1 week	10–12
MANGOES	YES	2–3 days	10–12
NECTARINES	YES	3–5 days	10–12
ORANGES	NO	10–14 days	10–12
PAPAYAS	YES	2 weeks	10–12
PEACHES	YES	3–5 days	10–12
PEARS	YES	3–5 days	10–12
PERSIAN MELONS	YES	7–10 days	10–12
PERSIMMONS	YES [2]	1–2 days	10–12
PINEAPPLES	YES	3–5 days	10–12

Type of Fruit	Ripen at Room Temperature	Time in Refrigerator	Months in Freezer
PLANTAINS	YES	3–5 days	N/R
PLUMS	YES	3–5 days	10–12
POMEGRANATES	NO	2–3 weeks	10–12
PRICKLY PEARS	YES	1–3 days	10–12
PRUNES	YES	3–5 days	10–12
QUINCES	YES	2–3 weeks	10–12
RASPBERRIES	NO	1–2 days	10–12
RHUBARB	NO	3–5 days	10–12
SAPOTE	YES	3–5 days	10–12
STAR FRUIT	YES	5–7 days	10–12
STRAWBERRIES	NO	3–5 days	10–12
TAMARIND PASTE	NO	2–3 weeks	N/R
UGLIFRUIT	NO	10–14 days	10–12
WATERMELONS	NO	1 week	10–12

[1] Lemon juice and lime juice can be frozen three to four months.
[2] You can also ripen a persimmon by wrapping it in foil and leaving it in the freezer overnight.

STORAGE TIMETABLE FOR PROCESSED FRUIT

Type of Fruit	Months in Pantry	Time in Refrigerator
CANNED FRUIT (except Citrus):		
Unopened	12	—
Opened	—	3–5 days
CANNED FRUIT, Citrus:		
Unopened	6	—
Opened	—	3–5 days
DRIED FRUIT	1	6 months
JELLIES, JAMS, AND PRESERVES:		
Unopened	12	---
Opened	1	6 months

NUTS AND SEEDS

Nuts and seeds are wholesome food, appealing because of their rich, slightly sweet flavor and their distinct texture, a happy balance of crisp and chewy.

They fit smartly into any course on the menu—in soups, sauces, salads, breads, and desserts, to name a few—but nuts and seeds, all by themselves, are the quintessential snack. Whether it's ball-park peanuts out of a bag or cashews out of a silver dish, nuts and seeds just taste good.

Buying Nuts and Seeds

Nuts and seeds come to market in a number of forms, depending on how much processing they have undergone.

Nuts in the shell are the least processed. Most nuts sold in the shell are not roasted, they simply come as is from harvest. The usual exceptions are peanuts, pistachios, and chestnuts, all commonly roasted in the shell.

Nuts in the shell should be clean, undamaged, and feel heavy for their size. Avoid nuts with splits, cracks, stains, or holes. *Do not buy or eat nuts with any signs of mold; they can be toxic.* A good test is to pick up a few nuts and shake them. If they make a rattling sound they are dry and possibly stale.

Nuts in the shell are a good buy because they store best and because they are almost always priced to yield nuts cheaper per edible portion. Of course, they leave you with the job of shelling. But fresh nuts you roast, season, and eat warm from your own oven make other nuts taste like salted erasers in comparison.

Shelled nuts may be raw or roasted. Roasted nuts are often, but not always, salted. Shelled nuts come packaged whole, in broken pieces, slivered, sliced, and finely ground.

You can judge shelled nuts by eye when they are in jars or cellophane bags. Only canned nuts must be bought on faith. Look for plump nuts, uniform in color and size. Dark shriveled specimens are probably stale.

When you buy shelled nuts in bulk, you have two extra measures of quality to apply. The nuts should feel crisp, not rubbery, and smell fresh, not rancid.

Storing Nuts and Seeds

Without protection against moisture, high temperatures, and air, nuts and seeds become rubbery and their natural oils turn rancid.

Shelled nuts will stale in a matter of days if left uncovered at room temperature. Nuts in the shell stay fresh longer but even they need careful handling to be stored successfully for more than about four weeks.

Nuts and seeds, in whatever form, should always be stored in tightly covered containers or in airtight plastic bags. The refrigerator is often the best place for them. (It is possible to freeze nuts and seeds, but they tend to lose flavor and crispness after thawing, so buy only what you can use within a reasonable time.)

Nuts purchased in cans or vacuum-packed jars can be stored on a cool, dry shelf until they are opened. The sealed packages keep nuts fresh for twelve to eighteen months. Jars should be placed away from direct sunlight. After opening, store cans and jars of nuts, tightly covered, up to one week on the shelf or up to several months in the refrigerator.

ALMONDS

An almond tree bears a fruit like a small green peach and the pit of the fruit is the almond, shell on the outside, edible kernel on the inside.

The familiar nut appears in markets in various states of undress: in the shell (fruit removed); out of the shell; and blanched (skin removed).

Some whole almonds are spiced or flavored; some blanched almonds are slivered.

A pound of almonds in the shell yields about 1¼ cups whole nuts.

Almonds in the shell are among the most durable nuts. If sealed in a tightly closed canister, they can be stored in a cool, dry place for up to a year but should be transferred to the refrigerator in warm, humid weather. Keep shelled almonds in a tightly covered container or plastic bag in the refrigerator up to nine months.

BRAZIL NUTS

Brazil nuts are harvested in the wild in South America. The rough brown shell is about the size and shape of an orange section.

Brazil nuts have a higher fat content than most other nuts and therefore are especially susceptible to rancidity. Keep both in-the-shell and shelled Brazil nuts in airtight containers or bags in the refrigerator or freezer up to nine months.

CASHEWS

You never see cashews in their shells because their natural housing contains caustic oils. Whether roasted or raw, the cashews you buy have been processed to remove the oils. The plump, curved kernels are ideal raw and in quick-cooking recipes but tend to get soggy when cooked in baked goods.

Keep cashews in an airtight container in the refrigerator up to six months.

CHESTNUTS

Chestnuts have a smooth, dark brown shell about the size and shape of a small plum. They may be eaten raw, but most people prefer them hot from the roaster, when their buttery-flavored meat has a soft, mealy texture.

Refrigerate chestnuts in the shell but not in an airtight container. Use a perforated plastic bag and they should keep four to six months.

For longer storage, shell chestnuts (an admittedly tedious job) and blanch the kernels. To simplify the peeling, slash a cross at the flat end of each shell with a sharp knife, then boil the nuts for about 15 minutes. After draining, the shells and skins should slip off easily. Pack the kernels —whole, chopped, or pureed—in an airtight container, label, date, and freeze. They will keep nine to twelve months at 0° F. Use in cooking without thawing.

COCONUTS

Lining the inside of the coconut's shaggy shell is chewy white meat and in the hollow core is a liquid called coconut water. (Coconut *milk* is a concoction made by soaking shredded coconut meat in the cloudy coconut water.)

Before buying a fresh coconut, shake it to be sure there is liquid inside —the more water you hear sloshing around, the fresher the coconut. Avoid ones with damp or moldy eyes or cracked shells.

A medium-size coconut, around 1 pound, yields about 3 cups of shredded coconut meat.

A fresh coconut in the shell keeps well up to one month in the refrigerator. Once opened, tightly cover fresh coconut meat and refrigerate it up to four or five days. (The technique for opening a coconut that requires the least finesse is as follows: pierce the shell twice with an icepick and drain off the water; put the coconut in a preheated, 400° F. oven for 15 minutes; take the coconut out and smash it with a hammer. Cut the meat away from the shell with a short, sharp knife and pare off the brown skin.) Use fresh coconut milk and coconut water within two days. You can freeze shredded coconut meat in an airtight plastic bag for six months at 0° F.

Shredded coconut also comes canned or packed in cellophane bags. The canned is moist and closer to fresh. It may or may not be sweetened. After opening, canned coconut keeps five to seven days and dried keeps three to four weeks, tightly covered in the refrigerator.

FILBERTS

Filberts, also called hazelnuts, are almost always sold in their round, light brown shells. When buying filberts, pick up a few at random and shake them—stale nuts will rattle inside the shell.

One pound of filberts in the shell yields about 1½ cups of whole nuts.

Store nuts in the shell inside a tightly closed container up to three months in a cool, dry place, or up to nine months in the refrigerator.

MACADAMIA NUTS

Macadamias are round, sweet, buttery nuts grown mainly in Hawaii. They are almost invariably sold shelled, roasted, and salted, but they can sometimes be found fresh in the shell. Their shells are 1 inch across and ⅛ inch thick, not easy to break into. They are expensive and therefore usually savored one at a time; however, they can be substituted for other nuts in cooking and baking.

Store macadamia nuts in an airtight container in the refrigerator up to six months.

PEANUTS

Peanuts are universally beloved and rightly so. They show up in exotic places—spicy oriental sauce and cream of peanut soup, for instance—and glue together every kid's favorite lunch, the famous peanut butter and jelly sandwich.

Peanuts sold in the shell are sometimes raw but most likely roasted and should be labeled as such. A pound of peanuts in the shell yields 2⅓ cups of whole nuts.

Shelled nuts, both the small round Spanish variety and the larger, oval Virginia peanut, are available raw or roasted; salted or plain; blanched or unblanched.

Peanut butter is a smooth puree of at least 90 percent roasted peanuts ("chunky style" has crunchy bits of peanut added). The other 10 percent may be made up of sweeteners and stabilizers. One hundred percent pure peanut butters with no additives are readily available. Because there are no stabilizers, natural oil rises to the top of pure peanut butter, but a quick stir will restore the product to its regular consistency.

Raw peanuts in the shell can be stored in a cool, dry place for up to two months. Roasted peanuts in the shell do not keep as well as raw. They can be stored at room temperature, but for no more than one month.

In the refrigerator, sealed airtight, peanuts in the shell keep up to six months. Refrigerate shelled peanuts for no more than three months in a closed container.

Unopened jars of peanut butter keep up to one year on a cool, dry shelf. After opening, refrigerate peanut butter, tightly lidded, three to four months. Store homemade peanut butter, well covered, no more than ten days, and turn it upside down from time to time to redistribute the oil.

PECANS

Pecans are a Southern specialty particularly identified with Louisiana cuisine, which gives us rich, sticky pecan pie and sugary pecan praline. The pecan, with its smooth, brown, watermelon-shaped shell, is a variety of hickory nut that is native to America.

Shake the pecans in the shell before buying. A dry rattle means the nut is stale. A pound of pecans in the shell yields about 2¼ cups of nuts. Pecans are also sold shelled, whole, and in pieces.

In the shell, in a tightly sealed container, pecans keep two to three months at room temperature or six months in the refrigerator. Shelled nuts keep six months in the refrigerator or twelve months in the freezer, well sealed.

PINE NUTS

Pine nuts are what they sound like, nuts from pine cones that grow on a special type of tree. The nuts may also be labeled pignolias, Indian nuts, or piñons, depending on their country of origin. They are usually sold shelled, but in the Southwest you can find pine nuts roasted in the shell.

Seldom eaten raw, pine nuts often enhance pesto sauce, stuffed grape leaves, and poultry stuffing. Since pine nuts are expensive and sometimes hard to find, slivered, blanched almonds are a good substitute.

Pine nuts go rancid more quickly than other nuts. Store them, tightly covered, in the refrigerator for one month or in the freezer for six months.

PISTACHIOS

The pistachio is a bright green kernel inside a small round shell. The nuts are always roasted for sale, and the shell, naturally grayish, may be dyed red or bleached white (entirely for the sake of decoration). The roasted shell is partially open, like a spherical clam. Pistachios may also be found shelled.

Keep pistachios, whether in the shell or out, in the refrigerator. They will stay fresh for three months in a tightly closed container. In the freezer they keep one year at 0° F. in an airtight container.

SEEDS

Many stores carry pumpkin seeds and sunflower seeds as snack foods. Pumpkin seeds are usually sold hulled, roasted, and salted but can be found raw and still in the shell. Raw or roasted, sunflower seeds are usually sold in the shell but may also be found hulled and labeled sunflower nuts. These are delicious raw and can be an inexpensive substitute for nuts in most recipes.

You can keep these seeds two or three months at room temperature. For longer storage, store in airtight containers in the refrigerator up to one year.

WALNUTS

The most common walnut is the English walnut, with its round, wrinkly tan-colored shell; California grows 90 percent of the world's supply. Another variety, seen only rarely, is the distinctive black walnut, which has a very hard, dark brown shell.

English walnuts may be found in the shell; a pound yields about 2 cups of nuts. They are also sold shelled, whole, and in pieces. Black walnuts are almost always sold shelled, in bags or cans.

In the shell, walnuts keep three months at room temperature. Shelled walnuts should be refrigerated in tightly closed containers. They will keep up to one year.

STORAGE TIMETABLE FOR NUTS

	Time in Pantry	Time in Refrigerator	Months in Freezer
ALMONDS:			
in the shell	12 months	12 months	12
shelled	—	9 months	12
BRAZIL NUTS:			
in the shell	—	9 months	9
shelled	—	9 months	9
CASHEWS:			
shelled	—	6 months	9
CHESTNUTS:			
in the shell	—	4–6 months	—
shelled	—	4–5 days	9–12
COCONUTS:			
in the shell	—	1 month	—
shelled	—	4–5 days	6
milk or water	—	2 days	—
canned, open	—	5–7 days	6
canned, unopened	12 months	—	—
dried, unopened	6 months	—	—
dried, open	—	3–4 weeks	6
FILBERTS:			
in the shell	3 months	9 months	12
MACADAMIA NUTS:			
shelled	—	6 months	9–12
PEANUTS:			
raw, in the shell	2 months	6 months	9–12
roasted, in the shell	1 month	6 months	9–12
shelled	—	3 months	6

	Time in Pantry	Time in Refrigerator	Months in Freezer
peanut butter, unopened jar	12 months	—	—
peanut butter, opened jar	1 month	3–4 months[1]	—
PECANS:			
in the shell	2–3 months	6 months	12
shelled	—	6 months	12
PINENUTS:			
shelled	—	1 month	6
PISTACHIOS:			
in the shell and shelled	—	3 months	12
PUMPKIN SEEDS AND SUNFLOWER SEEDS			
in the shell and shelled	2–3 months	12 months	12
WALNUTS:			
in the shell	2–3 months	12 months	12
shelled	—	12 months	12
MIXED NUTS:			
in the shell	—	9 months	12
shelled	—	9 months	12
CANNED NUTS:			
unopened	12 months	—	—
opened	—	6–12 months	—

[1] Store homemade peanut butter 10 days.

3

Milk, Cheese, and Eggs

MILK AND MILK PRODUCTS

Milk is a blend of elements—proteins, fats, sugar called lactose—loosely bonded in water. Take a bucket of milk fresh from the cow and leave it in the barn for a while and this is what happens: first, the fat in the milk clumps together and rises to the top as sweet, viscous cream. Then, after only a few hours, random bacteria from the cow, the air, the dust begin to grow in the milk. Bacteria are everywhere. Certain bacteria turn milk sugar into acid, and as it becomes more acidic the milk slowly changes from a liquid into a tart gel.

This bucket of milk only appears to be in ruins. In fact, what you have here is similar to sour cream, yogurt, and cheese—all products of beneficial bacterial action.

The souring milk's acidity acts against certain microbes that cause spoilage, but the bucket of milk is by no means immune to trouble. There are overwhelmingly good reasons not to drink milk that has been left in the corner of the barn. Cows and people can contaminate milk with microorganisms that cause tuberculosis, diphtheria, typhoid, scarlet fever, and undulant fever in humans.

This is where pasteurization comes in. Dairies pasteurize nearly 100 percent of the milk produced in the United States, by briefly heating it to very high temperatures that kill *all* pathogens in milk and many other bacteria.

Pasteurized milk is safe but not sterile. The few bacteria left after heating, and those accumulated accidentally in handling, eventually cause spoilage. Notice that pasteurized milk goes from fresh to spoiled without becoming a delicious dollop of sour cream along the way. To make yogurt, buttermilk, and sour cream, dairies reintroduce pure strains of bacteria into pasteurized milk.

Buying Milk and Milk Products

The best guide to freshness is a dating code. Readable dating codes on dairy products, while not universal, are common. Many states require either a "sell by" date or a "use by" date to be stamped on the product label.

Find out the exact meaning of the dating codes on the dairy products you buy. Freshness requirements vary widely from state to state and in many cases there is a hidden code you can learn to decipher. Ask the grocer.

The market where you buy dairy products is to a great extent responsible for how fresh they are. A store with high turnover in yogurt, for instance, renews its supply with fresh cartons two or three times a week. But turnover is not as important as how the market stores its milk products. When the temperature of milk products increases to 45° F. or 50° F. during handling or in the display case, shelf life is cut in half.

Beware of dairy products stacked high in overcrowded refrigerator cases. Milk, cream, butter—they all need to be surrounded with cold, circulating air. Any time you buy a milk product that spoils much sooner than the dating code predicts, even though you have refrigerated it promptly, take it back to where you bought it and ask for a refund.

Dairies that deliver milk and milk products to your home should provide you with an insulated box to hold the food until you bring it into your kitchen. Try not to leave dairy products in the box more than an hour.

Storing Milk and Milk Products

All fresh milk and fresh milk products must be stored in the refrigerator or, in the case of ice cream, in the freezer. Dried and canned milk products can be stored on the shelf but must be refrigerated after opening.

You must be careful to keep fresh dairy goods cold between the market and your kitchen. Fifteen to 20 minutes in a hot car is enough to raise the temperature of a carton of milk ten degrees.

Milk, butter, and other fresh dairy goods should be kept wrapped or covered to keep contaminants and refrigerator odors out.

Milk-based prepared foods are particularly perishable. Store fresh, homemade puddings, custards, and creams two to three days, tightly covered, in the refrigerator.

It is possible to freeze these desserts successfully if they contain whipped cream, whipped egg white, or gelatin. Wrapped airtight, they keep about one month in the freezer. Thaw them in the refrigerator and serve them within one day.

MILK

Milk is rich in nutrition and decidedly earns its place as a symbol of health and purity.

The milk we drink is wholesome and fresh because of the rigorously applied technology of the dairy industry. Milk does not come to you straight from the source. A strawberry plucked from a bush can be delivered to you pretty much as is, but untreated milk can be, at worst, positively dangerous.

The dairy farmer takes responsibility for preventing contamination of the raw milk as it comes from the cow. He must keep his milking machinery meticulously clean and wash the cows' udders (called bags) before each milking. The milk, warm from the cow, travels through hoses directly from the bag to a refrigerated farm tank.

A sanitized, insulated tank truck comes to the farm daily to pick up the stored milk. Before taking it on, the driver tests a sample for bacteria and for fat content. He checks to see if the farm tank is properly chilled and even sniffs the milk to make sure it is fresh. The truck delivers the milk to a central dairy where it is rechecked, then transferred to yet another refrigerated tank to wait for processing.

The raw milk first goes through a centrifuge, which forces out sediment and separates the cream from the milk. Next the dairy recombines the cream with the milk in standard proportions to make whole, lowfat, or skim milk.

It is at this point that vitamin D is added. Vitamin D helps the body absorb the calcium in milk and most of the milk produced in the United States contains it. Vitamin A is another common additive.

The dairy next pasteurizes the milk at 161° F. for 15 seconds. Pasteurization increases shelf life but reduces milk's thiamine and vitamin C.

Pasteurized milk is then homogenized: forced under pressure through an extremely fine screen that breaks up fat into minute particles that remain uniformly suspended in the milk. If it is left unhomogenized, all of the fat in milk separates out and rises to the top as heavy cream.

Finally the milk is cooled and either packaged for retail sale or processed into another dairy product (cottage cheese, buttermilk, and so on).

Under the best conditions, fresh milk stays fresh ten to twenty days after it comes out of the cow. The bacteria that eventually multiply in milk do not usually cause illness, but the taste and smell of spoiled milk are certainly nauseating. (Unpasteurized milk sours before it spoils and can be used in baking. Pasteurized milk simply spoils.) Milk develops an off flavor toward the end of its storage life.

Buying and Storing Fresh Fluid Milk. Fresh milk should have a sweet, clean taste. A cooked flavor comes from haphazard pasteurization and a grassy or garlic taint means the cows were milked too soon after eating. Bacterial growth causes acid or fruity tastes. Only well-run dairies and careful retailers provide good milk.

Pasteurized liquid milk is categorized by the amount of fat it contains: *whole milk* has from 3 percent to 3.8 percent; *lowfat milk* ranges from 0.5 percent to 2 percent; and *skim milk* has less than 0.5 percent.

Inspect the list of ingredients for the milk you buy regularly. Some dairies add stabilizers and emulsifiers to milk. "Protein fortified" on the label means that nonfat milk solids have been added.

Chocolate milk is whole milk with flavoring and sweeteners added. A label reading "Chocolate flavored skim milk" indicates all milkfat has been removed and some artificial flavorings have been added.

When fresh *goat's milk* is available, it is usually ultrapasteurized—heated at 280° F. for 2 seconds. The higher heat destroys more bacteria than regular pasteurization, so the milk has a longer shelf life.

The dating code most often used for milk indicates the last date of sale and high-quality milk can be stored for several days after the printed date.

Safe storage for fresh milk is cold, clean, and dark.

Milk keeps longest at 40° F. Higher temperatures, even for a short time, allow devastating bacterial growth. Buy only milk that is well chilled in the market. Use home-delivered milk only if you can take it right from the driver to your refrigerator, or have an insulated box to hold the milk cartons for a short span.

Look upon the milk bottle or carton as a *storage* container and not as a *serving* container. When you want to have milk sitting on the table for drinking, for coffee, for cereal, or for any other use, don't leave it in its storage container. Pour just the amount you will need into a pitcher and leave the rest of the milk in the refrigerator. This way your milk supply stays cold and, consequently, fresh.

Never pour milk from a serving pitcher back into the storage container.

Besides being warm, the milk that sat on the table probably picked up some microorganisms as it was handled. If you pour this "used" milk into the clean milk you will contaminate your whole supply. Either discard the milk or cover and refrigerate the pitcher separately.

Keep milk clean by keeping it covered. Always put the cap back on the bottle; always refold the top of the carton. (You also seal out odors from other foods by covering the milk.)

Resist the temptation to drink your milk icy cold right from the bottle, too. The germs in your mouth will slide straight in and start to spoil the milk as soon as your back is turned.

Store fresh fluid milk in the dark, because light alters the flavor of milk and destroys its vitamin B_2. Paper cartons and tinted glass bottles are enough protection; clear glass or plastic is not. Some people detect a paper taste from cartons, but what light does to milk is probably worse.

Plan to store fresh fluid milk one to five days beyond the "sell by" date. Laws regulating milk sales are vastly different from place to place. In New York City milk must be sold or destroyed no later than four days after it has been pasteurized. In many Midwestern states dairies can set their own expiration dates. Wherever you are, you can usually count on milk having its characteristic sweet taste up to the "sell by" date, but only experience can tell you how much longer it can last. Lowfat and skim milk generally do not keep as well as whole.

You can freeze fresh fluid milks at 0° F. up to three months. The milk sometimes gets a bit grainy after freezing, but it is perfectly all right for both drinking and cooking. Freeze milk in an airtight container (an unopened paper carton is best) at 0° F. Thaw it in the refrigerator.

Buying and Storing Canned Milk. *Evaporated milk* is made from fresh, unpasteurized whole milk. After 60 percent of the water is extracted, the concentrate is heated, homogenized, canned, and heated again to sterilize the contents. Evaporated milk always contains vitamin D and may have other nutrients and chemical stabilizers added. A mixture of one part water and one part evaporated milk will have about the same nutritional value as an equal amount of fresh milk. Out of the glass, it tastes flat—cooked—but diluted evaporated milk works in recipes calling for whole milk.

Evaporated milk does not usually carry a dating code.

Health food outlets carry canned, evaporated goat's milk, in a similar concentration.

Sweetened condensed milk goes through much less processing than evaporated milk. Pasteurized milk is combined with a sugar solution, then water is extracted until the mixture is less than half its original weight. Heating is not required because the high sugar content prevents spoilage. It is very high in calories, too: 980 in 8 ounces.

Sweetened condensed milk can be a coffee lightener and an ingredient in dessert recipes. It usually is not interchangeable with evaporated milk. The label has a stamped code, often small and difficult to spot, that indicates the last date for use.

Canned whole milk, also called UHT milk (stands for ultra-high temperature), is whole milk packaged in aseptic containers, either cans or laminated paper cartons. This milk has the same composition as fresh, whole milk, tastes something like it, and yet can be stored at room temperature because of special pasteurizing. The date on the label allows for an eight-month shelf life, but the flavor begins to stale sooner.

Store unopened cans of evaporated and canned whole milk on a cool, dry shelf up to six months. Check the date on sweetened condensed milk for maximum storage. After opening, store canned milk in a clean, opaque container in the refrigerator. Kept cold and covered, it lasts three to five days.

Sweetened condensed milk may thicken and darken as it ages, but it can still be used.

Buying and Storing Dry Milk. *Nonfat dry milk* is a powdered concentrate of pasteurized skim milk. About 3 tablespoons of nonfat dry milk added to 8 ounces of water make 1 cup of milk that you can drink or cook with just like fresh milk, with a notable sacrifice in flavor. Add the dry powder alone to baked goods, hot cereals, casseroles, and meat loaf as a nutrition booster. A U. S. Extra Grade shield on the label indicates the highest-quality product available.

Flavored nonfat dry milk may be found packaged as a low-calorie diet food (artificially sweetened) or, at the other end of the gamut, as malted milk or cocoa mix. The key ingredient is the dry milk, so buy and store these products accordingly.

Dry whole milk has a higher fat content and therefore a shorter shelf life than nonfat; otherwise, it can be used in exactly the same way. Dry whole milk is less widely available but can usually be found where camping supplies are sold.

You may well find a dating code on dry milk containers, often a "sell by" date.

An unopened package of nonfat dry milk can stay in the pantry for as long as a year. The same holds true for flavored nonfat dry milk and dry cocoa mixes. Heat shortens its shelf life, so if your kitchen gets warmer than 72° F., use the dry milk within a few months.

Unopened packages of dry whole milk last three months in a cool, dry place. It is better to refrigerate them than to leave them at room temperatures above 72° F.

After opening a package of dry milk, transfer the powder to a tightly covered glass or metal container (dry milk can pick up odors from plastic containers) and keep it in the refrigerator. Unsealed nonfat dry milk keeps for a few months; dry whole milk for a few weeks.

Dry milk can be left at room temperature in a sealed container but tends to lump and to absorb odors, especially in humid weather.

Combine the dry milk with water at least several hours before you plan to use it to give it time to dissolve fully and to develop a fresher flavor. Store reconstituted milk exactly as you would fresh fluid milk (see p. 134). Keep it refrigerated three to five days.

Buying and Storing Cultured Buttermilk. When special bacteria are added to fresh fluid skim milk, part of the natural milk sugar converts to acid and the result is thick, tangy buttermilk. Dairies may add salt, butter particles, citric acid, or other ingredients to enhance flavor. Use buttermilk in recipes calling for sour milk.

Like other fluid milk, buttermilk needs refrigeration. Keep it in a closed, clean container, pouring out only the amount you need and quickly returning the milk to the refrigerator.

The flavor of buttermilk is best the first week after purchase. Because its acidity discourages bacterial growth, buttermilk often keeps another five to seven days after that without spoiling.

If your buttermilk is not spoiled, but past its peak of flavor, you can still use it successfully in cooking.

Freezing buttermilk does not work very well because it tends to separate. If necessary, you can freeze it, thaw it in the refrigerator, and stir it thoroughly to remix the liquids and solids. The taste won't be quite the same as fresh buttermilk, but it will work fine for cooking.

Buying and Storing Eggnog. Many dairies package eggnog for holidays. It is a mixture of milk, eggs, sugar, and cream. Look for the dating code on cartons of fresh eggnog before you buy.

Commercial eggnog should still taste good three to five days after purchase. Keep it well chilled in the refrigerator in a closed container, following the instructions for fresh fluid milk (see p. 134).

Homemade eggnog should be made with fresh ingredients in small enough batches so there is no need to store a lot of leftovers. Keep it cold even as you serve it. If it is in a punch bowl, set the bowl on a bed of ice. It is difficult to plan down to the last cupful, so if you do have surplus eggnog, put it in a clean glass jar with a screw-on lid and refrigerate it for no more than one day.

You can freeze both homemade and commercial eggnog and keep it for up to six months at 0° F. Defrost eggnog in the refrigerator. If it has separated, you can stir it back together with a wire whisk or process it for a few seconds in a blender.

Buying and Storing Specialty Milks. *Certified milk* is raw (unpasteurized) milk produced according to government standards by a handful of licensed dairies. While the flavor is the best possible, there are risks. Even these tightly controlled producers occasionally slip up and package unsafe milk. It is reckless to consume any raw milk except that which is labeled "certified." *Multivitamin and multimineral milks* are fortified with various nutrients. The natural salt has been extracted from *low sodium milk* and replaced with potassium.

Acidophilus milk is another cultured milk. It has particular bacteria added that make lactose (milk sugar) more digestible. Before 1975 this type of milk had a soft curd and a very acidic taste. New techniques have made it possible to produce acidophilus milk with flavor and texture almost identical to those of fresh whole milk. You can find it in health food stores and larger supermarkets.

Lactose-reduced milk is whole milk treated with an enzyme that converts lactose into more easily digestible sugar compounds. It is ultrapasteurized and dated.

All of these specialty milks are fresh fluid milks that can be used and stored in the same way as ordinary fresh milk. If you keep them cold, clean, and dark, they keep well three to five days after purchase.

CREAM

Milk fresh from the cow has fat droplets evenly dispersed throughout the liquid. Because the butterfat weighs less than the water in milk, the droplets clump together and rise to the top as cream.

Cream, therefore, is simply milk with a very high fat content. Whole milk has around 3.25 percent butterfat content; cream has from 18 percent to 40 percent. Cream is thicker than milk and has a rich yellow tint that comes from the carotene in the fat.

Buying Cream. Fresh, fluid cream is classified by fat content. *Half-and-half* is the lightest cream, with only about 11 percent butterfat. *Light cream* (also called table or coffee cream) has at least 18 percent fat content. With around 30 percent butterfat, *light whipping cream* is the lightest cream that will whip into a stable foam. *Heavy cream* whips up the best; it has 36 percent or higher fat content. Heavy cream is often ultrapasteurized for a longer shelf life, but this type does not whip as well as ordinary heavy cream and has less flavor. It helps to shake the carton before whipping. "Heavy" cream is not actually heavier by weight but is usually more viscous than lower-fat creams.

Prepared whipped cream comes in aerosol cans and contains, in addition to cream and sweeteners, chemical stabilizers. Read the label so you don't confuse this product with "dessert toppings," which contain vegetable oil and other additives.

Buy cartons of cream that are well chilled and clearly dated.

Storing Cream. Cream, except ultrapasteurized, is slightly more perishable than whole milk, so apply at least as much caution in storing it. Measure the temperature in different parts of your refrigerator and keep cream in the very coldest.

Don't take out the whole carton of cream and let it sit on the table during breakfast. Pour out just the amount you need and return the rest of the cream to the refrigerator. Never return room-temperature cream to its original container. Refrigerate it separately in a tightly covered glass or ceramic container. Always store cream in a closed container, preferably one that keeps out light.

Pasteurized, fresh fluid cream will stay fresh one to four days after the "sell by" date. Ultrapasteurized cream will keep up to a month (about three to four days past the "sell by" date) as long as it remains sealed. You can store it one to four days after opening.

Cream sometimes forms solid white flecks when it is poured into coffee. As it ages, cream becomes more acidic. Coffee contains acid that along with its heat is enough to curdle the cream. The cream may not actually be spoiled, but when it doesn't dissolve well in coffee you know it is near the end of its shelf life. Try to use it up right away in your cooking.

You can easily recognize fully spoiled cream by its sour, repulsive smell.

You can freeze cream but it won't mix well with hot coffee after thawing. Thawed cream will still be good for cooking but not for whipping. Freeze it in an airtight container, filled to the top. Cream will keep four months at 0° F.

You can whip cream in advance and store it several hours before you serve it. Spoon the whipped cream into a wire strainer suspended over a bowl and refrigerate. Water will drain out of the cream, leaving it nice and thick.

Sweeten the cream by beating in powdered sugar (which gives a smoother texture than granulated), or flavor it by beating in a little vanilla, liqueur, mace, or cocoa.

A neat trick is to whip the cream for freezing. Whip light or heavy whipping cream into stiff peaks, then freeze it in scoops on a cookie sheet. Gently wrap the frozen blobs in plastic, stack them loosely inside a rigid container, and seal it airtight. The whipped cream will keep one to two months at 0° F. It can go direct from the freezer to the top of the dessert without thawing. You can also float it on top of hot coffee or cocoa.

Prepared whipped cream in aerosol cans keeps four to six weeks in the refrigerator. Do not freeze.

BUTTER

To make butter, you start with cream and whip it in a food processor with steel blades for 3 to 5 minutes or beat at high speed with an electric mixer for 5 to 7 minutes or stir it vigorously, by hand, for 45 minutes. The particles of butterfat in the cream eventually separate from the liquid, and when the liquid is drained off (this liquid is natural buttermilk, not the same as the cultured buttermilk at the grocery store), the remaining solids are pure butter. The butter must be washed in cold water and kneaded into a dense, cohesive lump. It takes 2 cups of cream to make 1/2 to 3/4 cup of butter.

It took a strong, tireless arm, but butter making used to be done at home in a wooden butter churn. Dairies now churn 8,000 pounds of butter in a single batch, using pasteurized cream but following the same procedure Great-grandmother did.

Buying Butter. The label rates butter for quality. AA is the best butter, smooth-textured, sweet, and fragrant. Grade AA butter always comes from top-quality fresh cream. Grade A butter, also from fresh cream, is fairly smooth with good flavor. Grade B may taste slightly tart because it is made with sour rather than fresh cream. Not all butter is graded, but the United States Department of Agriculture does put its stamp on many brands; the stamp simply tells you the butter comes from a producer who has met federal standards of sanitation.

By itself, the color of butter, from deep yellow to off white, varies depending on the breed of cow, her diet, and the time of year. Dairies often add natural coloring to butter but are not required to mention it on the label.

• SALTED BUTTER. Salt acts as a preservative as well as a flavoring and the amount added to butter differs from brand to brand. When using this butter in cooking, you should reduce the amount of salt you add accordingly.

• UNSALTED BUTTER. Also called sweet butter, this is frequently specified in recipes for the very reason that the saltiness of salted butter is unpredictable.

• WHIPPED BUTTER is soft because it is full of air bubbles. It spreads easily but is not suitable for cooking, since you must measure it by weight rather than by the customary volume.

Choose butter that is firm, cold, and in a clean, unbroken package. A good guide to both quality and freshness is the scent. Give the carton a good sniff and buy the butter with a sweet, full aroma.

Imported butters are usually sold frozen. Their higher water content makes them more perishable than domestic brands.

Storing Butter. Butter lasts for months, so how you store it is more important than how long you store it. Butter absorbs any passing odor—onion, cigarette smoke, detergent, fish—and takes on a funny taste. The paper wrappers on sticks of butter are enough to protect them for a while, but if you keep butter for more than a few weeks, overwrap it in plastic. Keep unwrapped sticks of butter in a covered dish.

Look for a date on the package of butter. It may be as much as three months ahead. If you keep butter refrigerated, it should stay fresh a few weeks beyond even that date. It loses flavor as it ages (most noticeable in unsalted butter) but remains usable.

When butter spoils, chemical changes take place that release an obnoxious odor. There is no mistaking butter that is rancid.

Butter at 40° F., the usual temperature in the refrigerator, is hard and difficult to spread. The butter compartment in the door of most refrigerators stays at higher temperatures—50 to 60° F. A week or two in this balmy little box won't hurt the butter, but it is a better idea to hold just one stick at a time there and to keep the rest of your supply colder.

If you use butter very quickly it is all right to store a stick at room temperature for a day or two, as long as it is in a covered dish.

If you have a supply of butter that is running out of storage life, you can clarify it to keep it usable a while longer. Clarifying removes the milk solids, the most perishable part of the butter. Heat butter very slowly in a bowl set over a pan of hot water until it is completely melted. Spoon out the clear yellow liquid and discard the white milk solids that have sunk to the bottom. Either strain the clear butter through a damp cheesecloth or refrigerate it and scrape off any residue that rises to the top as it hardens. Refrigerate clarified butter in a clean, tightly lidded jar for up to three months. Though it does not have the same intense flavor as whole butter, clarified butter blends neatly into sauces and won't burn as readily in the sauté pan.

To freeze butter, double-wrap it in plastic or foil so it won't absorb odors. It keeps flavor and freshness six to nine months at 0° F.

Store imported butter in the freezer, thaw it in the refrigerator, and store it thawed up to one month.

MARGARINE

Margarine is whipped up out of vegetable fats. To the palate it is a merely passable substitute for butter, but to anyone who must control cholesterol it is an essential substitute.

The American Heart Association recommends choosing a margarine that lists liquid oil as its first ingredient and that has twice as much polyunsaturated fat as saturated. All this information should be on the label. If it is not, choose another margarine.

Despite margarine's apparent edge over butter, it is certainly not a "health food." For cooking, pure vegetable oils high in polyunsaturates, such as corn or safflower oils, are healthier and contain none of the many artificial additives in margarine.

Buying Margarine. Manufacturers turn out margarine in any number of forms and formulas. *Regular margarine* contains no less than 80 percent fat, along with water, skim milk solids, salt, preservatives, emulsifiers, artificial color, and flavorings. The highest-quality margarines contain all vegetable fat, usually corn, soybean, or safflower oils. Regular margarine is formed into quarter-pound sticks and approximates the consistency of butter. A few products combine margarine and butter in various proportions. Some brands offer unsalted margarine. You can use regular margarine and the margarine/butter combination interchangeably with pure butter in almost any recipe, although the simpler the dish the more you will notice the substitution.

Soft margarines are all vegetable oil, with no milk solids. They contain salt, artificial flavor, vitamins A and D, and preservatives. Packed in tubs, soft margarine is easier to spread than regular. *Liquid margarine,* packaged in plastic squeeze bottles, does not harden even in the refrigerator.

Light or diet margarine has from 40 to 60 percent fat, and proportionally more air and water by volume. It is not a successful substitute for butter or regular margarine in baking or frying.

Buy margarines that are well refrigerated and check for a dating code. Margarine does not spoil readily, but poor refrigeration detracts noticeably from its flavor and texture.

Storing Margarine. Keep margarine well wrapped or tightly lidded, because it tends to absorb odors from nearby foods.

Most margarines store well for as long as four to five months in the refrigerator. Diet margarines should be held no more than two to three months.

Regular, soft, and liquid margarines freeze very well and will keep up to one year at 0° F. They should be double-wrapped to fend off odors. Diet margarine may appear curdled when thawed, but the eating quality is unaffected.

YOGURT

When certain friendly bacteria—*Lactobacillus bulgaricus* and *Streptococcus thermophilus* to be exact—convert the natural sugar in milk to lactic acid, milk thickens, acquires a pleasant tartness, and you have yogurt. Sour cream is a similar food that results when different strains of bacteria are introduced into fresh cream.

Buying yogurt. Most commercial yogurt starts with a mixture of fresh lowfat and nonfat dry milk. Whole milk and goat's milk yogurt are also available. Manufacturers often add sugar, fruit, colorings, and chemical stabilizers. Ingredients and calorie content are very different from one kind of yogurt to another, even when they are the same brand. Read the label for exact information. If you don't want unknown sweeteners and other additives, buy plain, lowfat yogurt and add fruit, jam or honey to your own liking.

Each carton of yogurt should have a date stamp. Usually it is the last date of sale, so you can expect the yogurt to store well for several days beyond that.

Pick clean, tightly sealed, well-chilled containers.

For information on frozen yogurt, see p. 146.

Storing Yogurt. Yogurt is fairly stable, so you can usually count on at least ten days of storage after purchase, unless the dating code indicates otherwise. Yogurt kept clean and cold will be safe to eat even longer than this but develops a harsher taste.

Keep yogurt in a sealed container in the refrigerator. Serve only the amount you need and return the rest to the refrigerator. Always keep it covered.

A bit of cloudy water may separate out from the yogurt during storage. This is not a sign of trouble. Just pour off the liquid or stir it back in. Another approach is to lay a square of cheesecloth on the surface of the yogurt to absorb the water. Replace the towel daily and after several days the yogurt will be as thick as sour cream. Use the thickened yogurt as a delicious, lower-calorie, lower-cholesterol substitute for sour cream.

You can freeze yogurt but the results will vary from brand to brand, from homemade batch to homemade batch, and from flavor to flavor. It may be almost unchanged when thawed or it may be too thin.

You can also use an ice cream maker to freeze yogurt, following the manufacturer's directions.

Freeze yogurt in its original, sealed carton or in an airtight rigid container. It keeps six weeks at 0° F. Thaw three hours at room temperature or one to two days in the refrigerator.

SOUR CREAM

Sour cream is a cultured food made from pasteurized, homogenized cream. The dairy injects a combination of bacteria, called souring agents, into cream. These agents act on the cream, creating its characteristic tangy flavor and stiff consistency.

Buying Sour Cream. According to federal labeling standards, *sour cream,* or *cultured sour cream,* or *acidified sour cream* has at least 18 percent milkfat. *Sour, cultured sour, or acidified sour half-and-half* contains 10.5 percent milkfat. Sour half-and-half is interchangeable with sour cream in recipes but has a thinner flavor. If the label calls it *dressing,* the sour cream contains other milk products, such as butter, but is essentially the same as regular sour cream.

Look at the label for two things: check for extra ingredients such as sweeteners and stabilizers; and check the dating code, for the fresher sour cream is the richer it tastes.

Buy only clean, sealed, chilled cartons of sour cream.

Storing Sour Cream. Keep sour cream in a covered container in the refrigerator. Sour cream lasts about four weeks after packing. You can store it ten to fourteen days beyond the "sell by" date if it is kept cold and clean.

Dip out the amount of sour cream you need and keep the rest chilled. Don't use the storage container as a serving container.

If some liquid separates from the sour cream, it is not spoiled. Pour off the liquid or stir until it is reabsorbed. The microorganisms that spoil sour cream usually declare themselves by growing a pink or green scum—a sure sign that it's time to throw out the carton.

Even the freshest sour cream curdles or separates if it gets too hot. Add it to cooked foods only after you have taken them off the heat or cook over very low heat, stirring constantly. First mixing in 1 tablespoon flour per 1/2 cup sour cream helps prevent separation, too.

Sour cream does not freeze well.

CRÈME FRAÎCHE

Crème fraîche is sweet, thick cultured cream with 30 percent butterfat content. Akin to sour cream but with its own distinct flavor, crème fraîche can be a dessert topping or a lavish ingredient in soups and sauces.

Buying Crème Fraîche. Many urban supermarkets have ultrapasteurized crème fraîche in small cartons in the refrigerator case. It is expensive, so take the trouble to track down a dating code. Like other fresh dairy goods, crème fraîche loses flavor sitting on the shelf.

Buy clean, fully chilled cartons.

If you cannot buy crème fraîche in your market, you can whip up a homemade version of it, using either buttermilk or a freeze-dried starter (available only from a few mail-order sources). For the buttermilk version, combine 1 cup cream with 2 tablespoons buttermilk in a large jar, cover, and store in a warm place 12 to 18 hours, until thick. Stir the mixture and refrigerate in a covered container. If you use heavy cream, the crème fraîche can be used like butter to baste broiled foods or to thicken sauce, off heat. Made with light cream, it is a perfect dessert topping but not suitable for cooking.

Storing Crème Fraîche. Commercial crème fraîche keeps up to 1 month in the refrigerator if tightly covered. Take out only the amount you need at the moment and return the container to the icebox quickly.

Store homemade crème fraîche seven to ten days.

Do not freeze crème fraîche; it separates badly when thawed.

ICE CREAM

Buying ice cream or the frozen desserts that resemble it takes a bit more care than choosing other milk products. Within fairly narrow limits milk is milk and butter is butter, but ice cream is made in as many different ways as there are dairies making it. Also, variations on ice cream, such as frozen yogurt and frozen pudding, abound.

Buying Ice Cream. There are standard labeling practices for most frozen desserts that help the buyer to figure out just what is in the package.

Ice cream, frozen custard, French ice cream, and French custard ice cream are the best quality. They have the highest milkfat and milk solids content. The words "French" and "custard" indicate a high percentage of egg yolk in the product, making the dessert richer and smoother.

Ice milk has less milkfat, protein, and food solids than ice cream, but often more sugar.

Sherbet has less fat and milk solids than ice cream. It has more sugar and usually contains fruit and fruit acid. Citrus sherbets have 2 percent fruit content, berry sherbets 6 percent, and other fruit sherbets must contain 10 percent. Some sherbets have nonfruit flavors such as coffee.

Water Ice contains no egg or milk products at all. It is basically water, sugar, and flavorings. An example is lemon ice.

Mellorine is a frozen dessert in which vegetable or animal fats are substituted for some of the milkfats.

Frozen yogurt has less milkfat than ice cream and less sugar than sherbet. It is made with cultured milk but often pasteurized after processing so it is no longer true yogurt.

Frozen pudding takes the form of an ice cream bar on a stick. There are no federal or state standards for frozen pudding or frozen yogurt, so you will have to deduce from the list of ingredients what their sugar, fat, milk, and chemical makeup might be.

Flavoring ingredients in these frozen desserts will be described in one of three ways. If it is called "strawberry ice cream" all of the flavoring is natural. "Strawberry flavored ice cream" indicates that some artificial flavoring is added, but less by weight than the natural flavoring. There is more chemical than natural flavoring in "artificially flavored strawberry ice cream."

Some manufacturers date their frozen desserts, so while you are reading the label, look for the code too.

Ice cream products are whipped during manufacture. The whipping incorporates air into the ice cream, making it light and smooth. A single pint of ice cream may hold anywhere from 12 percent to 50 percent air. Some air is necessary to keep the ice cream from turning into a rock, but the less air in the pint the more food you are getting and the finer the texture of the product. Twelve percent air tastes better than 50 percent air. The amount of air (called overrun) is not stated on the label and is hard to judge until you actually sit down and eat the ice cream. You might compare one brand with another by actually weighing each pint on a scale in the store. The heavier pint-sized carton has more ice cream and less air in it.

Your market should keep ice cream stored in a freezer with a door, or at least well below the frost line in an open chest. Buy cartons of ice cream that are completely solid and clean. A sticky or frosted package has probably partially thawed sometime in its history.

Get the ice cream home as quickly as you can. Have the store pack it in an insulated bag or, if that is not available, double-bag it.

Storing Ice Cream. Store ice cream and other frozen desserts in the freezer at the lowest possible temperatures. Ice cream can keep up to two months at 0° F. In an ice cube compartment, the ice cream can keep no more than a week.

During long storage ice cream tends to pick up odd odors and to lose its creamy texture. Ice cream is at its best no more than seven to ten days after it goes into your freezer.

Every time you take the carton out of the freezer the ice cream will suffer a setback, refreezing coarser and icier than before. You can't really prevent the change in texture but you can protect ice cream from drying and from off odors. After opening a carton, put plastic wrap over the exposed surface of the ice cream or slip the whole package into a plastic bag and seal it before you put it back in the freezer.

If by accident ice cream or any frozen dessert thaws completely, throw it away or, if it is still cold, eat it immediately over fruit or in a milk shake. Do not refreeze it. Complete thawing makes dangerous bacterial growth possible.

DAIRY SUBSTITUTES

For convenience, economy, or because of dietary restrictions, many people replace fresh dairy foods, particularly cream, with what are called nondairy products, made from vegetable oils, sweeteners, and a staggering list of additives such as sodium stearoyl-2-lactylate, polysorbate 60, and sodium silicoaluminate. In general, these substitutes have a much longer shelf life than their natural counterparts.

If the reason you are buying dairy substitutes is to avoid milk, look carefully at the package's list of ingredients: sodium caseinate, a common additive, is a derivative of milk protein. If you are concerned about fat and cholesterol, skim milk may be a cheaper, lower-calorie, healthier choice for your coffee, since the fats in dairy substitutes are usually partially saturated and their sugar content is high.

Buying and Storing Dairy Substitutes. There are two kinds of *nondairy creamers.* One is dry, in powder form. The other is sold as a frozen liquid. These lighten coffee and usually contain a high percentage of sweeteners, such as corn syrup. Unopened jars of powdered nondairy creamer keep up to two years. Kept dry, the contents will store one year after the jar is opened. Frozen liquid creamer keeps one to two years at 0° F., as long as it is in its original sealed carton. Thawed, the liquid keeps two to three weeks in the refrigerator.

Nondairy dessert toppings have the appearance of whipped cream. They are frozen or packed in aerosol cans for sale. Some brands include real milk products along with the usual vegetable oils. Dessert toppings sold in cartons also store satisfactorily for a year or more in the freezer. After thawing, you can keep them, covered, in the refrigerator for seven days, or you can refreeze them for another several months.

Nondairy dessert toppings in aerosol cans keep in the refrigerator two to three months. There is usually a readable dating code on the can to guide you. Do not freeze the aerosol can.

Filled milk and *imitation milk* are substitutes for whole milk that contain all or part vegetable oils, such as soybean oil, and sometimes artificial ingredients. These are most often sold canned. Filled and imitation milks have the same storage requirements as whole milk in the same form—fresh, fluid, canned, or dried. See "Milk," pp. 133–38.

INFANT FORMULA

Prepared infant formula is primarily water and nonfat cow's milk. Among other ingredients, it contains sweeteners: sometimes lactose, which is milk sugar; and sometimes corn syrup or other sugars. Coconut and soybean oils are common; vitamin and mineral supplements are universal. A few brands contain mono- and diglycerides, chemicals that keep the liquid from separating. Some formulas contain soy substitute and no milk at all.

Buying and Storing Infant Formula. Formula, whether liquid or dry, ready to eat or in concentrate, will be marked with a "use by" date. Observe the date and follow all the instructions for preparation and storage that appear on the package; infants are very susceptible to illness from improperly handled food. Store open containers of formula in the refrigerator for no more than forty-eight hours. Always discard any formula that remains in the bottle after a feeding.

For convenience, you can keep formula sealed in sterilized baby bottles in the refrigerator, but forty-eight hours is still the time limit.

STORAGE TIMETABLE FOR REFRIGERATED
MILK AND MILK PRODUCTS

Product	Days in Refrigerator	Months in Freezer
BUTTER	30–90	6–9
BUTTER, Clarified	60–90	6–9
BUTTERMILK, Fresh fluid	7–14	3
CREAM	1–4	4
CREAM, Ultrapasteurized, unopened carton	30	4
CREAM, WHIPPED:		
Commercial	30–40	1
Homemade	1	2
CRÈME FRAÎCHE:		
Commercial	20–30	N/R
Homemade	7–10	N/R
EGGNOG:		
Commercial	3–5	6
Homemade	1	6
HALF-AND-HALF	3–4	4
ICE CREAM AND FROZEN DESSERTS	N/A	1–2
MARGARINE:		
Regular and Soft	120–150	12
Diet	60–90	N/R
MILK, Fresh fluid	1–5	3
NONDAIRY CREAMER, Frozen liquid	14–21	12–24
MILK-BASED PREPARED FOODS (homemade custard, pudding)	2–3	1

Product	Days in Refrigerator	Months in Freezer
NONDAIRY DESSERT TOPPING:		
Carton	7	12
Aerosol can	60–90	[2]
SOUR CREAM	7–21	N/R
YOGURT	7–14	1–1½

[1] Desserts made with whipped cream, whipped egg white, or gelatin can be frozen up to one month.
[2] Do not freeze aerosol cans.

STORAGE TIMETABLE FOR PANTRY MILK PRODUCTS

Product	Shelf Time (unopened)	Refrigerator Time (opened)
INFANT FORMULA	12–18 months	48 hours
MILK, Canned:		
Evaporated and Whole	4–6 months	3–5 days
Sweetened Condensed	4 months	3–5 days
MILK, Nonfat Dry	1 year	2–3 months
MILK, Whole Dry	3 months	2–3 weeks
MILK, Dry, liquefied	0	3–5 days
NONDAIRY CREAMER, Powdered	2 years	1 year [1]

[1]Refrigeration not necessary.

CHEESE

Cheese is a delectable fabrication capable of preserving the goodness of milk for years. Cheesemakers warm milk, coagulate, cut, reheat, drain, knit, press, salt, and cure it. The character of the milk and the exact way these steps are carried out determine what the cheese will be.

Roquefort cheese, for example, is made from ewe's milk in the same tiny French village where it has been manufactured for a thousand years. The sheep's milk is warmed and left to curdle, a process speeded by the addition of rennet, a natural enzyme (from a calf's stomach) used in most cheesemaking. After about two hours the curd is cut and the remaining liquid, called whey, is drained off.

The mold that will eventually spread in blue-green veins through the cheese is *Penicillium roqueforti,* cultivated in rye bread. Loaves permeated with mold are dried and finely ground. As the drained curd is set into 7½-inch metal hoops, the powdered mold is sprinkled on each layer. After several days the rounds of curd go to curing rooms in the famous limestone caves near the village.

The caverns' gothic arched ceilings are perpetually damp: underground hot springs warm the caves and cause condensation, while cool currents of air flow through fissures in the rock walls. During their first week in the caves the curds' surfaces are repeatedly salted. Before they are set on racks to cure, the rounds are punctured so air can circulate through the cheese to nurture the mold. Finally the cheese is left to age for two to five months.

As any cheese ripens, natural enzymes and microorganisms mature the flavor and establish the texture: Parmesan cheese becomes harder and drier and its taste grows stronger; Brie softens and develops a mellow flavor. Roquefort is both creamy and slightly crumbly when it is aged; the cheese is shot through with a network of blue mold; and the taste is effervescent. Each fully cured round of Roquefort is wrapped in foil and stamped with the image of a red sheep—only cheese from these caves can be thus stamped and labeled Roquefort. Similar cheeses from other areas in France are called Bleu.

Hundreds of different cheeses are made in every part of the globe and each has a distinct life history and individual character.

Buying Cheese

You can't be serious about cheese in a supermarket. Though there are usually a few good cheeses in the average dairy aisle, there is nothing to compare with the scope and quality of a specialty shop.

A good store knows its cheeses and handles them all properly. However attractive it is to have food displayed on counters, all natural cheeses are better off in the refrigerator, and that is where good stores keep them. There will be no strong, musty odor at a good cheese shop, because the best dealers keep cheese well wrapped and discard any overaged or moldy cheese.

The greatest advantage in shopping in a specialty store is the opportunity to sample before you buy. Cheese, especially imported cheese, may be expensive and the only way to be sure of what you are getting is to take a taste from the very same piece you plan to purchase. You don't have to be an expert on type and country of origin to know whether that bite of cheese tastes good to you.

No matter where you shop for cheese, examine each piece closely before you buy. Mold appears as white, pink, green, or blue flecks, or sometimes furry patches. The blue cheeses, such as Roquefort, of course, have desirable mold throughout, but you can easily spot mold that doesn't belong. Avoid pieces with cracked and darkened edges, signs of drying. And don't buy cheeses that look greasy on the surface, because they may have been warmed and chilled too many times.

Prepackaged supermarket cheeses are usually stamped with a dating code that allows some storage time at home. Choose packages that are clean, tightly sealed, and not sticky.

The flavor of cheese ranges from bland to extremely pungent and only the expert knows exactly how a particular cheese *should* taste. Nevertheless, if you bring home a piece of cheese that seems foul rather than simply strong, take it back.

A number of unexpected ingredients can be present in cheese without being declared on the label. Natural and artificial color are common as are texturizers like calcium chloride. Preservatives such as sorbic acid and potassium sorbate are usually listed.

In deciding on the quantity of cheese to buy, think about its usual storage life. A rule of thumb is, the softer the cheese the sooner it spoils. Buy the very softest ripened cheeses no more than a few days before you plan to eat them. You can stock up on harder cheeses—some will last many weeks or months in your refrigerator.

When cheese is the centerpiece of a party, plan on serving about 3 to 4 ounces per person.

Storing Cheese

All cheeses are in a continuous state of change. Firm cheeses, such as Cheddar or the very hard Parmesan, have already aged for some time, and a few more months in your refrigerator will make very little difference in how they taste. Soft cheeses, on the other hand, change radically in a short space of time. In a matter of weeks Brie turns from a bland, chalky nonentity to the glossy smooth queen of cheeses. Wait a while longer and Brie becomes a soggy, bitter mess. The delicate tart flavor of young goat cheese becomes powerfully musty as it ages in your refrigerator.

Keep all cheeses moist and cold and completely protected from giving off or taking on odors. This means wrapping them once or even twice with meticulous care.

Keep all natural cheeses in the refrigerator. Except for blue cheeses (see pp. 160–61), wrap them in plastic or foil.

When cutting into cheese, try to keep its surfaces as flat as possible. You can then smooth and press the wrap nearly airtight around the cheese. Spreading butter over the cut surfaces of hard cheese before wrapping gives a little extra protection. For softer cheese, use the flat of a knife to shape a level surface without scoops and gouges.

Cheese tastes best when it is served at room temperature, but don't be tempted to set an entire round of cheese on the table unless you plan to eat the whole thing. Cheese that has been warmed and chilled repeatedly toughens and dries out. Slice off only the amount of cheese you need and keep the rest in the refrigerator.

Mold sprouts very readily on some cheeses. What to do with moldy cheese is a question without an easy answer. The usual advice, including a recommendation from the United States Department of Agriculture, is to cut away the moldy portion of the cheese and save the rest. The Food and Drug Administration, on the other hand, has indicated that mold presents a very small, but real, danger to human health. Certain molds can cause allergic reactions; some can cause infection. Cutting away visible mold does not eliminate the problem, because mold sends roots deep into the cheese. But again, the risk is small, since most molds are not harmful. A sensible compromise is to cut away a full 1/2 inch off the moldy part of a cheese and use the remainder.

Freezing Cheese

Freezing cheese is possible, but not without some sacrifice of quality. After freezing, cheese is mealy, brittle, and difficult to slice, but if it is not frozen for too long, it is perfectly good for cooking, and freezing it is better than throwing it out. The cheeses that freeze most successfully are brick, Camembert, Cheddar, Edam, mozzarella, Muenster, Parmesan, Port du Salut, provolone, Romano, and Swiss. Blue cheeses become crumbly but keep their flavor. Cream cheese, cottage cheese, and processed cheese suffer drastic changes in texture in the freezer.

To freeze cheese, double-wrap chunks no larger than 1/2 pound in plastic or heavy-duty foil and seal airtight. Cheese keeps well up to six months at 0° F.

FRESH CHEESES

Fresh cheeses are not aged or ripened. They are the simple products of a controlled curdling of milk. Like cottage cheese and cream cheese, most fresh cheeses are very moist and have relatively short storage lives.

Buying and Storing Fresh Cheeses. Unripened cheese tastes best when it is freshest. A number of cheeses in this group have dating codes on the label that will show exactly how fresh they are.

• COTTAGE CHEESE comes in several styles. *Dry curd* is just what it sounds like, the drained curds of skim milk formed by culturing with specific bacteria and, in some cases, rennet. It is similar to, but drier than, creamed cottage cheese. *Creamed cottage cheese* is dry curd mixed with a milk or cream dressing. When a label reads "cottage cheese" only, it is cream style. "Large curd" is lumpier than "small curd" cottage cheese and often tastes slightly more acidic. *Pot cheese* is another term for cottage cheese, but in some parts of the country the name denotes a slightly drier or whipped product.

Look for a "sell by" or "use by" date on the carton and, while you are at it, read the list of ingredients—there are usually stabilizers and other additives in cottage cheese you may want to avoid.

Making cottage cheese is a delicate and variable process like all cheese production, and one brand is not the duplicate of all others. Shop around for the cottage cheese with the freshest flavor.

You can store cottage cheese in a tightly lidded container in the refrigerator about one week. If you store the cheese upside down in its own carton it lasts two to three weeks and will lose all its flavor before it actually spoils.

• FARMER'S CHEESE is another bland, smooth cheese, but this one is pressed into a loaf you can slice. Tightly wrapped in plastic or foil, it keeps up to two weeks in the refrigerator.

• CREAM CHEESE is a white, soft, and spreadable fresh cheese that is sweet, but with a little tang to it. Prepackaged cream cheese is widely available, plain or flavored, regular or whipped, but most brands of cream cheese have stabilizers added, so read the label for ingredients. Cream cheese without additives is less gummy but doesn't keep as well. Look for a dating code because the fresher the tastier. *Neufchâtel* is like cream cheese, but the American variety is softer and has about two thirds the butterfat.

Store cream cheese and Neufchâtel, tightly wrapped, in the refrigerator up to two weeks.

• RICOTTA looks something like cottage cheese but is much smoother and sweeter. Ricotta should be snowy white; if it yellows it is too old. Refrigerate it up to one week in a tightly lidded container.

• MOZZARELLA in southern Italy is made from water buffalo milk, but of course cow's milk is used in the United States. Italian markets often manufacture mozzarella daily and sell the soft, hand-kneaded balls of cheese still juicy with fresh whey. The dry, prepackaged mozzarella in supermarkets has a rubbery texture and is meant for cooking. Both types are better when fresher. Moist mozzarella is good for two or three days in the refrigerator; the dry version keeps two weeks or more.

Freezing Fresh Cheeses. Most of these cheeses do not freeze well because they take on an odd texture. You can freeze them when they are combined with other ingredients in dips, spreads, and casseroles. Ricotta and farmer's cheese can be frozen, up to six months, if they are very well sealed. Dry mozzarella can also be frozen for six months and will still be suitable for cooking.

SOFT CHEESES

Most soft, ripened cheeses enjoy only a brief spell of perfection. You must learn to buy and store them so you can eat these cheeses at their peak.

Buying and Storing Soft Cheeses. The most widely available soft cheeses are:

• BRIE AND CAMEMBERT TYPES. During the cheesemaking process these cheeses are sprayed with special molds to form thin, flexible rinds that are white and velvety-looking. The cheeses ripen from the outside in, turning creamier and more flavorful as time goes by. In addition to Camembert and Brie, this category includes Carré de l'Est, Coulommiers, Limburger, and Liederkranz.

When sold whole, these cheeses are wrapped and packed in boxes. To judge the cheese, open the box and sniff. It will smell a little musty, but there should be no ammonia scent. Very gently prod the cheese with your finger—it should feel springy and bulging. A firm cheese is not yet ripe enough to serve. You can buy a whole cheese when it is underripe, then wait for it to come of age at home in your refrigerator, but if the cheese is very hard, or has a hard core, it may never ripen at all. Always leave it in its original package until it is ready. Once it is fully ripe, the cheese keeps for only two or three days, tightly wrapped, in the refrigerator.

You can usually buy wedges of Camembert or Brie cut from a large wheel. You may be able to taste the cheese but, barring that, take a good look at it. The interior of the slice should have a satiny sheen and appear ready to ooze.

Don't choose a piece of cheese that looks too runny because it is overaged and probably tastes bitter. Cheeses cut too soon show hard, white streaks in the center. Once sliced, these cheeses stop aging, so buy them ripe and use them as soon as possible. You can store the wedge at room temperature for up to twelve hours or refrigerate it two or three days.

Camembert and Brie are available canned, but you have no way of judging ripeness, which makes buying one a bit risky.

• CRÈMES are soft, smooth cheeses extravagantly enriched with cream. *Double crème* contains 60 percent butterfat; *triple crème* has 75 percent. Most of these cheeses have very light rinds and can be judged for ripeness in the same way as Brie. Some brands are sealed but stamped with a "use by" date. These cheeses should have a fresh scent, with no trace of ammonia. Their rinds should be bright-looking and well shaped. Among the *crèmes* available in the United States are Boursault, Brillat-Savarin, Caprice des Dieux, Crema Danica, Explorateur, and Supreme. Boursin is a fresh, unripened *triple crème* that has no rind. It usually bears a dating code.

These buttery, voluptuous *crèmes* tend to be less temperamental and longer lived than the Brie- and Camembert-type cheeses. In their original wrappings they will keep for a week (or more if clearly dated). Once opened, the cheese will last two to three days if well wrapped and kept chilled.

• GOAT CHEESE. Made, obviously, with goat's milk, these piquant cheeses are soft, airy, and mild when young but change character dramatically as they age. Fresh goat cheese should have a very fresh scent and look moist. Sample it, if you can, to be sure that it hasn't matured past your palate. Older goat cheese is very pungent, sometimes overpoweringly so. You can keep a young goat cheese mild and soft for at least a week to ten days. If you like the emphatic, earthy flavor of older goat cheese, you can keep the cheese for weeks and allow it to age in your refrigerator. Wrap it in plastic or foil and, if it is strong-smelling, seal it inside a jar. You can preserve the fresh taste of young goat cheese and stop it from ripening further by submerging it in pure olive oil. In this way, it can be stored for months at room temperature.

The goat cheeses you are most likely to find are Montrachet, Pyramide, Sainte-Maure, and Banon (which is wrapped in chestnut leaves).

• FETA is a soft Greek cheese with a bright, salty flavor. Greeks and Italians make feta with sheep's or goat's milk, but the American version uses cow's milk. It tastes best when fresh but will keep for months packed in a brine of salt water. Packing it in milk reduces the saltiness.

Freezing Soft Cheeses. A whole, uncut Camembert may survive freezing with just a little change in texture, but none of the other soft cheeses freeze very successfully.

SEMISOFT CHEESES

Semisoft cheeses are not nearly so mercurial as the soft cheeses, but still they change, ripen, and harden during storage. Though they tend to be bland, some semisoft cheeses, especially the monastery cheeses (so called because the method of cheesemaking originated in European monasteries), have earthy aromas and complex flavors. Monastery cheeses include Port du Salut, Beaumont, Bonbel, Saint-Paulin, Havarti, Pont L'Évêque, Tilsiter, Tomme de Savoie, Oka, and imported Muenster.

American-made semisoft cheeses—Monterey Jack, brick, Muenster, and beer cheese—can be rich and interesting or virtually tasteless, depending on the manufacturer. Danbo and Tybo are Danish semisofts; Bel Paese and Taleggio are Italian.

Buying and Storing Semisoft Cheeses. All of these cheeses should look plump and feel springy, not hard. While Monterey Jack is occasionally aged into a very hard grating cheese, the other semisoft cheeses are at their best only when young and moist.

If possible, examine the rind of the cheese. It should be clean and smooth with no cracks, bulges, or blemishes. The cheese should adhere closely to the rind.

In the case of monastery cheeses, their scent may be misleading—these cheeses often smell stronger than they taste.

You can refrigerate most semisoft cheeses for three to four weeks with no dramatic change in flavor or texture, but Bel Paese and Taleggio keep well for no more than two weeks in the refrigerator.

Wrap semisoft cheeses very tightly in plastic or foil and isolate the strong-smelling ones in their own lidded jars.

Freezing Semisoft Cheeses. You can freeze semisoft cheeses, but they won't be quite the same after the experience. The most noticeable change is that they become mealy. They are perfectly suitable for cooking after they have been frozen, however.

Wrap the cheese very securely in plastic or heavy-duty aluminum foil and seal inside an airtight plastic bag. The extra coverage prevents the cheese from exchanging odors with other food in the freezer.

Thaw the cheese in the refrigerator and do not refreeze.

FIRM CHEESES

Firm cheeses store very well. Their resilient textures hold up under months of refrigeration and their flavors remain virtually unaltered.

Buying and Storing Firm Cheeses. The textures of the different cheeses in this group actually vary considerably. Some Cheddars are dense and slightly flaky, some are lighter. The Dutch cheeses slice like semisoft cheese but store as well as firm.

• CHEDDAR-TYPE CHEESES. English Cheddar, American Cheddar, Canadian Cheddar, Herkimer, Colby, Cheshire, Derby, Double Gloucester, Leicester, and Tillamook are all members of the Cheddar group, the most beloved cheeses among the British and Americans. Mellow when young, they grow invigoratingly sharp as they age. Vast rounds of English Cheddar usually weigh 70 pounds; American, 20 to 30 pounds. The colors range from eggshell white to vibrating orange but have no bearing on flavor.

When choosing a Cheddar type, first look for uniform color. White streaks or spots indicate uneven ripening or mold. The texture should be fairly smooth, not grainy or crumbly. If there is a rind to examine, make sure it is clean and uncracked, with no bulges or darkening. Bitterness is an unpleasant taste in no way related to the sharpness of cheese. Bitter-tasting Cheddar has been badly manufactured, so return it to the store.

A whole, uncut Cheddar (there are 5-pound peewees) can keep for months in the refrigerator. It continues to age, and the flavor sharpens. Cut pieces keep one to two months if they are well wrapped and refrigerated. You can seal a piece of Cheddar-type cheese tightly in plastic or foil to store it but, for extra shelf life, wrap the cheese first in a piece of cloth soaked in a solution of two parts water and one part vinegar. Another method is to spread the cut side with butter before wrapping the cheese.

• SWISS-TYPE CHEESES. A Swiss-type cheese has a nutty, buttery flavor, but more importantly it has holes formed by gases that special microorganisms produce while the cheese cures. Among the Swiss-type cheeses are Appenzeller, Gruyère, raclette, and Emmenthal from Switzerland, Fontina from Italy, Jarlsberg from Norway, American Swiss, and Comté, a French Gruyère.

A look at the holes in Swiss-type cheeses gives a hint about their quality. The holes may be large or small depending on the variety, but they should be even and regular in shape. They should also have the moist, gleaming look that is a sign of full ripeness in the cheese.

Avoid Swiss-type cheeses with thickened or grayish rinds—they are too old.

Cut pieces of Swiss-type cheese keep well for one to two months when tightly wrapped in plastic or foil and stored in the refrigerator. A whole, uncut cheese keeps longer, up to several months in the refrigerator, gaining a stronger flavor as time passes.

• DUTCH CHEESES. Among the most reliable cheeses you can buy are the Dutch-made Edam and Gouda. There are American-made versions that can also be quite good.

Edam cheese comes as a flattened ball coated in bright red wax. Gouda has the same shape but may have a yellow wax coating.

The waxy coating gives you some clue to quality. When it is clean, smooth, showing no cracks, leaks, or bulges, the cheese inside is probably very good.

The cheese itself should look smooth, dense, and creamy.

Whole, uncut Edams and Goudas keep one to two years in the refrigerator, aging but never turning too sharp. Cut Dutch cheeses keep one to two months when well wrapped and stored in the refrigerator.

• PROVOLONE. This cheese comes in a variety of gourdlike shapes. It is wound around with cord and hangs temptingly in Italian markets and specialty cheese shops. Young or *dolce* provolone is smooth and sweet; the aged cheese, *piccante,* is flaky and has a bite.

The glossy rind should be completely smooth and unblemished. The interior of the cheese should be uniform, with no holes or signs of graininess.

Well wrapped, provolone keeps for several months in the refrigerator. You can store it in the rind for several weeks at room temperature.

Freezing Firm Cheese. Any of the firm cheeses can be frozen to extend shelf life, but not without giving up their characteristic textures. They will be crumbly and taste granular after thawing, but you can still use them for cooking without difficulty.

You can grate firm cheeses before you freeze them. It is most convenient to pack the grated cheese in plastic bags, in premeasured quantities of 1/2 cup, 1 cup, or whatever is appropriate for your recipes. Otherwise, double-wrap them in evenly sliced chunks under 1/2 pound. Thaw cheese in the refrigerator and do not refreeze.

BLUE CHEESES

The dark veins of mold that marble the blue cheeses account, at least in part, for the brilliant tang—as much a sensation as a flavor—that distinguishes this group. Roquefort is the best known, but there are many more, including English Stilton, and Gorgonzola from Italy. Danish blues include Danablu and Mycella. Among other French varieties are Bleu de Bresse, Fourme d'Ambert and Pipo Crème.

Buying and Storing Blue Cheeses. Blue cheeses should look moist and not too crumbly. French blues, apart from the veins, should be very white, rather than yellow or gray. Others may be slightly off white, but in either case the color should be even throughout the cheese and not darkened at the edges.

Tasting a sample is the best guide to buying a blue cheese. Don't buy one that is too salty or has a bitter taste. It is overaged.

Small chunks of blue cheese keep up to four weeks in the refrigerator. A large section of blue cheese will keep for months if you cut off pieces as you need them and keep the rest securely wrapped and refrigerated. Although you may simply store them wrapped in plastic, they do best with a little air circulation. You can allow for air without also drying out the cheese by wrapping it first in a damp cloth (soaked in a vinegar brine if you like), then depositing the piece in a lidded jar. A glass cheese bell works just as well as the jar, if you have one.

Freezing Blue Cheeses. You can freeze blue cheeses, but after thawing they will be much less creamy.

HARD CHEESES

The hard cheeses are intensely flavored, dense, and brittle. The well-known Parmesan is a hard cheese, as are Romano, Asiago, and Grana Padano.

Buying and Storing Hard Cheeses. Imported Italian hard cheeses are expensive but incomparably good. The very best is Parmigiano Reggiano (Parmesan), manufactured under strictly controlled conditions and aged at least fourteen months. True Parmigiano Reggiano has the name stenciled in red on the rind. Pecorino Romano, paler and sharper, may be made from goat's or cow's milk. Grana Padano is milder. Its production is less rigidly governed, so the quality is a little less reliable. Asiago, Cheddarlike when young, may be aged into a hard cheese, though it is less distinguished than the princely Parmigiano Reggiano. These cheeses should be pale yellow with only very few streaks of white.

The American versions of Parmesan, Romano, and Asiago, often sold grated, are waxy and dull in contrast to the Italian cheeses. Wedges of Stella brand Parmesan are superior to other domestics.

Refrigerate hard cheeses lightly wrapped and they should keep several months. The pregrated American cheeses last up to a year, either in the refrigerator or on a cool shelf.

Because they are so long-lived, there is really no need to freeze these cheeses unless you want to avoid the stronger flavor and drier texture that develops during long storage. Hard cheeses do come through freezing relatively unchanged.

PROCESSED CHEESES

All the cheeses discussed up to this point have been natural cheeses, either fresh or ripened. Processed cheeses are natural cheeses that undergo various treatments. Except for cold pack cheese, they are heated and pasteurized to stop natural ripening. Often, other ingredients are blended into the cheese for flavoring, softer texture, and longer shelf life.

Most of the cheese designated "American cheese" is a processed version of natural Cheddar.

Buying and Storing Processed Cheeses. There are standard designations for the different types of processed cheeses.

• COLD PACK CHEESE and COLD PACK CHEESE FOOD are blends of natural cheeses, mixed to achieve a smooth consistency. Cold pack cheese food contains additives such as sweeteners and flavorings. Unopened containers keep three to six months, up to the date stamped on the label. After opening, refrigerate these products in tightly lidded containers two to three weeks.

• PASTEURIZED PROCESS CHEESE consists of a blend of fresh and aged natural cheeses. Flavorings and bits of meat, fruit, or vegetable may be added. This type of cheese is long lasting but does require refrigeration. You can store it for months in its unopened package, up to the date stamped on the label. After opening, refrigerate the cheese, well wrapped, for three to four weeks.

• PASTEURIZED PROCESS CHEESE FOOD and PASTEURIZED PROCESS CHEESE SPREAD are made of fresh and aged natural cheeses, with milk solids and water added. These products contain less cheese, less milkfat, and more moisture than pasteurized process cheese.

The products called "cheese spread" also contain chemical stabilizers.

These products can be stored up to six months at room temperature or up to the date stamped on the label, whichever comes first. After opening, refrigerate the cheese, well wrapped or tightly lidded, three to four weeks.

• FRENCH PROCESSED CHEESES are pasteurized natural cheeses, all very soft. They come plain or elaborately spiced with almonds, peppers, orange, garlic, and many other flavorings. Common brand names are La Vache Qui Rit, Rambol, and Gourmandise.

The small, foil-wrapped bite-size pieces can be stored on the shelf for months. Larger pieces should be refrigerated, well wrapped. Unopened packages keep for four to six months. After opening refrigerate the cheese, well wrapped, three to four weeks.

Freezing Processed Cheeses. You can freeze cold pack cheeses in airtight containers up to six months and they will still be smooth and spreadable when thawed. Other processed cheeses become dry and brittle after freezing. They can be frozen, then used for cooking, or frozen as an ingredient in prepared foods.

STORAGE TIMETABLE FOR SELECTED CHEESES

Type of Cheese [1]	Weeks in Refrigerator	Months in Freezer
APPENZELLER	4	6
ASIAGO	8–24	6
BEL PAESE	3–4	6
BLUE TYPE	2–4	6
BONBEL [2]	3–4	6
BRICK	4–8	6
BRIE, Ripe	3–5 DAYS	6
CAMEMBERT AND		
CAMEMBERT TYPE, RIPE	3–5 DAYS	6
CHEDDAR [3]	4–8	6
CHESHIRE	4–8	6
COLBY	4–8	6
COLD PACK CHEESE	2–3	6
COTTAGE CHEESE	1	N/R
CREAM CHEESE	1–2	N/R
CRÈMES	2–3 DAYS	N/R
DERBY	4–8	6
DOUBLE GLOUCESTER	4–8	6
EDAM [4]	4–8	6
EMMENTHAL	4	6
FARMER'S	1–2	6
FIRM TYPE	4–8	6
FETA	8–12	N/R
FONTINA	4	6
GOAT	1–6	N/R
GORGONZOLA	2–4	6
GOUDA [5]	4–8	6
GRUYÈRE	4	6
HAVARTI	3–4	6
HERKIMER	4–8	6
JARLSBERG	4	6
LIEDERKRANZ, RIPE	3–5 DAYS	6
LIMBURGER	1–2	6
MONASTERY TYPE	2–4	6
MONTEREY JACK	2–4	6
MUENSTER	2–4	6
MOZZARELLA, FRESH	2–3 DAYS	N/R
MOZZARELLA, DRY	2–4	6
NEUFCHÂTEL	2	N/R
PARMESAN	1 YEAR	N/R

Type of Cheese [1]	Weeks in Refrigerator	Months in Freezer
PARMESAN, GRATED, AMERICAN	1 YEAR	N/R
PORT DU SALUT	2–4	6
POT CHEESE	1	N/R
PROCESSED CHEESE[6] (opened)	3–4	N/R
PROVOLONE	8–12	6
RICOTTA	1	6
ROQUEFORT	2–4	6
SEMISOFT TYPE	2–4	6
STILTON [7]	2–4	6
SWISS	4	6
TILLAMOOK	4–8	6
TILSITER	2–4	6

[1] All times shown apply to cut pieces of cheese.
[2] A whole Bonbel keeps 4–6 months in refrigerator.
[3] A whole Cheddar keeps 6–9 months in refrigerator.
[4] A whole Edam keeps 9–12 months in refrigerator.
[5] A whole Gouda keeps 9–12 months in refrigerator.
[6] Store unopened processed cheese up to the date stamped on the label.
[7] A whole Stilton keeps 9–12 months in refrigerator.

EGGS

The eggshell has a reputation for being Nature's perfect package. The prudent cook should keep in mind, however, that the package was designed to bring baby chicks, not breakfasts, into the world.

An egg is vulnerable, like all fresh foods. Despite the protective shell, time works mischief on an egg, blunting its full flavor, thinning its rich, fluid substance.

A good egg is a fresh egg. A newly laid egg has a compact yolk suspended in thick, cloudy white. Out of the shell, the orange-yellow yolk of a newly laid egg is a gleaming half dome sitting high and centered in dense white; only a small amount of watery fluid spreads out from the small area of cohesive white. A week-old egg has a thinner white that allows its somewhat deflated yolk to float off center. The amount of watery fluid about equals the amount of viscous white. The yolk of a three-week-old egg lies nearly flat and all the white is watery. This egg is completely edible and its nutritional components are intact, but its taste and appearance are very pallid compared to an egg still warm from the hen.

Several things have happened to the egg during the three weeks. The carbon dioxide that made the white cloudy when it was fresh has evaporated through the pores of the shell. This loss means a loss in flavor as well. The egg also has released moisture and absorbed air through the shell. The air has collected in a pocket at the egg's wide end, between membranes that separate the white and the shell.

Buying Eggs

The changes that take place in the egg as time goes on can be observed without actually opening it up. Inspectors judge the freshness of an egg by "candling" it, holding it up to an intense light to see what is inside. When the yolk is obscured by cloudy white, the egg is fresh; when the pocket of air is small, about the size of a dime, the egg is fresh. The grade of an egg is a measure of its freshness, based on candling. Grade AA eggs are the freshest; Grade A are slightly older, Grade B, most of which go to restaurant and institutional kitchens, are older and often slightly mis-shapen.

The grade is an indication but not a guarantee of how fresh an egg is. Eggs can take anywhere from two to thirty days to get from the hen to market. Any dating on the egg carton is based on the day eggs were packed rather than when they were laid.

USDA-graded eggs carry a three-figure code for the packing date that represents the day of the year, numbered consecutively from 001 (which is January 1). Translating that code, while you are standing in the super-market aisle, takes some doing, so just search out the carton with the highest number.

A number of states, by no means all, require a "sell by" date to appear on the package. The "sell by" date may allow as much as two weeks of shelf time from the packing date. There is no control, however, over the length of time between packing and when the hen actually laid the egg. It is even possible for a single graded, dated carton to contain eggs that range from very fresh to not so fresh.

Considering the world of difference between a fresh egg and a stale egg, you should shop around for the brand and the grocery store that most often delivers the very best eggs. You may even find it worth the time to search out a specialty shop that regularly sells Grade AA eggs. Many large stores carry only Grade A.

But you don't have to rely entirely on grading, dating, and comparison shopping to come up with the best. There are ways you can make an immediate assessment of the eggs themselves:

• NO CRACKS. Examine each egg closely and never buy one with hairline cracks or breaks of any kind in the shell. A broken shell permits salmo-nella and other undesirable microorganisms to invade the egg, making it unsafe to eat.

• CLEAN SHELL. When an egg is laid it has a natural coating, the cuticle, that seals the shell, preventing bacteria from penetrating and moisture from evaporating. In the United States, packers wash eggs with detergent to remove surface bacteria, and the cuticle is consequently stripped away. Some packers replace the cuticle with a film of oil on the shell, and some packers do not.

Unwashed eggs do have a few spots on them but, with cuticle intact, they stay fresh-tasting longer than washed eggs. You are taking no undue risks if you eat an unwashed egg right from the nest.

However, unless you do pluck it right from the nest, any egg you buy will be a washed egg, so there is no good reason for it to have dirt on the shell. Therefore, buy only clean eggs.

• QUIET AND HEAVY. These qualities are difficult to detect but you can learn with some experience. The freshest eggs feel heavier for their size because they have lost less moisture. And, because the amount of liquid inside the shell of a stale egg has decreased, it will slosh when you pick it up and shake it.

• SHAPE AND SIZE. These two factors are not measures of freshness but may affect the storage life of an egg.

In addition to being graded for freshness, eggs are classified by size. From the smallest to the largest, the sizes are Peewee, Small, Medium, Large, Extra Large, and Jumbo. (Recipes are generally based on the use of Large eggs.)

Older hens lay the larger eggs. Older hens also produce eggs with more fragile, less protective shells. Medium and Large eggs are more likely to have tough shells that hold in moisture.

Avoid misshapen eggs. In normal, oval eggs, the yolk is cushioned by the white and suspended at a distance from the air pocket. An oddly shaped shell may force the yolk out of this protected position.

• SHELL COLOR. The color of the shell, white or brown, makes absolutely no difference in the quality, flavor, texture, nutritional value, or freshness of the egg. Certain breeds of hen lay white eggs and others lay brown.

Of course, once you bring an egg home and open it up, you can judge its freshness by the character of the white (thick and cloudy when fresh) and the yolk (rounded and centered in a fresh egg).

In the white of the egg, you will also see strands of opaque material. These are supposed to be there. They are called chalaza and they help hold the yolk in place inside the egg.

You may occasionally find a blood spot in the egg yolk. It is safe to eat, but if it bothers you, dip it out with the edge of the broken shell. These blood spots become diluted as the egg ages, so a bright red spot is actually a clue that the egg is fresh.

Storing Eggs

Keep stored eggs chilled, below 40° F., at all times. One day at room temperature will age the egg as much as one week in the refrigerator.

Keep stored eggs covered. Eggs readily absorb odors from, and lose moisture to, the air circulating in the refrigerator. The best container for them is the carton you bought them in. Don't use the open egg racks in the door of the refrigerator. The rack is too warm and unprotected.

Store eggs with their broad, rounded ends up. This position helps the chalaza keep the yolk centered in the white, away from the air pocket where it might encounter unfriendly bacteria. The rounded end of the egg is also less likely to break when accidentally bumped.

Don't wash eggs before you store them. If they have been coated in oil, you would be washing away that valuable protection.

Eggs can be stored safely, in the refrigerator, four to five weeks after purchase, as long as they are clean and unbroken. Needless to say, after five weeks an egg will have lost a lot in the way of flavor and texture— fried, for instance, it will be flat, bland, and rubbery—but it will be perfectly edible nonetheless.

During the first week of storage, the still very fresh eggs are ideal for frying, poaching, or any kind of use in which the appearance of the egg is important. Fresh egg whites whip most easily, too.

After one week the eggs are best suited to hard-cooking, because the white has shrunk away from the shell slightly, making it easier to peel. You can also soft-cook week-old eggs. You can still fry or poach them but they will not be as pretty and the yolks tend to break more easily.

An egg that has spent over two weeks in your refrigerator is best for scrambling or for using as an ingredient in other cooked foods.

Eating eggs raw or partially cooked—as in eggnog, hollandaise sauce, Caeser dressing, or even soft-boiled eggs—entails only a slight risk of food poisoning from salmonella (see p. 362). Healthy people can afford to take the chance, since less than one percent of all fresh eggs carry the bacteria, but salmonella infections can be fatal in infants, the elderly and the chronically ill. These people should eat only thoroughly cooked eggs.

Eggs that are accidentally cracked while in storage should be taken out of the shell and put in a covered glass container in the refrigerator. They can be stored for two to three days out of the shell. If you don't know when the shell was broken, store them for no more than twenty-four hours and use them only in baked goods or other thoroughly cooked foods.

When your recipe calls for only whites or only yolks, you can save what is left of the egg in covered glass containers. Whites will keep well for a week to ten days in the refrigerator, and they freeze very well. When storing yolks alone, cover them with water. They will keep no more than two or three days.

You can keep yolks longer if they are hard-cooked. Slip the raw yolks gently into a pan of cold water and bring the water to a boil. Take the pan off the heat and let it stand, covered, for 10 to 15 minutes. Remove the yolks with a slotted spoon and store them in an airtight container for four or five days. Crumble the hard-cooked yolks and sprinkle them over the foods of your choice.

An entire hard-cooked egg in its shell can be stored in the refrigerator two to three weeks. A peeled hard-cooked egg can be refrigerated for one week, as long as it is kept in a sealed container.

Some strange and alarming substances have been applied to eggs in the attempt to preserve them. People have tried sealing the egg up with the likes of cactus juice, soap, clay, salt, and ashes, all with varying degrees of success. The Chinese delicacy, the thousand-year-old egg, is a duck egg that has been coated with, or buried in the ground with, a mixture of ashes, salt, and lime and left for about three months (not a thousand years). Inside the shell, the egg turns into a dark jelly but can be eaten without fear of harm. (You can store a thousand-year-old egg one to two weeks at room temperature.)

Waterglassing—storing eggs in a solution of sodium silicate—is a time-honored method. (Sodium silicate is also used for sealing concrete floors.) As long as the eggs are completely covered with the syrupy waterglass and kept chilled (40° F.), they will keep for three months. But there is no need for most households to resort to this cumbersome technique, for eggs can be stored up to a year in the freezer.

Freezing Eggs

Eggs can't go into the freezer in their shells because they expand and crack at those temperatures. Open the eggs and freeze them, separated or whole, in large quantities or small, depending on how you plan to use them.

Freeze only clean, fresh eggs that had no breaks in the shell.

• WHOLE EGGS. Break the eggs into a rigid container. About 5 Large eggs equal 1 cup. With a fork, pierce the yolks and very slowly stir the eggs to mix them. Don't whip up any foam as you stir because air bubbles will dry out eggs as they freeze.

Yolks, when frozen, turn into a gel that stays thick and sticky after thawing. Adding either salt or sugar to the egg mix keeps the yolks closer to their natural consistency. Stir in 1/8 teaspoon salt for every 2 eggs in the batch if you intend to use the eggs for main-dish cooking. For every 2 eggs going into baking or desserts, add 1 teaspoon sugar or corn syrup.

Seal the container, leaving no more than 1/2 inch head space. Write on the label how many eggs, whether salted or sweetened, the date packed, and the "use by" date. Freeze at 0° F. and keep up to nine months.

For extra convenience, pour the whole-egg mixture, very thoroughly blended, into an ice cube tray to freeze it. As soon as they are hard, transfer the egg cubes to a plastic bag, label, and seal. A single cube thaws quickly and will equal 1 egg in your recipes.

• EGG YOLKS. Break one egg at a time, separating the white into a cup and slipping each yolk into a rigid container. Six or seven yolks equal about 1/2 cup. Pierce each yolk and stir in either 1/4 teaspoon salt, 1 teaspoon sugar, or 1 teaspoon corn syrup per 1/2 cup.

Seal the container, leaving no more than 1/2 inch head space. Write on the label how many yolks, whether salted or sweetened, the date packed, and the "use by" date. Freeze at 0° F. and keep up to nine months.

Hard-cooked yolks can be frozen too. Sealed in a plastic bag and held at 0° F., they will keep two to three months.

• EGG WHITES. Break one egg at a time into a cup, separating each white so that no yolk at all mixes in. Pour the whites into a rigid container, seal, leaving 1/2 inch head space. Label with number of egg whites, date packed, and "use by" date. At 0° F., egg whites keep up to one year. Thawed egg whites can be whipped. For increased volume, allow them to sit at room temperature for half an hour beforehand.

Hard-cooked egg whites become very rubbery after freezing.

Thaw eggs in the refrigerator or under cold running water. Never refreeze them.

EGG SUBSTITUTES

The egg substitute in the frozen food section of the supermarket is almost entirely egg whites with vegetable oil, emulsifiers, colorings, and nutrients added. The product can be stored up to one year at 0° F. After thawing, refrigerate it up to seven days.

STORAGE TIMETABLE FOR EGGS

	Days in Refrigerator	Months in Freezer
WHOLE EGGS:		
In shell	28–35	N/A
Out of shell	7	9[1]
EGG WHITES ONLY	7–10	12
EGG YOLKS ONLY	1–2[2]	9[1]
WHOLE EGGS, HARD-COOKED:		
in the shell	14–21	N/R
peeled	7	N/R
EGG YOLKS, HARD-COOKED	4–5	2–3
EGG SUBSTITUTES	7	12

[1] Mixed with salt or sugar as a stabilizer.
[2] Covered with water.

4

Meat, Poultry, and Fish

FRESH MEAT

The butcher shops of sixteenth-century Europe overflowed with fresh mutton, goat, beef, poultry, pork, rabbit, partridge, roe deer, boar, lark, heron, woodcock, and all other manner of game. And when fat, tender haunches of sheep were within reach of even a humble peasant, the rich and well placed flaunted tables set with an amazing overabundance of meat dishes. Songbirds, sweetbreads, pork tongues, ham, geese, roast veal, calves' feet, sausage, breast of goat, rabbit, and pigeon would be served in a single gala meal.

Europeans brought their limitless appetite for meat to North America where settlers hunted bear, moose, porcupine, raccoon, turkey, and the doomed passenger pigeon. As cities on both continents grew, drawing an ever increasing population away from the land, fresh meat became scarcer and more expensive but remained the staple food of Western civilization, always in demand. (These early epicures, by the way, had only the vaguest guesses about meat's contribution to good health. Their taste for meat derived purely from the pleasure of it. Not until the mid-1800s did scientists identify specific nutrients and begin to understand how different components of food—or lack of them—affected the body.)

Beginning in the nineteenth century, new methods of feeding and transporting cattle, hogs, and other food animals once again put fresh meat in the average man's daily diet, though not, to be sure, with the intoxicating variety of Renaissance cuisine. These same methods, further refined, bring beef, pork, lamb, and other meat to current markets with a regularity which the American cook takes utterly for granted. A complex set of laws and regulations, administered by the Department of Agriculture, assure the buyer that, to a very great extent, fresh meat is pure and safe to eat.

Fresh, unspoiled meat is not difficult to recognize—it is moist, satiny red or pink, and has almost no scent. Spoiled meat is equally easy to spot. Rank odor, spongy, dull gray flesh, and discolored patches on the surface are its unmistakable signs.

Quite a number of things happen to cause meat to change so distinctly. Within the cell structure of the meat itself enzymes by the millions cause chemical reactions that break down tissues. At the same time the fats in meat can react with air and water in a way that changes their chemical makeup. The transformed fats are rancid: they have a strong, unpleasant taste and odor. Eventually, even at freezer temperatures, rancidity will spoil meat.

Most devastating to fresh meat are microbes, invisible yeasts, molds, and bacteria that attack and eventually destroy the meat. Slime, strange colors, furry growths, phosphorescence, and reeking odors are evidence of the microbes' activity. A few common microbes contaminate meat and grow without leaving any clues that they are there, and many of these can cause illness.

It usually takes from seven to ten days for meat to get from the slaughterhouse to the market. Because it is packed, shipped, butchered, and displayed under constant refrigeration, meat stays fresh. The industry also goes to great lengths to keep meat clean: microbes can come from the animals' own intestinal tracts, from the air, from surfaces in the slaughterhouse and butcher shop, and from hands that touch the meat. Carcasses are frequently vacuum-packed in heavy plastic bags for protection.

Government agencies, federal and local, regularly inspect meat, all meat animals, and packing plants. Their standards for cleanliness are extremely high, though the issue of how well the standards are applied is controversial. Wholesome meat, from healthy animals slaughtered and packed under very sanitary conditions, earns, literally, a stamp of approval. A circular symbol, stamped in edible purple dye on the carcass, means the meat has passed government inspection.

The laws that regulate the meat industry were originally passed to stop the widespread practice of slaughtering and selling diseased animals under filthy conditions. Questions raised today about the purity of America's meat center not on cleanliness but on substances, such as antibiotics, hormones, and heavy metals. When these substances are ingested by the living animal, as they commonly are, traces remain in the meat after slaughter. (The use of drugs and hormones on food animals is, of course, among the techniques that keep meat cheap and abundant.) The USDA does test meat and poultry for pesticides, herbicides, drugs, mercury, and hormones. There are tests for a hundred separate chemicals and metals. Legal standards either set maximum amounts allowable or ban the substance altogether. Once the slaughtered animal has been approved, inspectors frequently spray the carcass with a very dilute solution of chlorine, considered perfectly safe, to disinfect the surface of the meat.

If you follow in the footsteps of the industry and keep the meat you buy cold and clean, you can store it very successfully at home. Many foods lose flavor and even nutrients during storage—a fresh egg is never as good later as it is right out of the nest; orange juice tastes best the moment it is squeezed from the fruit—but meat, properly handled, is delicious for all of its storage life.

Buying Meat

Except in small grocery stores and independent butcher shops, most meat you buy is precut, wrapped in plastic, and stuck with an encoded label. The first word on the package label (according to the national standard) identifies the kind of meat inside, pork for instance.

Next the label states where the cut comes from on the animal. For example, "chuck" is the name for beef shoulder; "round" is beef or veal leg; "sirloin" is the hip area of beef, veal, pork, and lamb.

The specific retail cut—roast, steak, chop, and so on—is in third position on the label. If the meat has been boned, cubed, larded, smoked, or otherwise processed, that information appears last.

Meat is graded according to the age of the animal, the amount of marbling (meaning the streaks of fat within the lean—the more fat the more flavor), and the color and texture of the lean. As these factors vary, so does the palatability of the meat.

Prime is the highest grade for beef, veal, and lamb. The second highest rating is Choice. Most of the meat you find in supermarkets and butcher shops is Choice. Prime cuts appear much less often. Meat graded Good (the grade right below Choice) is carried regularly in some grocery chains and rarely in others, but is virtually never labeled Good. Instead, markets use a house brand name for the meat or refer to it as "generic."

There is much less variance in quality in pork carcasses than in other meats, so fresh pork is not usually graded for the consumer.

We are accustomed to fluctuations in the fresh vegetable section—one day we see robust green heads of lettuce, then the next day's shipment arrives all pale and limp. The same abrupt changes can happen at the meat counter, but they are not so easy to see. Uniform labeling and grading do not guarantee that every "Veal Boneless Rump Roast (Choice)" you ever pick up is going to be the same. The Choice grade, for example, will be applied to meat whose general condition falls into a fairly broad category. While the points on which the carcass is rated—marbling, texture, maturity, color—suggest that the meat will taste good, they don't promise it. Meat quality will vary from store to store and even from week to week at the same store.

Still, it's possible to make nearly infallible judgments about the freshness of the meat you buy. Many stores help you by putting dating codes on meat labels. The date on the package may show the date the cut was wrapped or the last date of sale. Use the date to find the freshest meat and to figure out for yourself how much longer you can store it at home.

Pick up the package and look it over. Fresh meat is springy to the touch, moist, well colored, and nearly odor free. Sure signs of spoilage are: drying at the edges; darkened, mottled patches; brown or dull gray color; or strong odor.

The package itself should be clean and unbroken with little or no juice in the container. Excess drippings often mean the meat has been frozen and thawed. Take only well-chilled packages, stored deep in the refrigerator case.

Don't resign yourself to grabbing whatever is left at the meat counter and making do. Call on the butcher to cut meat to order, to trim precut meats, and to answer your questions about storing and cooking the meat. In fact, you will get extra storage life at home from meats cut or ground to order the same day you buy them.

As a rule of thumb, buy 1/4 pound per serving of boneless and ground meat. Allow 1/2 pound per serving of meat with the bone in. Very bony meats, such as ribs, call for 3/4 to 1 pound per serving. Well-trimmed meat cuts, ones with less fat and bone left on, are more expensive per pound but have a higher proportion of edible meat per pound than less closely trimmed cuts.

Once you have chosen fresh meat, get it home as fast as you can. At temperatures higher than 40° F., all the processes of deterioration accelerate.

Storing Fresh Meat

A rise of a few degrees in temperature can mean many days of storage lost. Use a thermometer to find the coldest part of your refrigerator. That is where you should keep the meat. At 36° F. to 40° F., the range of most refrigerators, meat will keep for the minimum times listed on the storage timetable later. Some appliances have a special meat compartment that stays between 29° F. and 34° F. At these temperatures, meat keeps for the maximum number of days shown on the chart and even longer.

You want a little air circulation around stored fresh meat to dry off the excess surface moisture, which encourages bacterial growth. The plastic film on prepackaged meat is porous enough to allow some air to pass through, so you can leave supermarket cuts in their original wrap. Butcher paper works well too—just loosen the ends enough to let some air inside the package. Loose plastic or paper covering is necessary because you don't want the meat to get so dry on the surface that its juices begin to evaporate.

Handle meat as little as possible before you store it. Don't do any slicing or grinding until you are ready to cook. Wash your hands and work surfaces both before and after contact with raw meat. Keep raw meat juices, a common source of food poisoning bacteria, away from all other foods in the refrigerator: place possibly leaky packages of meat on a plate or in a separate compartment; and immediately wipe up drippings from refrigerator surfaces.

Freezing Fresh Meat

There are some good reasons to store meat in the freezer. You can buy large quantities at sale prices and freeze the meat for future use. You can reduce the number of trips to the supermarket by buying and freezing enough meat for a week or more at a time. When your dinner plans change, you can stash the meat in the freezer and save it from spoiling.

There are also reasons *not* to freeze meat. Freezing, to some degree, will desiccate and soften meat and sap its flavor. You can't avoid it but you can reduce this loss of quality with careful handling. You must wrap meat airtight, freeze it fast, and keep it very, very cold.

Freezing occurs when the water that is a natural component of all meats turns to solid ice crystals. Just as water in an ice cube tray expands as it freezes, so does the water in meat. The sharp-edged crystals push into the surrounding tissue, leaving a trail of ruptured cells. When thawed, meat will have lost some of its natural springiness.

The water in meat that is *outside* cell walls freezes first. As it does, it leaches water from *inside* the cell walls. When meat thaws, the original balance does not return to normal. The water released during freezing seeps out of the thawing meat. The meat is less juicy and loses flavor and nutrients as well.

The faster meat freezes the smaller the ice crystals will be, and of course smaller ice crystals will do less damage. You can speed freezing in a couple of ways: adjust your freezer to the lowest possible temperature; package meat in small portions; place packages directly on freezer surfaces, not on top of other frozen foods.

All of the natural processes of deterioration that go on in meat at higher temperatures do not stop at 0° F., they merely slow down. Microbe activity stops in the freezer, but most bacteria, yeasts, and molds survive and as soon as the meat begins to thaw they once again become a threat.

For the best results, it is absolutely necessary to keep frozen meat 0° F. or below. Storing it at even 10° F. will cut storage life in half.

Freezing meat fast and keeping it cold are important but still not enough. Meat must be sealed in airtight packages before going into the freezer. Plastic-backed freezer paper, sealed with freezer tape, is generally the best wrapping material. The paper molds easily to the shape of the cut and resists tears; the plastic backing keeps air out.

An alternative is plastic bags, but these are easily pierced by protruding bones. If you use a plastic bag it should be the type clearly labeled "freezer bag." The freezer bags used with home heat-sealing machines work particularly well because they can be closed more securely than others.

Heavy-duty aluminum foil molds well and keeps out air, but it too is susceptible to punctures. Clear plastic freezer wrap tends to spring open unless the package is heavily taped.

There are two basic wrapping techniques for frozen meat.

• For large cuts, use the "butcher wrap." Spread a sheet of freezer paper on the counter, plastic side up. The sheet should be much longer and wider than the food to be wrapped. Place the meat diagonally across one corner of the paper. Take the point of that corner and pull it up, over, and around the meat to create a sort of tube. Fold both ends of the tube inward, over the top of the meat, then roll the meat forward to the far corner of the sheet. Pull that last corner tight and tape it to the package.

• The "drugstore wrap" works best for ground meat, a stack of chops, stew meat, liver, and other variety meats. Lay out a sheet of freezer paper that is twice as long and twice as wide as the cut to be wrapped. Place the meat in the center. Pull two opposite edges of the paper up until they align over the meat. Keeping them together, turn the edges down. Turn this fold over on itself again and again until it fits closely across the top of the food. Crease the edge of the fold to hold it in place. This will leave you with an open-ended tube. Press the loose paper at each end of the tube into a V. Fold both Vs over and tape them securely on the top of the package. For extra protection, use the tape to seal the fold across the top of the meat.

For refrigerator storage, you want to handle the meat as little as possible until it is time to do the cooking. But for freezing it is best to trim, cut, or grind meat to suit your cooking needs before packaging it for the freezer.

Ground meat freezes best shaped into hamburger-size patties for fast freezing, even if you will be cooking it in another form. Stack patties, placing a folded piece of freezer paper (shiny side out) on top of each one so they won't stick together. Center two or three short stacks on a large sheet of freezer paper and finish off the package with the drugstore wrap.

If you plan to take out a few patties at a time and leave the rest frozen, stack them in a plastic freezer bag, squeeze out all the air inside, and seal it tightly. Repeat the squeezing and sealing every time you open the bag. You can season ground meat before freezing but do it with a light touch. The flavor of garlic, onion, and black pepper, to name just a few spices, will intensify during storage.

Trim excess fat off *steaks* and *chops* before freezing. Mold aluminum foil over sharp, protruding bones—this will prevent tears in the wrapping. Steaks and chops should be stacked with plastic wrap or freezer paper between them to keep them from sticking together, then packaged using the drugstore wrap.

Stew meat should be trimmed to remove excess fat and any tough, elastic connective tissue before you cut it into cubes. Wrap the cubes first in clear plastic or aluminum foil no more than one layer deep. This first wrap keeps the meat from bunching up. Then seal the packaged meat into freezer paper, using the drugstore wrap.

Save all your meat trimmings. Put bones and meat scraps into a plastic freezer bag. Force all the air out of the bag, seal it, and keep it in the freezer. These odds and ends are the essentials of a savory stockpot.

Thawing Meat

There are two perfectly good ways to deal with frozen meat. You can thaw it slowly in the refrigerator or start cooking it while it is still frozen. Both methods keep the meat fairly juicy.

Putting frozen meat directly into the oven makes a big difference in the cooking time required. Large *frozen roasts* take up to 50 percent longer to cook. The internal temperature of the meat is the best measure of doneness, so insert a meat thermometer into the center of the roast after it has thawed in the heat of the oven. Wait to season the roast until it is thawed, otherwise the seasonings will slide right off. The oven should be on a low to moderate setting, 300° F. to 350° F., to ensure that the meat is evenly cooked on the inside.

Frozen pot roasts and *stew meats* take about 25 percent longer to braise in liquid than fresh meat. However, only thawed meat will brown well prior to moist cooking.

Frozen ground meat, steaks, and *chops* can easily overcook on the outside. To avoid this problem, lower the usual cooking temperature. In the broiler, position the meat at least six inches from the source of heat. Add about 25 percent to the normal cooking time for these cuts.

Pan frying frozen meat results in a lot of troublesome splatter as the moisture from the ice crystals hits the hot fat. It is best to thaw cuts for frying, but if that is not possible, cook the meat at a low to medium setting with the pan covered until the meat thaws completely.

For *deep frying* you must thaw the meat before cooking.

Thawing meat slowly in the refrigerator allows time for the meat to reabsorb some of the moisture lost during freezing. Leave the thawing meat in freezer wrap but loosen it to allow some air flow. Put the package in the refrigerator on a plate to catch any drippings.

In the refrigerator, *large roasts* take four to seven hours per pound to thaw. *Small roasts* need three to five hours per pound. *Steaks* and *chops* 1 inch thick defrost in twelve to fourteen hours. *Ground meat* and *stew meat* thaw in eight to twelve hours.

You can cook meat thawed in the refrigerator as though it were fresh, but do it without delay. Thawed meat stored more than one day, even in the refrigerator, tends to develop a high bacterial content.

Thawing meat at room temperature is never a good idea. The outside of the meat thaws quickly and is likely to be at room temperature for hours. After four hours, bacteria can multiply dangerously.

One safe way to defrost meat quickly is to submerge it in cold water (never higher than 50° F.) after you have sealed the meat inside a plastic bag. Cook it immediately after thawing.

A microwave oven will thaw meat in minutes, but you should watch over the process very carefully. Follow the directions in your oven's instruction book. Reposition the cut once during the defrost cycle, especially if it is large. Separate steaks and chops as soon as they have softened slightly. Taking these steps will insure that the meat thaws evenly throughout. If you have one, insert a specially designed microwave thermometer into the meat to see that the center has reached at least 32° F. Another way to check the meat is to slip a sharp knife into the center and probe for hard ice crystals. Always allow a few minutes' standing time between thawing and cooking to allow juices to be reabsorbed into the meat.

You can refreeze partially thawed meat as long as there are still ice crystals in the flesh, but you will be sacrificing so much flavor and texture that you should think twice about doing it. Cooking thawed meat and then refreezing it will give you better results.

Storing Cooked Meat

Refrigerate cooked meat immediately after the meal—don't wait for it to cool to room temperature. Wrap the warm meat loosely until it is thoroughly chilled, then seal it in heavy-duty aluminum foil, using the drugstore wrap described on p. 180 (check for punctures and double-wrap the meat if you find any). The meat won't dry out so quickly if it is left in large pieces, but remove as much bone as you can before wrapping. You can also place cooked meat in a shallow container—plastic, metal, or crockery will do—and cover loosely with plastic wrap until the meat is completely chilled. Once cold, cover the dish with a snug lid or tightly molded foil. Refrigerate for up to four days.

Meat gravy and broth are much more perishable than the meat itself. Since broth and gravy keep no more than one or two days in the refrigerator, it is usually wise to pour them into an airtight rigid container and freeze them right away.

If you have broth or gravy in the refrigerator for two days, you can boil it for a few minutes to kill any bacteria, return it to a clean container, and keep it another two days. By boiling it every two days you can keep broth or gravy safe for up to two weeks, but freezing is more sensible.

If you won't be using your leftovers within a few days you should freeze the cooked meat after it has cooled in the refrigerator. Cooked meat, well wrapped, keeps two to three months at 0° F. and should be reheated without any thawing or be thawed in the refrigerator.

Cooked meat frozen in broth or gravy tastes better than dry-frozen because liquid protects the meat from air exposure. Remove bones and freeze the cooked meat in its own liquid or in gravy, using an airtight rigid container. The container should be the right size: you want about 1/2 inch air space between the food and the lid, but not much more. Casseroles, soups, and stews should be packed in the same fashion.

Store combination dishes no longer than two to three months in a 0° F. freezer. Heat them directly from the frozen state, as described in "Cooking for the Freezer" on pp. 335–37.

Commercially Prepared Meat Dishes

Whether it's enchiladas or Yankee pot roast, rules for storing commercially prepared foods are few and simple.

Frozen entrees can be stored in a home freezer, at 0° F., for two to three months. When you buy them in large quantities, write the last date for use in waterproof ink right on the box to help you keep track. Never thaw the entrees—keep them frozen and carefully follow directions for heating. Leftovers should be covered and refrigerated for no longer than three to four days. Never refreeze them.

Store canned entrees in a cool, dry place. If you are likely to lose track of it, date the can with the last date for use, which should be no longer than one year from the date of purchase and no more than six months if your storage area is frequently above 72° F. After one year the canned entree will still be safe to eat but it will have lost a noticeable amount of taste and texture. Once opened, the canned food can be kept for three to four days in the refrigerator in a covered container.

Keep dehydrated entrees no more than six months on a cool, dry shelf.

Carry-out entrees from restaurants and delicatessens should be eaten the same day they are purchased. Wrap any leftovers and refrigerate them as soon as you can. Keep them for no more than one day. Do not use plastic carry-out containers to reheat food in the microwave.

Marinating Fresh Meat

A marinade softens meat's tough connective tissues and imparts flavor to the meat at the same time. An acid ingredient—usually wine, vinegar, or lemon juice—in concert with meat's own enzymes does the tenderizing. Oil in the marinade infuses the meat with the seasonings, which can be anything from garlic to rose petals.

Meat may be coated with the marinade or may soak in the liquids for several hours or days. Since the acid and any salt in the mixture discourage microbe activity, marinating meat actually extends its storage life and in fact was traditionally used as a method of preserving meat.

You can marinate meat at room temperature for four hours if you plan to cook it immediately. For longer periods, always refrigerate. Meat steeped in marinade keeps up to a week in the refrigerator. Wipe marinated meat dry with a paper towel before you cook it—dry meat browns more easily.

BEEF

While it is true that cattle were first domesticated about 10,000 B.C. somewhere around the eastern Mediterranean, the United States has adopted beef so ardently that a steer might join the eagle as the national symbol. Americans eat about 125 pounds of beef per person per year (though this figure is declining), which is more than any other nation. The pioneers cleared the Great Plains of an estimated 60 million buffalo and of the Indian tribes that depended on them, making way for a huge cattle industry, backdrop for the quintessential American: the cowboy. In the nineteenth century great cattle drives moved the animals the long miles from their grazing land to railroads, which took the animals to the feed-lots and packinghouses of Chicago and Kansas City. Today, cattle are slaughtered where they are raised and shipped to market vacuum-sealed in plastic, the symbol of the twentieth century.

Beef varies in quality more than any other meat. While veal, pork, and lamb come from very young animals, beef cattle are raised to full maturity —about eighteen months—before slaughter. The animal's age and its diet directly affect the taste and tenderness of the meat. Most people prefer the flavor of grain-fed beef. About half of American beef is fat-tened on grain, the other half is grass-fed.

In many states, you can find "natural" beef from cows raised without antibiotics, artificial hormones, or feed additives. The beef is govern-ment approved but not usually graded. The meat, fresh or flash frozen, costs more but you can expect very high quality and you know exactly what you are paying for.

Buying Beef. When you are buying a beef cut, the first thing you want to know is its grade.

The highest grade, Prime, goes on beef that is the tenderest, juiciest, and most flavorful. When you find it fresh in retail stores, which won't be every day because restaurants buy up almost all Prime beef, it will be much more expensive than other grades.

Prime beef is often "aged." The traditional way to age beef takes as long as six weeks. The carcass hangs on a hook in a refrigerated locker with controlled humidity; and as the meat ages its natural enzymes soften the connective tissue that makes meat tough. Beef shrinks as it ages; the meat darkens; and the flavor intensifies. Despite cold temperatures, mold grows on the surface of the meat and must be cut away when aging is finished. A modern, speedier way to age beef is to hang it at higher temperatures (60° F. to 70° F.) surrounded by ultraviolet light that slows bacterial growth. Beef cuts will also age if simply left refrigerated in their vacuum-sealed, oxygen-free polyethylene bags—bacteria can't multiply without oxygen, but the enzymes can still do their work. Some connoisseurs complain that meat aged in plastic takes on a faint sourness. If you buy aged beef, ask the butcher which method has been used and compare the effect of the different methods for yourself. Of course, aged beef is a delicacy that you must pay a premium price for—you cannot age beef successfully at home.

Freezing prime beef at home diminishes the very taste and texture that you pay a dear price for. When you buy prime beef, plan to use it within a few days.

On the other hand, mail-order outlets and specialty stores offer commercially frozen prime beef processed rapidly at such low temperatures that the quality is nearly indistinguishable from that of fresh meat. You must store it at 0° F. or below to retain this high quality.

Choice is the grade of beef most commonly available in supermarkets and butcher shops. Choice beef is very high-quality meat that will be tasty and tender if stored and cooked properly. Good-grade beef, usually sold as a house brand, is equally nutritious but lacks the tenderness and full flavor of the higher grades.

As for freshness, it is fairly simple to recognize. Beef should be bright red on the outside and dark purple on the inside. The red color, called bloom, develops naturally after the meat has been cut and exposed to air. The surface will turn from red to brown after it has been exposed to oxygen for a number of days. The brown color does not necessarily mean that the beef is spoiled, but it does mean that the meat is nearing the end of its storage life.

The meat should feel springy, not mushy or waterlogged.

The fat of the beef should be creamy white. Yellowish fat is associated with older or grass-fed animals whose meat is less desirable.

Storing Beef. Beef has a high proportion of saturated fats and therefore has a longer storage life than meats with less stable fats, such as pork. Beef cuts, except ground beef, stored loosely covered in the refrigerator, last up to five days at 35° F. to 40° F. In a meat compartment that maintains a temperature of 29° F. to 34° F., beef can keep without spoiling for as long as eight days.

Beef freezes as successfully as any meat can. Beef will keep well at 0° F. for six to twelve months. It must be trimmed of excess fat, sealed in airtight packages, and frozen rapidly to maintain high quality in the freezer.

GROUND BEEF

Retail meat outlets grind their own beef fresh every day, using trimmings from other cuts and adding a certain amount of fat to arrive at the right proportion. The grade or cut of meat that goes into ground beef is not really important—tough or tender, marbled or lean, the final product has uniform texture and fat content.

When beef goes through the grinder, microbes, ever present on its surface, end up distributed throughout the meat. Ground beef spoils much more quickly than other cuts of meat and should be bought and stored with that in mind.

Buying Ground Beef. Meat counters usually offer at least four varieties of ground beef:

- Ground Beef. This label identifies meat with the maximum amount of fat allowed, 30 percent.
- Ground Chuck. From the shoulder area of the carcass, this meat will contain anywhere from 15 to 27 percent fat and cost a little more than Ground Beef.
- Ground Round. Leaner than Ground Chuck, this cut comes from the hip area. It will be priced higher than chuck.
- Ground Sirloin. The most expensive cut and the leanest, this comes from the back area.

A number of chain stores label ground beef with the exact proportion of fat to lean. The higher the percentage of lean the higher the price will be.

Occasionally a butcher uses his own terminology on ground beef, calling it "super lean" for instance. You can sometimes judge the fat content of the meat just by looking at it, but if you want to be sure you should ask.

Even though leaner ground beef is more expensive, it is not necessarily better. It's the fat in meat that has most of the flavor. Seventy percent lean ground beef is good for hamburger patties, 80 percent lean, or chuck, is good for meat loaf. In most cases, you can drain off extra fat after cooking. Ninety percent lean works well in combination dishes and is the best choice for people cutting down on calories and fats.

If at all possible, cook ground beef the same day you buy it. It should have a bright red color and no detectable odor.

If you have the butcher grind an appropriate cut to order, such as round steak or sirloin tip, you can be sure of getting the maximum storage life from the meat.

You can buy cuts to grind or chop at home as well. Beef chopped in a home food processor will have a firmer texture, which you are likely to prefer to the mushier store-bought meat. Be sure to keep at least 10 percent fat by volume with the meat.

Supermarkets frequently package preformed ground beef patties for hamburger sandwiches. Be sure you know what you are buying before you pick these up. They may contain additives and extenders such as soy protein. Ground meat containing soy protein does not keep as well as ground meat alone but does make juicier hamburgers.

Storing Ground Beef. Store ground beef in the refrigerator in its store wrappings. It keeps no more than two days from the time it was ground, so you should plan on using it the same day you buy it or the next.

Refrigerated ground beef will turn brownish, which is the effect of exposure to oxygen. The meat is not necessarily spoiled when it is brown, but it is a good idea to cook it immediately and store it cooked if you want to keep it longer. Throw out ground meat that has developed an unusual smell.

Form ground beef into patties for quick freezing (see p. 181 for wrapping instructions). You can also season and shape the beef into a meat loaf for freezing as long as the loaf is not too thick—a flat loaf will freeze more rapidly. Ground beef will keep well at 0° F. for two to four months.

Another convenient way to store ground beef is to make meatballs. Arrange the cooked meatballs on a tray and put them inside an airtight plastic bag. They won't stick together, so you can remove as many as you need at any one time and leave the rest in the freezer. You can reheat the meatballs in sauce without thawing.

PORK

Pigs are remarkably efficient at converting fodder to meat. They produce 20 pounds of edible meat for every 100 pounds of food they consume, which is three times the rate for beef cattle. One of the factors in the pig's superior equation is that every part of the animal except the whiskers is fit, in one form or another, to fill a platter.

In China farmers caught on very early to the advantages of pigs as livestock. Around 2000 B.C. the average Chinese household included a few pigs of a breed small enough to be kept indoors and raised on table scraps. Archaeological findings show that these pigs were slaughtered at under one year of age.

Pork still comes from very young animals, though today's hogs are huge and bred for lean meat. Because the eating quality of such short-lived pigs is relatively consistent, most pork is sold without a USDA grading. (It is, of course, always inspected for wholesomeness by a federal, state, or city agency.)

Unfortunately inspection cannot guarantee that pork is free of *trichina*, the microorganism that causes trichinosis. Although this disease-causing organism is not as common as it once was, heat all fresh pork thoroughly enough to destroy any possible trichinae. This means the meat must be cooked to reach an internal temperature of no less than 140° F. Many cookbooks specify 170° F. to 185° F. for pork, but these high temperatures dry out the meat and are not really necessary for safety. A cautious compromise of 160° F. insures that all parts of the pork have been heated completely and that the meat will still be juicy.

About one third of a pork carcass reaches the consumer as fresh meat. The remaining two thirds are cured, smoked, or otherwise processed, which is why there are fewer fresh pork than beef cuts available.

Buying Pork. Good-quality pork has grayish-pink, fine-grained flesh, not too heavily marbled. The fat is creamy white and firm, the bones soft and tinged with red. An exception is pork tenderloin: this lean, boneless cut is deep red.

Pork's storage life is generally shorter than that of other meats. Look at the dating code and use your own judgment to be absolutely sure the pork you are buying is as fresh as possible. Don't settle for meat with discoloration or dry-looking edges.

Storing Pork. If it is kept loosely wrapped in the coldest part of the refrigerator, pork lasts from two to four days. The fat in pork is fairly unstable. It will begin to go rancid even under the best freezer conditions. Pork, wrapped in an airtight package and frozen at 0° F., will keep its high quality for no more than six months (see pp. 179-81 for specific wrapping and freezing techniques). At higher or fluctuating temperatures, cut the storage time in half.

Ground Pork and Fresh Sausage. Unseasoned ground pork should be purchased the same day it is ground and be cooked that day or frozen. Fresh breakfast sausage (seasoned ground pork), whether packed loose or in links, should also be used within one day. It is essential that you check the dating codes on these meats to make sure they are fresh at the time of purchase. Any ground meat is especially vulnerable to bacterial contamination.

If you buy ground pork or breakfast sausage already frozen, double-wrap the package and keep it in the freezer for no more than three months. You can freeze ground pork and fresh sausage at home but, again, keep it for no more than three months.

LAMB

Sheep, one of the earliest domesticated animals, from the beginning served as much more than meat on the hoof. A sheep produced wool and milk and, at the end of a long life of trekking over rocky hills and fields, would be roasted for dinner. Its pungent, sinewy flesh required the most industrious tenderizing.

However, the sheep that are now raised in the American West are bred to produce only one thing: lamb. Lamb is the meat of a sheep one year old or younger and it is sweet, musky, tender, and delicious, especially when served pink: rare or medium rare.

Buying Lamb. The lamb you see at the supermarket is labeled one of three ways. "Lamb," "Genuine Lamb," and "Spring Lamb" are all guaranteed to be from animals under one year old. The average age of lamb going to market is five to seven months. The term "Spring Lamb" is applied only to meat sold from March through the first week of October, but otherwise is the same as Lamb or Genuine Lamb. Some people taste a difference in Spring Lamb and prefer it to the others, but these choices are not very widely available.

Lamb meat should be pinkish red and fine-grained. A cross section of the bone should be red and porous. The fat should be white, firm, and not too thick. In general the darker the meat the older the animal it came from and the more distinct the flavor. If you can find it, lamb from two- to four-month-old animals, which is pale and extremely tender, is the very best.

Mutton is from sheep over two years old. Practically no mutton is sold in the United States, but you may encounter Yearling Lamb—from animals between one and two years old. Both mutton and yearling lamb have comparatively strong flavors.

Specialty shops sell very young lamb that may be labeled Baby Lamb, Milk-fed Lamb, or Hothouse Lamb. These terms all refer to meat from animals six to ten weeks old that have been raised on an all-milk diet. The meat should be pinkish white and have a very soft texture. The flavor is extremely delicate.

Fresh vacuum-packed New Zealand lamb is consistently excellent. Its frozen counterpart is tasty but can be too dry for roasting. Another imported variety, Pré-salé (presalted) Lamb, comes from herds that graze in salt marshes in France. It is considered the most delicious.

There are USDA grades that apply to lamb, and most cuts you find in markets are either Prime or Choice. Prime has more generous marbling, but Choice is completely satisfactory lamb, tender and juicy.

Storing Lamb. Large cuts of lamb will keep well two to four days in the coldest part of the refrigerator, still in its store wrappings. Ground lamb and lamb cubed for stew meat lasts only one or two days. Be sure to remove the tough outer layer, called the fell, on a leg of lamb, if the butcher hasn't.

Larger cuts of lamb can be stored in the freezer for six to nine months at 0° F. Ground lamb, thin chops, and stew meat will keep no more than three to four months (see pp. 179–81 for specific wrapping and freezing techniques).

VEAL

Veal is a kind of blank canvas for the culinary artist. The very light texture of veal and its sweet, neutral flavor blend happily with thick tomato sauce or piquant wine sauce, with a splash of lemon or a sprinkle of aromatic herbs. It is highly favored in continental cooking for reasons not only of taste but also of economics. In Europe, where beef is not as much in demand and where grazing land is scarcer, farmers slaughter very young male cattle for veal rather than undertaking the expense of raising them to maturity. (The females are spared, since they grow up to produce milk.) High-quality veal is, in consequence, widely available there. That is not true in the United States.

Buying Veal. The most desirable veal comes from two- to three-month-old calves that have been fed only milk, either mother's milk or formula. Milk-fed veal is velvety pale pink, almost white, with no marbling. It has very little border fat and what there is should be creamy white. The flesh is firm and springy, not mushy; the bones are soft and moist with a reddish tinge to the marrow.

Older, grass-fed calves produce reddish meat, the color coming from iron in their feed. Deep pink color is not the only indication that veal is not milk-fed: the flesh is coarser than younger animals'; the fat may be yellowish; and the bones are whiter.

Older calves, particularly those over six months of age, yield tougher meat, lacking the mild flavor of better veal. Between six and eleven months of age, veal calves should be identified as baby beef. Veal is expensive and there is no point in paying the premium price for baby beef. The meat of calves six to eleven months old has none of the delicacy of veal and none of the full flavor of mature beef.

It may be difficult for you to find young, milk-fed veal unless you search out either a fine butcher shop or an Italian specialty market.

Some of the veal you will find at retailers will be graded by USDA standards either Prime or Choice. The meat is judged by the same characteristics of firmness, fine grain, and color that you can see for yourself at the meat counter.

Storing Veal. Large cuts of veal, such as roasts, will keep in the refrigerator three to five days. Veal is most often trimmed down to cutlets and thin scallops, and these, along with ground veal, are very perishable. Refrigerate them for no longer than two days.

Veal can be frozen in airtight packages and stored at 0° F. up to nine months for large cuts, and up to four months for smaller cuts.

BONES AND STOCK

Making stock, the rich liquid essence of meat, is the most honored technique for extracting all the flavor and all the nutrients from meat. The bones and scraps trimmed from the meat you prepare can be combined with aromatic vegetables, covered with water, and left to simmer slowly for hours. There are no unbreakable rules for making stock. You can use any meat and any variety of herbs and vegetables.

The strained broth adds an incomparable flavor to sauces, stews, and soups; as a substitute for plain water, it enhances recipes calling for poaching, braising, and even steaming.

In some cases you will want to make a pure stock, using only one kind of meat and carefully chosen vegetables. Veal stock, for instance, has an especially delicate character desirable in white sauces. Beef makes a heartier stock frequently called for in soups and stews. Pork and lamb, which are a bit too strong in flavor to be used in a basic stock, are perfectly appropriate in broths for pork and lamb recipes.

Buying Bones. When buying meat bones for stock, look for leg bones, called shank bones, because they contain lots of the substance called collagen that acts as a natural thickener. Shank bones are often prepackaged at the meat counter but you will probably have to ask the butcher for other types of bones. Get a variety—they are inexpensive and each type has its own flavor quality to add to the stock. Have the butcher crack the bones for you so that all possible flavor will be released into the stock.

Choice- and Good-graded meats are equally acceptable for stock. Bones should be red, porous, and moist. Reject meat and bones with an off odor or off color.

Beef and veal bones, in addition to being ideal for stock, also yield marrow, a smooth, fatty inner core that can be cooked and used as a garnish or spread on toast. For marrow, buy only heavy leg bones that have been sawed into 2- to 3-inch-long sections. Once you have scooped out the marrow, the bones can still be used for stock.

Storing Bones. Raw bones and marrow are very perishable. Refrigerate them for no more than two days. In the freezer, bones will keep well for six months, marrow for no more than one month.

As you accumulate raw bones and meat trimmings from your daily meals, place them in a plastic bag, force the air out of the bag, tie it securely, and put it in the freezer. Hold the bones and scraps until you have several pounds' worth, then start up your stockpot.

You can follow the same procedure for saving cooked bones and meat scraps, but store them in a separate bag. The cooked trimmings should be stored in the freezer no longer than two to three months. They can go directly from freezer to stove without thawing.

Cooked bones and marrow, well wrapped, can be held in the refrigerator for three to four days. As with raw bones, crack them before they go into the stockpot. Cooked marrow does not freeze well.

Storing Stock. In the refrigerator, stock will keep safely for one to two days. You can extend its refrigerator life by taking it out after two days and bringing it to a boil to destroy any bacterial growth. The layer of fat that rises to the top of the stock will seal and protect it quite effectively, so wait to remove the fat until you are ready to use the stock.

To freeze stock, pack it in small, recipe-size quantities (around 1 cup) in rigid containers. Leave about 1/2 inch air space at the top of the container, but no more than 1 inch. You can also freeze it in ice cube trays for smaller amounts. At 0° F., stock will keep for four to six months. Remove fat from the stock before you freeze it because the fat tends to turn rancid even at 0° F.

VARIETY MEATS

Organ meats, such as liver, brains, and sweetbreads, are nutritionally rich and relatively inexpensive. Many people are put off by the idea of variety meats and that is too bad—when fresh and well prepared, these meats are as subtle and pleasing as any dish can be.

Buying Variety Meats. Organ meats are extremely perishable, more so than muscle meats because of their softer texture and higher bacterial content. You must be alert to any signs of drying or decay in the variety meats you buy.

All variety meats should be bright in color, moist, and plump. They should have little or no odor. Avoid organ meats that show signs of drying or darkening at the edges.

• LIVER. Calf's liver is reddish brown and virtually odor-free. Baby beef and beef liver is dark red and has a slightly beefy smell. Lamb liver is very similar to calf's liver in appearance and even in flavor. Pork liver is rarely found at retail butcher shops—its stronger taste is not as appealing to most consumers. Fresh liver has a smooth, glossy surface with no hint of dullness or slime.

• KIDNEYS. Beef, calf, lamb, and pork kidneys are all generally available. They should be moist and firm, with no soft spots. Avoid kidneys that have any trace of objectionable odor. The fat surrounding beef and veal kidneys, called suet, is light and delicate. It can be saved for use in pastries and steamed puddings.

• BRAINS. They should be clean, pale pink, and glossy.

• SWEETBREADS. These are the thymus glands from calves, baby beef, and lambs. Sweetbreads are beloved by connoisseurs of fine food and their price usually reflects the fact. They are best when they are very light in color and have a fresh odor.

• HEARTS. Hearts should be bright red with no tinge of gray.

• TONGUE. The best fresh tongues are beef and veal tongues under 3 pounds. The surface should have a fine texture.

• TRIPE. Tripe is the stomach lining of a beef or veal animal. It is very light-tasting and pleasantly chewy. Tripe should be uniformly white and have a honeycombed texture.

Storing Variety Meats. Refrigerate fresh organ meats, loosely wrapped, for no longer than two days. They are very susceptible to bacterial decay.

Freezing does a lot of damage to the flavor and texture of all organ meats, so avoid it if you can. If freezing is necessary, trim all excess fat from the meat, wrap it in an airtight package, and freeze it rapidly. Brains, tongue, and sweetbreads should be kept in the freezer only up to two months. Liver, heart, tripe, and kidneys will retain their quality up to four months in the freezer.

RABBIT

Rabbit is a mild meat, tasting a lot like chicken. A few specialty stores and urban supermarkets sell fresh domestic rabbit, but you are more likely to find it frozen.

Buying Rabbit. Rabbits are best when young, so look for one that weighs no more than 2½ pounds.

The rabbit should look plump and pink. Frozen rabbit should be well wrapped, feel hard, and have no signs of freezer burn—discolored patches on the surface.

Storing Rabbit. Fresh rabbit keeps one to two days in the refrigerator. You can store frozen rabbit six to nine months at 0° F.

GAME

Buffalo and venison are lean red meats that can taste as good or better than prime beef. Some specialty stores regularly carry them (and other less common game, such as boar and bear) but they are most readily available through mail-order purveyors. Most commercial game meats come from animals raised on farms where they are allowed to range for food. They are usually free of the antibiotics and synthetic hormones common in beef cattle.

Game that you purchase fresh or flash frozen should be stored according to the recommendations for prime beef.

Hunters should go into the field equipped with enough information and the proper tools to clean and store wild game. The meat of wild game is very perishable to begin with, and the chances for bacterial contamination are great.

Game animals should be bled, cleanly gutted, and cooled immediately after they are killed.

The cleaned carcass can be hung for several days in a cold place (around 35° F.) or butchered and refrigerated for one to two days. Be sure to remove all excess fat before storing the meat either way, because it quickly goes rancid.

The practice of strapping large game to the fender of the car and driving home is particularly wrongheaded. The heat from the engine can ruin the meat in a matter of hours.

Once game has been hung or refrigerated and the flesh is no longer rigid, it should be cooked or frozen as soon as possible. There are professional butchers who can cut up, package, and quick-freeze your game for you.

Always cook game thoroughly; never serve it rare. Wild game sometimes carries disease and parasites. The parasite that causes trichinosis is sometimes found in bear meat, for instance.

If you freeze game yourself, start by trimming off all the fat you can. Wrap the meat in airtight packages and freeze it rapidly. Game can be kept for six to nine months at 0° F.

How to Use the Storage Timetable

Meat stored in the refrigerator will eventually spoil, becoming discolored and rank-smelling. How long it lasts before spoiling depends on a number of factors. If the meat is perfectly fresh when purchased, if it spends little time at room temperature between the market and home, and if it is kept very clean during handling, the meat should keep well for the maximum times listed on the table.

Meat in the freezer is not likely to spoil but only gradually to lose so much of its quality that it will become very dry and tasteless. Meat will store satisfactorily up to the maximum times listed on the table only if it is wrapped properly and the temperature of the freezer rarely fluctuates above 0° F. The temperature of your freezer or freezing compartment is critical. The type of freezer unit that is not separate from the rest of the refrigerator is designed for ice cube trays only. It is not cold enough to keep meat from spoiling, even if meat appears to be frozen hard. If your freezer maintains a temperature of 10° F. above zero, cut the maximum time limit in half.

STORAGE TIMETABLE FOR MEAT

Type of meat	Days in Refrigerator (35–40° F.)[1]	Months in Freezer (0° F.)
BEEF:		
Roasts, Steaks, Ribs	3–5	6–12
Ground, Cut for Stew	1–2	2–4
Liver, Kidney, Heart, Tripe	1–2	2–4
Brains, Sweetbreads, Tongue	1–2	1–2

Type of meat	Days in Refrigerator (35–40° F.)[1]	Months in Freezer (0° F.)
VEAL:		
Roasts, Chops, Ribs	3–5	6–9
Ground, Cutlet, Stew	1–2	3–4
Liver, Kidney	1–2	2–4
Brains, Sweetbreads	1–2	1–2
PORK:		
Roast, Chops, Ribs	2–4	3–6
Ground, Fresh Sausage	1–2	1–2
Liver, Kidney, Heart, Chitterlings	1–2	2–4
LAMB:		
Roasts, Chops, Ribs	2–4	6–9
Ground, Stew	1–2	3–4
Liver, Kidney, Heart	1–2	2–4
Brains, Sweetbreads	1–2	1–2
BONES	1–2	6
RABBIT	1–2	6–9
GAME	1–2	6–9
COOKED MEAT:		
Beef, Veal, Pork, Lamb, combination dishes	3–4	2–3
COOKED MEAT:[2]		
Gravy	1–2	2–3
STOCK	1–2	4–6
FROZEN PREPARED ENTREES	3–4	2–3
CARRY-OUT ENTREES	1	N/R

[1] Meat refrigerated at lower temperatures (29–34° F.) will keep slightly longer.
[2] Meat coated in broth or gravy should be stored according to these time limits.

CURED AND PROCESSED MEATS

Toward the end of each year, as the days grew shorter and the nights colder, the sixteenth-century European housewife faced the task of stocking her larder against the uncertainties of the coming winter. Farmers slaughtered livestock in the fall rather than feed and house the animals through the harsh weather. The abundant fresh meat bought in October and November would be packed in salt to preserve it until spring. Since salt could itself be quite expensive, the shrewd buyer chose only meaty, tender cuts for curing. A tough, stringy leg of mutton or pork was "not worth its salt."

When meat is heavily salted, the normal chemical composition of the flesh changes and the new chemistry inhibits decay. Meat that has been completely dried can last through not just one winter but many. In damp climates smoking meat has been the surest way to dry it; wood smoke that penetrates meat leaves behind substances that further curtail bacteria.

Salting and smoking are still the essential elements in curing meats. Processors add sugar, seasonings, and spices for flavor; and they use nitrites and nitrates to give meat a pink color and to prevent the growth of the organism that causes botulism. Unlike the preserved meats of earlier times, today's ham, bacon, bologna, and corned beef are most often mild-cured, for flavor rather than longevity. They are therefore moist, not hard, and quite perishable.

Many processed meats in the delicatessen are not cured at all but have been prepared from fresh meat in some other way. Certain cold cuts are simply cooked meat. Luncheon meat is often a blend of seasoned cooked pork and beef formed into a loaf to be sliced for sandwiches. Liverwurst is a blend of finely ground cooked pork and livers which may include some smoked meat as well.

Buying and Storing Cured and Processed Meats

There is one general rule about buying and storing cured and processed meat, fish, and poultry products that applies to every single variety: *read the label.* The label will tell you whether the meat is uncooked, partially cooked, or cooked and ready to eat.

Look for storage instructions on product labels, such as "keep under refrigeration." Many cured meats also come with essential dating information.

When there is no label, ask the grocer about storage and cooking requirements for the food. A bratwurst from one source may be fresh, while another brand may be sold fully cooked. There is really no way to figure this out for yourself because every manufacturer has unique recipes for preparing even the most common cured and processed meats.

BACON

Bacon comes from the layered slabs of lean and fat on the belly of a hog. The meat is cured in salt brine, and it may be smoked or not.

Buying and Storing Bacon. Most bacon you find is presliced and vacuum-packaged. Look first for a date stamp that shows the last date of sale. The meat itself should be firm, fine-grained, and well colored, not dull or slimy.

As long as it is still sealed in its vacuum package, bacon keeps several weeks in the refrigerator, up to one week beyond the last date of sale on the label. Once you have broken the seal, wrap the package tightly in plastic or foil. An opened package stays fresh for no more than seven days in the refrigerator.

Some butchers and specialty shops sell bacon in slabs. It should look moist and smell fresh. A slab of bacon keeps several weeks in the refrigerator. Wrap it very securely in foil or plastic and slice off the ends as they dry and darken.

Vacuum-sealed bacon freezes well, but results are less predictable with frozen slab bacon or loose slices. Use heavy-duty aluminum foil or moisture-vaporproof freezer paper to wrap loose bacon airtight (see drugstore wrap, p. 180). You can keep bacon frozen at 0° F. up to one month. After that, the fat will begin to go rancid.

Defrost bacon completely in the refrigerator to eliminate a lot of unnecessary spatter and shrinkage during cooking.

You can save time and add storage life to bacon by cooking it first, then freezing it. Fry, broil, or microwave a whole package of sliced bacon until it is almost, but not quite, done to your liking. Drain the slices on paper towels or a paper bag. When they are cool, place the number of slices you normally serve at breakfast on a fresh paper towel and fold over, put down another portion of bacon, fold over, and so on. Seal the cushioned bacon strips in an airtight plastic bag and freeze. To serve the bacon, take out what you need and reheat it for a minute or less in a frying pan, broiler, or microwave oven. Or crumble the cooked bacon while it is still frozen and use it as a garnish.

Frozen cooked bacon keeps four to six weeks at 0° F.

Canadian bacon is a leaner cut of cured, smoked pork from the loin area of the hog. Sliced, it will keep for three to four days in the refrigerator—keep it tightly wrapped in plastic or foil. A chunk of Canadian bacon lasts up to seven days at refrigerator temperatures. Because it contains less fat than regular bacon, a chunk of Canadian bacon will keep in the freezer a little longer, for six to eight weeks. Vacuum-packed slices can stay frozen for six weeks, but the maximum for hand-wrapped slices is three weeks.

Fully cooked Canadian bacon should be stored the same as cold cuts (see p. 204).

HAM

Strictly speaking, ham is the word applied to the hind leg of a hog, fresh or cured. In practice, the word "ham" also refers to cured cuts of pork other than the leg, such as shoulders, shoulder blades, and Boston butts. The shoulder is also called a picnic ham and usually costs less than a ham from the leg. A fresh ham is simply a leg of pork.

One method for curing ham, called dry curing, is to coat it with salt, sugar, and sodium nitrate or nitrite and marinate it for a week or more. But most ham is cured with brine: the meat may be soaked in it or injected with it. Processors add flavorings, either natural or artificial, to some hams and they may finish the cure by smoking the meat.

Buying Ham. One brand of ham will look and taste entirely different than another, depending on the quality of pork used and on the processor's unique recipe for curing, seasoning, and smoking. The word "premium" on the label has no standard meaning—it is no more than advertising—so you must experiment to find the ham that suits you. The phrase "water added" refers to the brine used for curing.

Hams come to market whole, in halves, or in slices, bone in, or partially deboned. Boneless hams are neatly reformed inside casings. The best-tasting hams have been smoked, but these are more expensive and less available. And the best cut is the butt half, the meaty upper part of the leg, with the bone in. The shank half has good flavor but more tough tendons. A loin ham (not a true ham) is another good cut; shoulder and Boston butt hams, comparable in flavor, are much fattier.

Hams are nearly impossible to judge for freshness unless they are sliced so the flesh is visible. When you can see the meat, look for fine texture and pink color. Press the meat. It should be firm, not spongy. Don't be troubled by iridescence on the surface. It is hardly ever a sign of spoilage and is most often a result of pigment changes caused by exposure to air and light, changes specific to hams containing nitrates. When you cannot see the meat, you have to rely on dating codes.

Ham is sold partially or fully cooked. A label that says "cook before eating" means it must be cooked as thoroughly as fresh pork before it is safe to eat. The phrases "ready to eat" and "fully cooked" mean the ham is safe to eat as is. If you buy a ham with no labeling, or with confusing directions, either ask the butcher to explain or treat it as uncooked meat.

• CANNED HAM. A few canned hams are one piece of meat but most have been deboned, cured, and formed under pressure into their characteristic shape. Though they are fully cooked inside the vacuum-sealed cans, canned hams are still perishable and must be kept under refrigeration.

A few canned ham products are sterilized at very high temperatures and may be stored safely on the shelf. Read the label, but when in doubt refrigerate.

• COUNTRY HAMS. Country ham is an American delicacy made in a number of different regions from hogs specially fed on nuts and fruit. The most famous country ham is the Smithfield from Virginia. The meat is dry-cured in salt, smoked for weeks at low temperatures, then aged for up to two years. These hams have deep red, dry, pungent meat, nothing like a soft, bland brine-cured ham's. Country hams from Kentucky, Tennessee, and Missouri aren't aged as long, so the meat is not as dry.

After aging, the surface of the ham develops an extremely unattractive layer of mold (you scrape it off before you cook the meat). This is the way it is supposed to look. There is no way to choose a country ham except to depend on the reliability of the processor and the retailer—usually only whole hams are sold, and these in cloth bags. The exceptions are authentic Chinese grocery stores, which often sell country ham by the pound.

• RAW HAMS. A number of European-style hams are dry-cured and air-dried but not smoked. Prosciutto and Westphalian are well-known examples found in specialty shops and delicatessens. Prosciutto should be more gold than pink and moist-looking with distinctly white fat. Famous Parma prosciutto cannot be imported into the United States, so look for Swiss or Canadian versions, which are better than overpreserved American prosciutto. Germany's Westphalian ham has dark flesh and should also look moist but not watery.

Storing Ham. Hams keep for about seven days in the refrigerator. If they are vacuum-packed or sealed in heavy plastic (not the standard meat-counter wrapping), leave them in their original package. Otherwise, wrap the meat tightly in plastic or foil. Hams stamped with a last date of sale code can be stored, unopened, for seven days beyond the date. After cooking, ham can go back into the refrigerator, tightly wrapped, for another three to five days.

A whole ham keeps in the freezer for three months at 0° F. Sliced ham should not be frozen for more than one month. Ham left too long in the freezer loses its rosy color and turns brownish gray. The meat will do best if it is in a vacuum-sealed package. Otherwise, wrap the ham very tightly in heavy-duty foil or freezer paper, or in a freezer bag with all the air squeezed out. Thaw in the refrigerator before cooking.

• CANNED HAM. Unopened canned ham keeps for six to twelve months in the refrigerator. Once the can is open, the ham keeps for up to seven days in the refrigerator, providing it is well wrapped in plastic or foil.

If you choose to bake the ham, the leftovers can be stored five to seven days after cooking.

Freezing ham in the can is pointless—it will not extend the ham's storage life. It is, on the other hand, sensible to freeze leftover canned ham for longer storage life. Seal the meat airtight in a plastic bag or in freezer paper. It keeps for one month at 0° F. The meat can be thawed in the refrigerator or reheated directly from the freezer.

• COUNTRY AND DRY-CURED HAMS. A whole, uncut country ham can be kept indefinitely in a cool, dry place, at around 72° F. Eventually, however, it becomes too dry. Considering the expense of the meat, you should probably plan to store it no more than six months to a year.

Once cut or prepared, the country ham is perishable and must be refrigerated. It will keep well for three to five days, wrapped in plastic or foil.

Raw, dry-cured ham should be kept, well wrapped, in the refrigerator for no more than three to five days.

Leftover country and raw ham can be frozen, well wrapped, for one month at 0° F.

SAUSAGES

There is a deceptively simple definition of the word "sausage": any meat that is ground up and stuffed into a casing is a sausage. But sausages come in confounding variety, including the humble frankfurter, spicy hard salami, spreadable liverwurst, and countless others.

Sausages are surely the last word in food conservation. Processors use every edible scrap of meat off a carcass to make sausages. But you probably don't eat knackwurst because it is practical, you eat it because it is juicy, garlicky, and delicious.

Buying and Storing Sausages. Thrill-seekers should make their way to Italian, or German, or Latin American, or Chinese, or any other ethnic markets and sample the sausages each culture prizes. It is wise to taste first and ask questions later, for the ingredients may be pig's feet, sheep stomach, pork blood, or other unfamiliar animal parts. Never mind. All but the dullest palates should be won over by these fat, spicy sausages.

Wherever you buy them, the best, most distinctive sausages are sold whole or in bulk. The meat should smell fresh; the casing should be dry and free of any trace of mold.

Supermarkets carry presliced sausages sealed in vacuum packages. These are virtually always stamped with a date, usually the last date of sale, which allows for about a week's storage time at home.

The meat inside the package should look fresh and moist but not wet. Avoid meat that shows drying at the edges or looks moldy. The package itself should be unbroken. Do not buy a package that is misshapen or bulging.

Sausages may be raw or cooked; they may be smoked, cured, and dried in any combination. You must know whether or not the meat is ready to eat or needs cooking. If you don't find the information on the label, ask the grocer.

• FRESH, UNCOOKED SAUSAGES. Common examples of fresh sausage are bratwurst, kielbasa, Polish sausage, Italian sausage, linguisa, pork sausage, country sausage, bockwurst, and weisswurst. Basically these are fresh ground meats, seasoned and packed into sausage skins.

In most cases, you can store these sausages, tightly wrapped in plastic or foil, for up to seven days in the coldest part of the refrigerator, but pork breakfast sausage (links, patties, and in bulk), bockwurst, and weisswurst are very vulnerable to bacterial growth and should be kept for no more than one day at refrigerator temperatures. Vacuum-sealed sausage has a freshness date on the label as a guide to storage time.

If you plan to make fresh link sausage at home, you need casings. A butcher can supply you with the necessary cleaned animal entrails packed in salt, which keep one to two weeks at room temperature.

Fresh, uncooked sausages can be frozen but keep well for no more than two months. Seal them airtight in heavy-duty aluminum or freezer paper, using the drugstore wrap (see p. 180) and freeze at 0° F.

• COOKED, AND COOKED, SMOKED SAUSAGES. Among fully cooked sausages are frankfurters, bologna, blood sausage, braunschweiger, knackwurst, liver loaf, liverwurst, and German mortadella. Supermarkets usually sell these in vacuum packs; delicatessens sell them in bulk. Sealed in vacuum packages, these sausages keep up to seven days beyond the last date of sale. Once opened, they will keep four to seven days in the refrigerator, as long as they are well wrapped in plastic or foil. The same is true of cooked sausage bought in bulk. Sealed in plastic or foil, cooked sausage keeps for two months at 0° F.

• DRY AND SEMIDRY SAUSAGES. Dry and semidry sausages have a slightly chewier texture. Among them are salami, summer sausage, Lebanon bologna, Lyons sausage, Italian mortadella, and pepperoni, to name the most common varieties. You can find whole dry and semidry sausages in delicatessens. Whole sausages keep best, but not if the skin is broken or moldy. Whole dry and semidry sausages should be wrapped in plastic or foil and stored in the refrigerator. A dry sausage keeps four to six weeks, a semidry sausage two to three weeks. A salty, white film may develop on the outside of the sausage. It is harmless and can be cut off when you are ready to eat the meat. Keep unopened, vacuum-packed slices in the refrigerator up to seven days beyond the last date of sale. Opened packages can be refrigerated up to seven days.

Sealed airtight in plastic or foil, dry and semidry sausages keep up to one month in the freezer.

COLD CUTS

Cold cuts like sliced boiled ham, headcheese, pepper loaf, cappicola, and jellied tongue loaf are cooked meats prepared for easy slicing. Most are formed into loaves, often with a gelatin base, sometimes with seasoning. Cold cuts are packaged and marketed very much like sausages.

Buying and Storing Cold Cuts. Because they are especially susceptible to bacterial growth, cooked meat products—both loaves and sliced cooked meat such as boiled ham or corned beef—should be bought in small quantities unless they are vacuum-packed and dated. Handle them as little as possible and keep them cold. Wrap cold cuts tightly in plastic or foil and plan to refrigerate them no more than three days.

Cold cuts in unopened vacuum packages keep seven days beyond the last date of sale stamped on the label. Once opened, store them up to three days. You can freeze cold cuts, tightly wrapped, up to one month at 0° F.

CORNED BEEF

Corned beef is a boneless cut of beef, often the brisket, that has been cured in a salty, seasoned brine. The word "corned" is a colloquialism that began when a very coarse salt, with grains the size of corn kernels, was used for curing.

Buying and Storing Corned Beef. Most of the corned beef you find is vacuum-sealed in a heavy plastic bag, with a date stamped on the label. There is the rare butcher who cures corned beef in the shop and sells it right out of the brine.

Refrigerate vacuum-packed meat up to seven days beyond the last date of sale indicated on the label. Conventionally wrapped and homemade corned beef keeps five to seven days in the refrigerator, as long as it is well wrapped in plastic or foil.

Once the meat is cooked, you can refrigerate it safely for three to five more days, wrapped tightly in plastic or foil.

Corned beef can be frozen successfully as long as it is tightly wrapped, either in its vacuum pack or sealed in plastic or heavy-duty foil. Uncooked, it will keep for one month at 0° F. Cooked corned beef lasts a little longer in the freezer—two months.

There is no need to defrost uncooked corned beef. It can go right into a pot of simmering water. If you mince cooked corned beef before you freeze it, the meat is ready for hashmaking right out of the freezer. Otherwise, defrost it in the refrigerator.

OTHER PROCESSED MEATS

- DRIED MEAT. Beef jerky is a salty, chewy dried meat good for snacking and carrying on hikes. Jerky can stay at room temperature safely for days but, for long-term storage, refrigerate it. It will keep well for two to three months.

Camping supply stores sell dried meats and dried meat entrees. Vacuum-sealed in foil, they are lightweight and can be stored without refrigeration. Keep packages in a cool, dry place up to a year.

- SALT PORK. Salt pork is a dry-cured cut of meat, very fatty, and lightly streaked with lean. It will keep for one month in the refrigerator. The high-fat, high-salt content makes freezing inadvisable.

- CANNED MEAT. Store canned meat (except canned ham—see p. 201) in a cool, dry place for up to one year. Extremes of heat and cold do affect canned meat, altering its flavor, texture, and nutritional value. If your pantry is warmer than 72° F. for weeks at a time, store canned meat no longer than six months.

Once the can is opened, treat the meat as you would cooked meat, storing it in the refrigerator three to five days or in the freezer two to three months.

How to Use the Storage Timetable

The storage time limits suggested below apply to foods that are hand-wrapped in plastic wrap or aluminum foil. (The only exception is the listing for Canned Ham.)

Products sold in vacuum-sealed packages can be refrigerated in their unopened containers for as long as the dating code on the label indicates. Once the vacuum seal is broken, the time limits on this chart go into effect. For instance, if you open a vacuum pack of bacon on the final date shown on the label, the meat will keep well for seven more days in the refrigerator.

You can freeze meat sealed inside a vacuum pack with good results, but freezing makes the plastic brittle, so handle it with care.

Almost all cured meats contain small amounts of nitrites or nitrates, which add flavor and a reddish bloom. Nitrites and nitrates also control bacteria growth, particularly botulism toxin, thereby extending the product's shelf life.

There are nitrite-free processed meats on the market, such as bacon, frankfurters, and bologna, for people concerned about the health hazards of the chemical. It is essential that these meats be kept frozen right up to the time they are prepared for the table.

STORAGE TIMETABLE FOR CURED AND PROCESSED MEATS

Type of Food	Days in Refrigerator	Months in Freezer
BACON:		
Sliced, raw	7	1
Slab, raw	14–21	1
Cooked	3–4	1–1½
BOLOGNA	4–7	2
CANADIAN BACON:		
Sliced	3–4	1
Chunk	7	1–2
CANNED MEAT:		
Opened can	3–4	2–3
COLD CUTS	3–4	1
CORNED BEEF:		
Raw	7	1
Cooked	3–5	2
FRANKFURTERS	4–7	2
HAM, Noncanned:		
Whole	7	3
Sliced	7	1
Cooked	3–5	1
HAM, Canned:		
Unopened	(6–12 months)	—
Opened	5–7	1
HAM, Dry-cured:		
Whole[1]	—	—
Cooked	3–5	1
JERKY	60–90	—

Type of Food	Days in Refrigerator	Months in Freezer
LIVERWURST	4–7	2
SALAMI:		
Whole	(2–6 weeks)	—
Sliced	21	1
SALT PORK	30	—
SAUSAGE, Fresh[2]	7	2
SAUSAGE, Cooked and		
Cooked, Smoked	4–7	2
SAUSAGE, Dry and Semidry:		
Whole	(2–6 weeks)	—
Sliced	7	1

[1] Store whole dry-cured hams six to twelve months in a cool, dry place.
[2] Fresh pork breakfast sausage (links, patties, and bulk), bockwurst, and weisswurst should be stored no more than one day in the refrigerator.

POULTRY

The finest poultry you can buy comes from freshly killed birds that have grown up scratching in the yard, eating a variety of foods. Many stores now offer this kind of poultry—chickens, turkeys, ducks, geese, pheasant, and quail—labeled "natural," "organic," or "free-range." The standard supermarket sells barn-raised poultry fed on a controlled diet that fattens the birds at an early age. Their meat is tender but often terribly bland and the insubstantial flavor is usually further compromised by prolonged storage at near freezing temperatures. All this bad news notwithstanding, there is some mass-produced poultry that is very good and you only have to find the brand and the retailer or butcher who supplies it in your area.

Buying Poultry

The flavor of poultry, except in the case of wild game, is delicate. Poultry tends to dry out in storage and, long before it is spoiled in any other sense of the word, poultry can taste flat and dull. Plan to keep it for the shortest practical time.

All chickens, including capons and Rock Cornish game hens, all turkeys, ducks, geese, and guinea fowl, are inspected individually for wholesomeness by the United States Department of Agriculture. Some poultry is tagged with the USDA inspection stamp. (This is the same inspection stamp that is printed with purple ink directly onto meat carcasses. Poultry skin won't hold the dye, so paper or metal tags must be attached.) The tag may also carry a USDA grade.

Poultry earns its grade by its general appearance alone: eating quality is not a consideration. Grade A birds are meaty and have unbroken skin; Grade B poultry is bonier and has less perfect skin. (Grade B is not often printed on labels.) Grade C poultry is quite scrawny-looking and unlikely to be sold at retail. Grade A is almost always the best buy because there will be more meat in proportion to bone. Grade B can be equally tasty and is economical for soup and stock making.

The taste and texture of poultry have a lot to do with its age. Younger birds are more tender than older ones. They can be roasted, fried, barbecued, sautéed, or broiled. Older birds, though tougher, have enough flavor to outlast stewing, braising, and poaching.

There should be some clue to the bird's age on its label. For example, a young duck would be called "duckling"; young chickens are labeled "broilers" and "fryers"; older turkeys are sometimes tagged "yearling." If you have the chance, you can test the age of a fresh, whole bird by pressing its breastbone—it will be quite pliable if the bird is young. For more details on which birds are young and which are not, read the section on each type of poultry.

Fresh poultry should be cleanly dressed, with no feathers, no bruises, and no broken skin. A plump, rounded bird is probably juicier and yields more servings per pound. Sniff the bird, even through the plastic wrap. It should have a fresh aroma. Because of an increased risk of food poisoning, do not buy whole birds already stuffed for roasting.

Specialty poultry like turkey parts or duckling are sometimes thawed for sale as fresh meat, and chicken shipped at very cold temperatures may accidentally freeze. Not only will these be less tasty, they will also be much more perishable than truly fresh poultry. A lot of liquid in the poultry package often means the bird has been defrosted.

In many cases the only available Rock Cornish game hens, ducks, geese, and squab are frozen. Frozen poultry is never as good as fresh, but commercial techniques of vacuum packaging and quick processing preserve flavor and juiciness remarkably well.

Frozen birds can only be tasty, though, if they have been properly handled in the market. Look for evidence that frozen poultry has been held at a steady 0° F. It should be rock hard and stored well below the frost line in the freezer case. Examine the package: frozen liquid inside is a sign that the bird has been thawed and refrozen; heavy frost means it has been stored too long. Don't choose a bird with powdery, discolored patches on the skin. You can't always see the skin because of opaque wrappers, and frozen poultry usually bears no dating code. In this case you have to rely on brand names for quality and on your grocer for conscientious storage.

When buying either fresh or frozen poultry, plan on about one serving for every pound. But the smaller the bird the more you need. A Rock Cornish game hen, for example, may weigh close to 2 pounds, but you should buy 1 bird per person. Ducks and geese have more fat, so 1 1/2 pounds per person should be allotted. Large turkeys, over 12 pounds, are so meaty, you can count on 1/2 to 3/4 pound per serving.

Refrigerating Poultry

For best results, take your bird out of the store's package and rewrap it. First, remove giblets (neck, liver, gizzard, heart) and store them separately, because giblets spoil sooner than muscle meat. Wrap the giblets in plastic or foil, and plan to cook them or freeze them within twenty-four hours.

Second, rinse the meat in cold water and pat it dry with paper towels. The rinsing will cleanse the bird of contaminants picked up during cutting and handling in the store. And because too much surface moisture encourages microorganisms that spoil, poultry stays fresher when its skin stays dry. Wrap the fowl loosely in plastic, foil, or waxed paper and refrigerate. If you want a crisp skin, leave it unwrapped.

A fairly large percentage of the poultry you buy, even frozen poultry, carries a small number of salmonella bacteria. These won't harm the meat and they won't harm you unless the bird is stored and handled carelessly. A few precautions can protect against food poisoning. Keep the rinsed, raw poultry thoroughly chilled until you cook it. Contaminated poultry left at 45° F. for four hours can cause illness (see "Food Poisoning," pp. 361–69). Remember that a bird warmed to 70° F. because it has been in the grocery bag for a couple of hours won't immediately plummet below 45° F. the minute you put it in the refrigerator. Another hour or so may pass before the bird is safely chilled. During this time, even in the refrigerator, bacteria have a chance to grow.

Keep fresh poultry drippings away from other stored foods and off refrigerator surfaces. If bacteria in the raw meat juices find their way into cooked foods, in particular, they can multiply to dangerous levels.

Always store poultry and stuffing separately. Warm stuffing inside a refrigerated bird, raw or cooked, takes hours to chill. During that time a few bacteria, which grow exceedingly well in moist bread products, can increase to millions. Stuff the bird just before it goes into the oven, never sooner.

Wash your hands after handling raw poultry and immediately wash all surfaces and utensils that have come into contact with the raw poultry in detergent and hot water.

High temperatures destroy salmonella bacteria, so thoroughly cooking poultry to at least 170° F. is your best protection against this type of food poisoning.

Freezing Poultry

Any air that reaches poultry while it is in the freezer can dehydrate the meat and begin to turn the fat rancid. Poultry must be packaged for the freezer in an airtight, moistureproof package.

To prepare it for freezing, take the bird out of its original store wrapping. Rinse it in cold water and dry it thoroughly with paper towels. Remove all traces of viscera that may adhere to its parts or to the inside of the body cavity. Pluck out any overlooked feathers or hairs on the skin. Trim off excess fat and skin. If you plan some special preparation for the bird, such as boning or skinning, do it now, before you freeze it.

When preparing a whole bird for the freezer, truss it to achieve a compact shape. Take a large sheet of vaporproof freezer paper or aluminum foil and follow instructions for the drugstore wrap on p. 180. With waterproof ink, label the package with the contents, the date frozen, and the "use by" date.

Another technique is to put the bird into a plastic bag which you then lower, bottom first, into a sinkful of cold water. The water will drive air out of the bag. Before the water level reaches the top of the bag, seal it tightly. Use masking tape to label the package.

Poultry parts are so oddly shaped, they require extra coverage to protect them from air in the freezer. Begin by folding each piece tightly in plastic wrap. Stack the parts onto freezer paper and use the drugstore wrap. Or you can slide them into a plastic bag and use the dunking method. When parts are stored in a plastic bag, they can be taken out one at a time, if that is most convenient.

The faster the meat freezes the less damage will be done, and small packages freeze quickest. Wrap no more than 4 pounds of parts in any one bundle and don't try to freeze large, whole birds at home, because they just won't taste right. Any bird over 7 pounds cannot freeze fast enough, even in the best home freezer. Cut these larger birds into parts before freezing.

Place packages directly on shelves, not on top of other frozen foods, and leave some air space between items.

Freeze poultry and stuffing separately. Stuffing inside a bird takes so long to freeze and to thaw that there may be enough time for bacteria to grow. Poultry that has been commercially stuffed and frozen is safe as long as you follow all the instructions on the label.

Thawing Poultry

If you are stewing, boiling, or poaching poultry, no thawing is necessary: move the fowl directly from the freezer to the stove. But, before roasting or frying, you should thaw frozen poultry completely. Thawed birds cook more evenly and turn a tantalizing brown. Furthermore, it is next to impossible to get breading or flour coatings to adhere to frozen meat. As for commercially stuffed and frozen birds, prepare them precisely according to the package directions for safety as well as for flavor.

To thaw poultry, place it in the refrigerator, still in its freezer wrap, on top of a dish or paper towel that can catch drips. Thawed slow and cold, the meat holds in moisture and nutrients. It takes about two hours per pound for poultry to thaw in the refrigerator. Large whole birds take three hours per pound. A big turkey can take several days.

It is not safe to thaw poultry at room temperature because bacteria can grow luxuriantly on the skin while the interior is still frozen. But there is a reliable technique for thawing poultry quickly. If the poultry is in a waterproof package, submerge it in cold water (about 50° F.). Change the water frequently to keep it cold. With this method, poultry thaws at the rate of about half an hour per pound.

Another possibility is a microwave oven; thaw according to the manufacturer's directions. A whole bird should be turned at least once during the defrosting to make sure that it thaws uniformly. Parts should be separated as soon as they are slightly softened.

Check for ice crystals in thawed poultry. Pinch individual pieces and run your hand around inside whole birds. If there are still traces of ice, increase the usual cooking time to make sure the poultry ends up heated to at least 170° F.

Always remove giblets from commercially frozen birds as soon as possible and either thaw them separately in the refrigerator or cook them immediately.

Cook thawed poultry within twenty-four hours.

Storing Cooked Poultry

Whisk cooked poultry into the refrigerator or freezer as soon as you can. Cooked poultry should never be left at room temperature for very long. Don't wait for it to cool first—chill cooked poultry immediately.

The most convenient way to store a whole bird is to remove all the meat from the bone. Wrap the meat loosely until it is completely chilled, then seal it in plastic bags or plastic wrap. The bones can be placed in their own bag and saved for making stock. Poultry parts can be wrapped bone in.

If, after the usual excesses of food and drink on Thanksgiving, you do not have the spunk to bone your 12 pounds of leftover turkey, refrigerate it right away under a loose umbrella of foil. But be sure to remove all the stuffing first. Bone the bird Friday morning and wrap the meat up tight for longer storage.

Always store stuffing separately.

Cooked poultry will keep well for three to four days in the refrigerator. Gravy is safe for only two days. You can extend the storage life of the gravy by bringing it to a full boil after two days, then returning it to a clean, tightly lidded container. Stuffing should be refrigerated no longer than two days—freeze it if you want to keep it longer; it keeps one month at 0° F.

Before you serve the leftovers, boil gravy for several minutes and reheat stuffing to at least 165° F.

Freezing poultry leftovers works best when you combine the meat with sauce or gravy. Slice the meat off the bone, then arrange it inside a small- or medium-size casserole that you have lined with a large sheet of aluminum foil. Pour the sauce or gravy on top, then fold and seal the foil. Place the casserole in the freezer and keep it there until the poultry becomes hard. You can then stack the foil packet in the freezer and put the dish back on the shelf. When it is time to cook the leftovers, slip the frozen package back into the casserole and put it right in the oven to reheat. Poultry coated with gravy will keep in the freezer for six months.

To freeze poultry meat without gravy, wrap it in foil sealed airtight. "Dry" poultry meat should only be held in the freezer for one month. Reheat it in the foil or use it in soup or stew without thawing. To use the meat in cold dishes, thaw it in the refrigerator.

Fried chicken with a thick bread coating can be left on the bone and frozen successfully for up to four months.

Poultry bones should be stored in a sealed plastic bag. They will keep in the refrigerator for two days and in the freezer for three months. These bones can be combined with other meat bones or simmered on their own to make stock (see p. 192).

Smoked Poultry.

Specialty shops and delicatessens carry smoked poultry, usually an expensive delicacy. Be sure to find out if the meat is fully cooked or requires more heating at home.

Fully cooked smoked turkey belongs in the refrigerator, where it should keep well up to one week. Smoked chicken and uncooked smoked turkey won't last quite that long—plan to keep it only two to three days.

Both will freeze well, but take even more care than usual to wrap the meat airtight. The smoked poultry is already quite dry and freezer air can make it very papery. It keeps up to six months at 0° F.

CHICKEN

French cuisine, Indian, African, Szechuan, Italian, Cajun—the list goes on and on—all hold chicken in high esteem, marinating it, dousing it in sauces, baking, broiling, and barbecuing it. And because it is high in protein, low in fat, and easy to digest, chicken, simmered and skinned, has a humbler role as a simple food for the very young and the calorie conscious. Chicken soup, of course, is a legendary (and effective) home remedy.

Buying Chicken. You will generally find five different types of chicken available:

- A *broiler/fryer* is a seven- to nine-week-old bird that weighs between 2 and 4 pounds. The best have tender, delicately flavored meat that is delicious broiled, fried, sautéed, or roasted. Most broiler/fryers at the meat counter will be cut up—halved, quartered, or divided into drumsticks, thighs, breasts, and wings. Chicken parts are convenient but always more expensive than whole birds.

- A *roaster,* or *pullet,* is three to seven months old and weighs between 3 and 7 pounds. This chicken is meatier and has more fat under the skin, which keeps it juicy during roasting.

- A *capon* is a castrated male bird that consequently grows very large and plump. It is four to five months old and weighs be-

tween 6 and 9 pounds. It is particularly well suited to roasting because its extra fat moistens the meat.
• A *stewing chicken* is the oldest chicken on the market. It is at least a year old and weighs 3 to 7 pounds. Only slow, moist cooking will tenderize a stewing chicken, but its flavor in soups, stews, and fricassees is unparalleled.
• A *Rock Cornish game hen* is a special breed of chicken that is very small but meaty and moist enough for roasting. It is five or six weeks old and weighs 1 to 2 pounds. They are usually sold frozen, but fresh ones can be found in many urban areas.

Choose whole and cut-up chickens that look moist and rounded. The breast of a whole chicken should be very full in relation to the legs.

You can't judge the quality of a chicken by its color—they range from white to yellow, the exact hue depending on the bird's breed and diet; the color has no bearing on taste. Whatever the color, the skin should not appear transparent or mottled. The skin should be unbroken.

Some chickens are labeled "fresh" or "fresh-killed." This usually means that the birds have been shipped under ordinary refrigeration, in an ice pack or a CO_2 "snow" pack, rather than under the colder temperatures of what is called deep chilling. Deep-chilled birds are stored and shipped at around 28° F. At that temperature the water in the carcass is frozen hard but, technically, the meat itself is not frozen. Deep-chilled chickens may be held for many days before reaching the consumer. "Fresh" chicken is more desirable than deep-chilled—it will most likely have juicier flesh. If you want to seek out the very freshest chicken, look for poultry shops and kosher butchers.

You should be aware that chickens, like red-meat animals, are heavily dosed with antibiotics. The maximum levels of these drugs are prescribed and presumably policed by the government. Inspection agencies also monitor feed and chickens for pesticide residue, so contaminated fowl is very rare. Hormones are not used at all for broiler/fryers. Larger chickens may have received hormones to fatten them, but again under the supervision of the USDA, which maintains that, theoretically, they are not dangerous to consumers. Buying and eating organically grown chickens should help you to avoid most of these chemicals.

Storing and Freezing Chicken. You can leave chicken in its supermarket plastic wrap or butcher paper if you plan to cook it the same day you buy it. Otherwise, take it out of the package, rinse, dry, and rewrap it loosely in plastic, foil, or waxed paper. Some brands of chicken come vacuum-sealed in very thick plastic. Chicken parts with no giblets can be stored in these packages for several days.

A whole chicken will keep well for one to two days in the refrigerator.

Plan to refrigerate cut-up chicken no more than twenty-four hours. If your refrigerator has a meat drawer that stays between 32° F. and 35° F., you can add a day to these storage times.

The giblets and neck, whether they are loose or in a paper packet, should be removed from the chicken and stored separately. Refrigerate them no longer than one day.

To freeze a whole chicken, first remove neck and giblets, rinse the bird inside and out, and pat it dry. Truss the bird to make it compact, then wrap it in an airtight package. A plastic bag is best, as long as you expel *all* the air from the bag and seal it securely. Chicken parts should be wrapped individually, then bagged for the freezer.

Under optimum conditions, a whole chicken retains good quality in the freezer up to twelve months. If the bird is not well wrapped, or if the freezer temperature fluctuates above 0° F., the chicken should be stored only half that time. Chicken parts and Rock Cornish game hens deteriorate more quickly. They will keep well for six to nine months in the freezer.

Cooked chicken can be refrigerated for three to four days. It should be well wrapped or stored in a tightly lidded container. If the cooked chicken is covered with gravy or sauce, don't keep it in the refrigerator more than two days; freeze it instead. Chicken in gravy freezes well for up to six months (see pp. 333–35 for suggestions on freezing cooked chicken dishes). Fried chicken, tightly wrapped in foil, lasts four months in the freezer. Cooked chicken frozen without sauce or gravy should be kept no longer than one month at 0° F.

TURKEY

The turkey is a native American bird. Early Spanish explorers graciously brought the chicken to the Western Hemisphere and returned to Europe with the turkey. The Old World was enthusiastic about the new fowl but, because of the usual muddle concerning exactly what piece of land these sailors had found, named it variously "bird of India" and "Calcutta hen." America's first settlers found wild turkeys as plentiful as pigeons in Central Park. After a couple of centuries, wild ones were hard to find and domestic turkeys made an appearance only during football season, those first gray months of winter. Now, just as football games sprawl across the calendar, turkey abounds year round.

Buying Turkey. Nearly all turkey meat for sale is marketed under brand names. This is true for fresh and frozen birds, whole birds and parts.

After comparison shopping, you can safely stick with your favorite brand since the quality will be very consistent. You still have to be vigilant about inspecting the poultry, however.

Fresh turkey should look moist and smell good. The skin should be pearly white, not bluish. Fresh turkeys are frequently deep-chilled (see p. 216) for shipping. Avoid the deep-chilled birds and ask for the fresher, moister turkeys that have been stored in ice.

As for frozen, it is hard to tell what you are getting—packages are usually opaque and undated. The best you can do is to buy from a reliable market and question the butcher if you have reason to think the bird has been frozen too long.

Read the label for information on additives. Turkeys are sometimes treated with sodium, sugar, artificial flavor, coloring, and other unexpected ingredients.

The kinds of turkey you are most likely to find at the supermarket are:

• WHOLE YOUNG TURKEYS. These weigh between 8 and 20 pounds. You will find them fresh and frozen, self-basting (meaning vegetable oil and/or butter has been injected into the bird) or plain. These turkeys may be hens or toms. All are tender because of their age (about four months old), and all are suited to traditional roasting methods. Carefully read the label on self-basting birds, because you may be buying ingredients such as palm oil and extra sodium that are less than desirable.

• TURKEY PARTS are packaged fresh and frozen, in a number of combinations. Wings, thighs, and drumsticks can be baked, fried, or braised; hindquarter roasts and breasts roast well. Turkey steaks and cutlets are boneless slices of breast meat that, for instance, are inexpensive substitutes for veal scallops.

• ROLLED TURKEY ROASTS are made up of boned, tied turkey meat, either all white or dark and white mixed. A similar-looking product is made up of cooked, processed turkey meat, so read the label to make sure you know what you are getting.

• OTHER TYPES OF TURKEY are much less common. There are small fryer/roaster turkeys that are 5 to 8 pounds. They are generally more expensive per pound. Mature toms are older birds, over 20 pounds (not every bird that weighs that much is a mature tom, though). They are too tough to roast and should be simmered.

Markets that monitor their frozen food sections will often thaw turkeys and parts to sell once the poultry has reached its time limit in the freezer. The thawed poultry, which should be labeled as such, is perfectly acceptable but has a very short storage life. You should cook it within twenty-four hours.

Storing and Freezing Turkey. Fresh turkey can be stored one to two days in the refrigerator in its original wrap. Plan to cook thawed turkey the same day you buy it.

A purchased frozen turkey will retain its quality for up to twelve months from the date it was originally processed, as long as it is kept at 0° F. Needless to say, it is difficult to determine this date. Furthermore, stores often put frozen turkeys on sale—the likeliest time for you to buy one to keep for a while—for the very reason that the birds have been in the freezer long enough. Unless you know otherwise, it is probably safe to assume a frozen turkey, particularly one bought outside the holiday season, has already logged several months of freezer time. Taking into account these months, plus the added strain on the bird of fluctuating temperatures in home freezers, plan to store a commercially frozen turkey for no more than six months at 0° F., and half that time at 10° F.

Commercially frozen turkey parts have a shorter storage life than whole birds. They will be best if held no longer than six months from the day they were packed. Using the same logic applied to whole frozen turkeys, keep frozen turkey parts in your home freezer for a maximum of three months.

Don't try to freeze a whole turkey yourself. Because of its size, the bird will freeze much too slowly. What you can do is to cut up the turkey for freezing. The butcher can help with the cutting, dividing off the breasts and other parts. Wrap each turkey part in plastic or foil, then pack the parts into airtight plastic bags in bundles no larger than about 4 pounds. Keep giblets separate. Force all the air out of the bags and seal them tightly. Home-frozen turkey parts keep their quality for three to six months at 0° F.

There are only two safe ways to thaw turkey. You can thaw it in the refrigerator, in its original plastic wrapping, on a tray to catch the drippings. This method is the safest and it results in the juiciest meat. Allow about three hours per pound of thawing time. This adds up to one day for an 8-pound bird, two days for a 14-pounder, and close to three days for birds over 22 pounds. A 4-pound package of turkey parts thaws in about one day.

The second thawing method is only safe for birds 10 pounds or less. Submerge the wrapped frozen turkey in cold (50° F.) water. Change the water regularly to keep the temperature low. Allow half an hour per pound thawing time.

Thaw commercially stuffed frozen birds according to package directions.

When the thawing time has elapsed, unwrap the turkey and remove the neck and giblets. Feel the body cavity for ice crystals. If the interior still seems to be frozen, refrigerate the bird until it is completely defrosted or cook it longer than normal to make sure it is heated throughout to 170° F. (or to 175° F. if the poultry is stuffed).

Cook turkey within twenty-four hours of thawing it. You can roast a thawed breast and braise or poach other parts.

Remove cooked turkey from the bone for storage. You can refrigerate the meat for three to four days. The bones can be bagged and stored for stock making—two days in the refrigerator, three months in the freezer. Sealed in foil, turkey meat lasts only about one month in the freezer. If it is frozen in gravy or sauce, it will keep for six months.

Thaw cooked turkey meat in the refrigerator and use it within one day.

DUCKS AND GEESE

Domestic ducks and geese are rich, fat birds, not as meaty as chicken or turkey, but with a more complicated flavor.

Buying Ducks and Geese. Most ducks and geese on the market are sold as commercially frozen whole birds. You can use brand names as guides to quality, trying different ones to find your favorite.

Ducks are usually prepared for market at two months of age and are therefore correctly identified as duckling. The Long Island Pekin breed of duck, weighing 4 to 5 pounds, is the most commonly available and, fortunately, delicious. A few Long Island ducks are raised on their name-sake island in New York State, but most come from producers in the Midwest. Pekin duckling is occasionally marketed fresh in large Eastern and Midwestern cities and San Francisco's Chinatown has fresh ducks daily. In all cases the fowl is shipped within a day of slaughter and is therefore very, very fresh. Mallard and Muscovy ducks are, if anything, more flavorful than the Pekin but are in much shorter supply. The Mallard tends to be smaller—under 3 pounds—and the Muscovy male can weigh up to 10 pounds. These breeds are virtually always frozen for retail sale.

The geese you find weigh between 5 and 14 pounds. Generally speaking, the weight is an indication of age. Under 9 pounds, the goose can be considered a gosling and therefore more tender than older birds. Geese are slaughtered, frozen, and shipped from July to December, so that by spring they are scarce as hen's teeth.

As with other kinds of frozen poultry, look for well-rounded, plump-breasted birds wrapped in airtight packages that show no signs of freezer damage. The skin should be clean and unbroken with no signs of mottling and no pin feathers. Any bruises or tears will cause the meat to discolor during cooking. Ducks and geese should have a fresh, not a gamy odor.

Storing Ducks and Geese. Commercially frozen ducks and geese can be stored at 0° F. for as long as six months. Ask how long it has already been in the freezer and subtract that time from the six-month maximum. Fortunately, some frozen birds carry an expiration date on the label.

Thaw ducks and geese in the refrigerator for best results. Leave the poultry in its original wrap but set it on a dish to catch drippings. A duck will take around twenty-four hours to thaw; a goose should be left for at least thirty hours. If you use the less preferable cold-water bath to defrost, allow three hours for a duck and four hours for a goose. Cook the birds immediately after thawing.

The thick layers of fat on fresh ducks and geese will serve to preserve the meat during storage. You can refrigerate the fresh poultry two to three days without losing too much flavor. However, home freezing is so damaging, it is not recommended.

That very plentiful fat encasing a goose or duck can be a great asset in future recipes, so don't just throw it out. Rendered and filtered, the fat can be stored in the refrigerator for months. One way to do it is to remove fat and fatty skin from the uncooked bird and puree it. Put the puree, along with 1/4 cup water per pound of fat, into a saucepan and melt it over a very low heat. Next, pour the liquid fat through a cheesecloth and set it in the refrigerator to congeal. When the fat is set, it separates from the water. Spoon the fat into a covered container and refrigerate. You can accomplish the same thing by siphoning fat from the roasting pan as the bird cooks and filtering it through cheesecloth. Use the rich, tasty fat for sautéing and for seasoning cooked vegetables.

SQUAB

A squab is a pigeon no more than a month old. These tiny birds are specially bred and raised for market.

Choose squabs, which are sold frozen except in specialty markets in urban areas, that weigh no more than 1 pound. If you find them fresh, buy birds with the palest skin and the plumpest form.

Commercially frozen squab should be stored at home for no more than four to six months at 0° F. Use fresh squab within one day.

GAME BIRDS

Game birds will spoil very quickly unless they are handled properly immediately after the kill.

Dressing Fresh-killed Game Birds. Bleed the bird first by slicing its jugular vein and hanging it head downward. Once it is bled, cut off the head and eviscerate the carcass. Scoop out clumps of fat and blood. Find the oil sac attached to the back near the tail and slice that out. Rinse the body cavity and pat it very dry.

Although the practice is out of favor, you can choose to hang the bird at this point. Hanging is an aging process that tenderizes the meat and intensifies its flavor. The place game hangs should be cool (below 45° F.), dry, dimly lit, and open to air circulation. Depending on the size of the bird and the temperature during hanging, time limits will vary from one to seven days. One test of "doneness" is that tail feathers pull out easily. A strong odor means it has hung too long.

Instead of hanging the bird, you can simply store it for one day in the refrigerator. This allows a period of aging that will tenderize the meat.

Plucking need not be done in the field: do it right before you refrigerate or cook the bird. Grasp small patches of feathers and pull them out toward the tail. The process is easier if the bird is dipped several times in scalding water, but dipping detracts from the keeping quality of the game. Singe off any remaining feathers and hairs with a candle flame.

A dressed, plucked game bird keeps three to four days in the refrigerator. You can freeze game birds and keep them six to nine months.

Buying Game Birds. Pheasant, quail, guinea hens, partridge, and wild turkeys are raised for retail sale in some parts of the United States. They can be found fresh or frozen. The usual poultry standards apply to selecting these game birds: fresh birds should be clean, moist-looking with unbroken skin and a fresh scent; frozen birds should be well wrapped, free of freezer burn, and should show no signs of premature thawing. Younger game will be more tender. A soft, flexible breastbone in fresh poultry indicates a young bird. As for frozen game, the younger birds will be the ones smaller in size.

Store fresh birds no more than one or two days. Frozen birds can be held up to nine months at 0° F.

GIBLETS

Invariably, whole birds and even cut-up chickens are sold with giblets and necks tucked inside. Edible poultry giblets are the heart, cleaned gizzard, and the liver. These variety meats also come packaged separately in meal-size portions.

Buying Giblets. You can't judge the quality of giblets that are passengers inside supermarket poultry, but you can decide if they are good enough to cook. Pull out the giblets from fresh poultry as soon as you get it home and out of frozen poultry as soon as it is sufficiently thawed. Giblets should have a fresh aroma and bright, uniform color.

When buying giblets separately, look for the same signs of freshness. Livers should be very red, with no traces of green or yellow. All types of giblets should be moist-looking.

Storing Giblets. Giblets have a very short storage life. Try to cook or freeze them the same day they are purchased, and always store them apart from poultry meat.

Cooking is the best choice, since giblets don't hold up well in the freezer. If you must freeze them, place them in an airtight freezer bag and keep them at 0° F. only two to three months.

STORAGE TIMETABLE FOR POULTRY

Type of Poultry	Days in Refrigerator	Months in Freezer
CHICKEN:		
Whole	1–2	6–12
Parts	1	6–9
ROCK CORNISH GAME HEN	1	6–9
TURKEY:		
Whole	1–2	6–12
Parts	1–2	3–6
DUCK	2–3	6
GOOSE	2–3	4–6
SQUAB	1	4–6

Type of Poultry	Days in Refrigerator	Months in Freezer
GAME BIRDS	1–2	6–9
GIBLETS	1	2–3
COOKED POULTRY:		
In gravy or sauce	1–2	6
Plain meat	3–4	1
Breaded, fried	3–4	4
STUFFING	1–2	1
GRAVY	1–2	6
STOCK	1–2	4–6
RENDERED FAT	(6 months)	N/A
SMOKED CHICKEN	2–3	6
SMOKED TURKEY	7	6

FISH AND SHELLFISH

In the best of all possible worlds the fish you buy would look as if caught in a wave of crushed ice in mid-swim; the clams would be so robustly alive they would clack their shells like castanets; aggressive blue crabs and heavy-clawed crayfish would thrash around in sparkling holding tanks at every neighborhood supermarket.

Reality presents another picture. Except in coastal towns and in urban specialty shops, fish is mostly sold in paper-wrapped bricks like books stacked on the freezer shelf, or, if fresh, skinned and presliced.

The closer seafood is to looking alive, the closer it is to *being* alive, the sweeter, more delicate and delicious it will be. Even the most reliable retailers sometimes sell fish that have been out of the water for a long time. There are inevitable delays in getting fish to markets: fishing boats often stay at sea for two weeks before they deliver their catch to wholesalers; and fish are often trucked in from long distances. A fish packed in ice stays edible for weeks but tastes great for only a few days.

Purchase fresh seafood, if you can, from a retailer who specializes. A genuine fish house will know how to store what it has and will be moving fresh fish in and out quickly.

If your only access to fresh fish is at the local supermarket, you should use a little caution. For instance, don't buy fish on Sunday—purveyors deliver Monday through Friday, so anything left on ice after Saturday has already been kept too long. Ask about your store's delivery dates and plan to buy fish and shellfish on those days only.

Frozen fish is second best but can be perfectly satisfying in the right recipe. Don't hesitate to buy frozen fish but don't expect to store it for a long time. It should be double-wrapped, because even at 0° F. the fishy smell can creep into other foods. It should be used within a week.

Even when you hook your own fish (or gather your own shellfish), there are techniques for short-term storage that protect the quality of the catch until you can get it sizzling in the pan.

FISH

Salmon, perch, trout, sole, and scrod are fish that find their way into markets all over the country, but hundreds of other species rarely leave their near shores. Fresh pompano and mullet in Florida, mahi-mahi and corbina in California, buffalofish in the South, and black sea bass in the Northeast—they are all reason enough to travel to distant places.

Buying Fish. Fish are not graded or labeled uniformly. In most states and municipalities the seller's only obligation is to identify fish that has been previously frozen, a procedure not always strictly followed. Always buy fish the same day you plan to cook it. If this is impossible, use the utmost care in storing it at home.

Fish are presented for retail sale in a number of different forms:

• LIVE FISH, held in glass tanks, are, of course, as fresh as it is possible to be. Nevertheless, take the time to pick out a particular fish that is lively and that has no cuts or bruises on the skin. The fishmonger will clean it, then cut it to order.

• WHOLE, DRESSED FISH look very much as they did when they were still swimming. The fins and viscera will have been removed already and the fishmonger will further prepare the fish to suit your needs. Be sure to have the scales removed, so you won't be faced with that nasty job at home.

Because fish stay moist and flavorful longer with their skin and bones intact, whole fish are a good buy, and it is fairly simple to check a whole fish for freshness. First, look it in the eye: fresh fish have bright, clear, rounded eyes with no trace of blood or milkiness. The fish's gills should be deep red, neither pink nor brownish. The skin should be shiny, free of slime, and show no red patches along the belly. Get close enough to the fish to sniff it. There will be a scent, but a strong, fishy odor means the fish is no longer fresh. The flesh should be firm and should spring back when pressed.

When buying whole fish, plan on 1 pound of fish per serving.

• FISH FILLETS are boneless slabs cut lengthwise off the side of the fish.

They may or may not be skinned. Fresh fillets are moist and translucent and the flesh is firm with a mild, not fishy, odor. If the fillet is mushy or discolored, it has been held too long. Fillets are the most perishable cut of fish. A fletch is half a fillet cut from large flat fish such as halibut.

A fillet is all fish, no bone. Plan on 1/2 to 3/4 pound of fillet per serving.

• FISH STEAKS are cross sections of large, firm-fleshed fish such as salmon. Steaks are sold with the backbone in and the skin on. Again, check the texture and odor of the fish. It should be resilient to the touch and mild to your nose.

Buy 1/2 to 3/4 pound of fish steak per serving.

If you have the choice, pick a whole fish and have it cut into fillets or steaks to order. Whole fish are more likely to be fresh and their freshness is easiest to identify. Very large fish such as salmon, swordfish, and tuna are not likely to be sold whole, but fortunately large fish, even precut, are better than average keepers anyway.

Fresh fish for sale should be displayed on a bed of ice. Precut, pre-wrapped fish sometimes show up at the meat counter. Unless you are certain that the package has been there for only a few hours, don't buy it. Meat-counter temperatures are not low enough to keep fish fresh-tasting.

Frozen fish usually come in the form of fillets or steaks. Examine the fish closely for the discolored patches that signal freezer burn. Buy packages that are tightly covered with vaporproof wrapping, frozen hard and stored well below the freezer line in the grocer's case. Frozen fish should have no fishy odor.

Fresh-caught Fish. Even fish pulled right out of the water can become smelly and off-tasting quite soon if they are not treated right. Take crushed ice or an ice pack on the fishing trip. If you have time, gut the fish as soon as you catch it, remove its gills, rinse it out, and lay it on the ice. Otherwise, lay the fish as is on the ice and clean it as soon as you can. The flesh of the fish is so unstable that the animal's own digestive system will start breaking down its tissues soon after it leaves the water.

An alternative is to keep the fish alive as long as possible, by putting it into a live well or attaching it to a stringer underwater. These methods only work on live fish and many fish don't survive long in these conditions. Dangling a dead fish on a line in a lake or ocean will do little to prevent spoilage unless the water is icy cold.

If ice is not available, gut the fish, remove its gills, and keep it cool. Fishermen who are hip deep in a mountain stream may have to use a creel packed with grass or mud to cool a fish. If this is your only choice, keep the fish out of the sun and get it on ice within two or three hours.

Storing Fish. Transport fresh fish directly from store to home without delay. The truly determined fish lover can pack ice in a plastic bag beforehand and carry fish at the right temperature from beginning to end.

Pat fish dry, wrap it tightly in foil or plastic, and seal it in an airtight plastic bag. Store it in the coldest part of your refrigerator. Fish held at 31° F. will last literally twice as long as fish stored at 37° F. If your refrigerator gets no colder than 37° F., you can bring down the temperature of the fish by storing it on a pan of crushed ice. Keep a layer of waxed paper between the fish and the ice. Use ice whenever you need to keep the fish overnight before cooking it. At normal refrigerator temperatures (35° F. to 40° F.), keep fish for no more than a few hours.

If you must store fish more than twenty-four hours, rinse it in salt water or lightly coat the fish with soy sauce before you wrap it for the refrigerator.

You can add piquancy to fresh fish and one or two days of storage time by soaking it in a marinade. Remarkably, after marinating several hours, fresh fish and shellfish alike turn white and flaky as though they were cooked and need no further preparation.

Freezing Fish. Fish do not fare well in a home freezer. Temperatures of minus 20° F. and lower will preserve fish with some success, but most home freezers are just not cold enough. Fish bought in a store is already a few days out of the water and should not be frozen unless it can't be avoided. Fresh-caught fish (often overabundant) may be frozen but only when necessary.

Before freezing, the fish should be scaled, gutted, and rinsed clean of all blood and viscera. Always remove the gills. Smaller fish (2 pounds or less after cleaning) should be left whole. Larger fish can be sliced into steaks or fillets. You can choose to freeze a large fish whole to suit a particular recipe, but the larger the fish the longer it will take to freeze and slow freezing does a lot of damage to the flesh. If you have a *very* cold freezer you can probably freeze a 4-pound fish whole with some success. Use the following methods to prepare fish for the freezer.

• DIPPING. Dipping fresh fish in the appropriate solution for half a minute before freezing will firm the flesh and reduce moisture loss. Dipping works on whole fish, fillets, and steaks. For fatty fish, such as trout, whitefish, swordfish, tuna, and mackerel, use a solution of 1 tablespoon ascorbic acid crystals (available at drugstores) in 1 quart very cold water. Lean fish, such as sole, snapper, flounder, perch, and bass, benefit from a dip in a solution of 1/4 cup salt in 1 quart cold water. (If you don't have ascorbic acid, dip fatty fish in salt solution, too.)

After dipping, the fish, still wet, should be sealed in two layers of freezer paper or heavy-duty aluminum foil and frozen; or, for best results, ice-glazed or frozen in an ice block (see following paragraphs).

• ICE GLAZING. A thick coating of ice on whole or cut fish will seal it against the drying effects of freezer air. First dip the fish in either a salt or ascorbic acid solution. Wrap the fish loosely in plastic wrap and lay it in a single layer on a baking sheet in the freezer. After four hours the fish should be frozen solid. Remove it from the freezer and take off the wrap. Submerge the fish in a pan of ice water for five seconds, then hold it in the air for five seconds. Repeat this two or three times. Each dunking will coat the fish in a thin layer of ice. Now cover the fish tightly in freezer paper or heavy-duty aluminum foil, using the drugstore wrap (see p. 180). Whole fish should be wrapped singly. Steaks and fillets should be interleaved with sheets of freezer paper to keep them separate. Wrap the package in a second layer of vaporproof paper, tape it, and label it with the date and contents.

• ICE BLOCK. Small whole fish can be frozen inside a block of ice, a method that will, like glazing, protect them against drying. Fill a tall, thin container like a coffee can or milk carton (you can also use a loaf pan) half full of cold water. Drop in one or more fish and set in the freezer. When the ice begins to harden, pour more water into the container to completely cover the fish, leaving 1 inch head space. Cover the container tightly in heavy-duty foil and put it in the freezer.

Adjust your freezer's thermostat to the lowest possible setting before putting in the fish. Place the double-wrapped, tightly sealed bundles directly on freezer surfaces—don't stack them—leaving a little air space around each package. Keep the freezer door closed for four hours after the fish has gone in.

If these procedures are followed carefully, fatty fish (see list on storage timetable) will keep their high quality in the freezer for three months and lean fish (see list) will keep well for six months.

Thawing Frozen Fish. Frozen whole fish, over 1/2 pound, should be thawed before cooking. Thaw them in their freezer wrap, on a tray in the refrigerator. Allow twenty to twenty-four hours' thawing time for whole fish 4 pounds and over, and twelve hours for 2-pounders. If the fish is in waterproof packaging it can be thawed under cold running water—one hour for 4-pound fish and half an hour for 2 pounds. Steaks, fillets, and very small whole fish are juiciest when they are cooked directly from the frozen state. If you intend to use a bread coating or to fry the cut fish, you should partially defrost them first (two hours in the refrigerator or ten minutes in cold running water). Use lower temperatures than normal to cook frozen fish to insure even heating. Fish is just too delicate to thaw in the microwave; you will most likely half-cook the fish if you try it.

Cooked Fish. Cooked fish will keep for three to four days in tightly lidded containers in the refrigerator. Don't try to reheat cooked fish; it will become much too dry. Serve it cold in salad or with a sauce.

Smoked Fish. Fish is first packed in salt or brine, then either hot-smoked

(110° F. to 180° F.) or cold-smoked (60° F. to 110° F.). Any fish can be smoked but, at retail, you will commonly find only herring, trout, whitefish, sturgeon, cod, and, most popular of all, salmon. The finest smoked salmon comes from Scotland and Nova Scotia. Lox is a very salty version of smoked salmon that has been soaked in brine rather than dry-cured.

A whole smoked fish or a large chunk can be stored, tightly wrapped, in the refrigerator up to two weeks. Slices of smoked fish deteriorate much more quickly. Wrap slices in plastic or foil and refrigerate them no more than a few days.

You can freeze smoked fish to extend its storage life, but the flesh will be soft when thawed. Double-wrapped and sealed in heavy-duty foil or freezer paper, smoked fish can be kept at 0° F. for two to three months.

Canned Fish. Canned tuna and canned salmon are staples in most American kitchens and canned sardines, kippered herring, crab meat, minced clams, and smoked oysters are on most grocers' shelves.

Store canned fish on a cool, dry shelf up to one year. In a pantry that is warmer than 72° F. for any length of time, the canned fish will lose some of its quality. Stored too long, the fish is safe to eat but acquires a strong, salty odor and flavor. In the case of sardines, this briny, intense taste can be looked upon as an asset. The French like sardines aged in the can as long as twelve years.

After canned fish is opened, store it in a tightly lidded container in the refrigerator for three to five days (fish salads in dressing last about three days). In the freezer, plain or in prepared dishes, well-wrapped fish from cans keeps two to three months.

Pickled Fish. Pickled herring, packed in sour cream or mustard sauce or vinegar or wine sauce, is a popular appetizer. The sweetest, most tender pickled herring is imported from Scandinavia. All varieties are in the refrigerator case in glass jars that may or may not bear a dating code. Unopened jars must be kept chilled, either up to the date on the label or no more than three months. After opening, keep the herring refrigerated up to three weeks.

Dried Fish. Fish that has been dried to the point of being leathery or papery is available in specialty shops and ethnic markets. Dried shrimp, for instance, is an ingredient in Japanese and Chinese cooking; dried salted cod is used in South American dishes.

Keep dried fish in an airtight container on a cool, dry shelf up to six months. In warm, humid weather, store the container in the refrigerator.

VARIETY FISH

If you shop in a specialty fish store you may have the opportunity to sample some less familiar seafood.

• EELS. Eels can be bought live in East Coast and some Midwest markets, mostly at Christmastime. Ask the fishmonger to clean, skin, and dress them for you. Uncooked, eel can be refrigerated, well wrapped, for one day. Freeze cleaned, dressed eel in the form of steaks. Treat with an ascorbic acid dip as you would a fatty fish (see p. 240). Cooked eel will keep for three to four days in the refrigerator.

• SQUID AND OCTOPUS. These baggy-bodied sea creatures are not as popular as they deserve to be—if you are lucky enough to find them fresh or frozen at the fish store, give them a try. If whole and fresh, look for clear eyes and a fresh scent. Have the fishmonger clean and dress them for you. Uncooked, they will keep for one or two days wrapped on ice in the refrigerator. Store cooked squid and octopus two to three days in the refrigerator in a covered container.

• FROGS' LEGS. Frogs' legs should be dry, not slimy, plump and pinkish-white. Fresh legs will be springy to the touch. Frozen, they should be solid and free of discoloration. Fresh or frozen, they should have a fresh scent.

Store fresh frogs' legs for no more than one day, well covered, on ice, in the refrigerator. Frozen, they will keep up to three months. Refrigerate cooked frogs' legs, well wrapped, up to two days in the refrigerator.

CAVIAR

True caviar is sturgeon roe: small, round, firm eggs colored blue-gray, deep black, dark green, or pale brown. *Beluga* caviar is considered the finest (and priced accordingly). Hardly less sensational are *osietr* and *sevruga* caviar, each from a different species of sturgeon. Pressed caviar, or *payusnaya,* is smaller, softer, or damaged roe sifted out in the grading process. The texture is like thick jam, but the flavor is still excellent. Salmon, or red, caviar is the most acceptable substitute for the real thing. Lumpfish caviar is popular in Scandinavia but most often reaches Americans highly preserved and heavily dyed—nothing like its fresh counterpart.

Fresh caviar is never vacuum-sealed; it is potted in tin or glass and stored at very cold temperatures. It should be labeled *malossol,* meaning "light salt" in Russian. Before you buy, ask the retailer to open the caviar's container: eggs should be whole and have a light, clear coat of fat. Vacuum-packed caviar is pasteurized and therefore chewier and less delicious than fresh.

Store unopened fresh caviar at a temperature between 29° F. and 32° F. —in a meat cooler with a separate temperature control—and it should keep one to four weeks. If you freeze it, as some purveyors recommend, you risk the eggs bursting into caviar mush. Once caviar is opened, you must use it within a day or two. Pasteurized caviar can be stored on the shelf two to four months, then one to two weeks in the refrigerator after opening.

CLAMS

The waters off the East Coast yield many varieties, but clam lovers concentrate on two: soft-shell clams, or steamers, usually steamed or fried; and hard-shell clams, also called quahogs (rhymes with snow dogs), consumed in raw splendor. The youngest, smallest quahogs are known as little necks; cherrystones are larger and about five years old. Bigger quahogs are tough and unwieldy on the half shell and are therefore called chowder clams.

The Pacific Coast claims its own soft-shell clams: razor clams (shaped like, and nearly as sharp as, a barber's straight razor), and two types of littleneck (no relation to little neck), which are the most tender. The famous pismo clam is superb but disappearing.

Buying Clams. Clams are on the market year round but the best are found from fall through early spring. You will find them in the shell, shucked, canned, or frozen. Clams in the shell must be alive when you buy them. They are alive if the shell is shut tight or clamps down when you tap it. Geoduck and soft-shell clams behave differently: their long, ridged siphons normally protrude from the shell. A healthy individual will retract the siphon abruptly when touched. Don't buy clams with gaping shells, broken shells, or a strong odor.

Shucked shellfish should be packed in their natural liquor. The liquid should be clear and smell fresh, and the clams should be plump.

Frozen minced clams are available. Buy only odorless packages that are frozen solid.

Clam Digging. Clam digging may be easy—you can pry little necks up by wiggling your toes in the sand—or difficult—to catch a giant West Coast geoduck, you need one pair of hands to throttle the yard-long neck and another to dig—but the gustatory rewards are great. Be aware, however, that clams are readily contaminated and have on occasion been the cause of hepatitis and paralytic poisoning. Check with the local fish and game department to find out if clams are in season and safe to eat.

Before you eat fresh-caught clams, put them in a bucket of clean seawater or brine (1/3 cup salt to 1 gallon water) combined with 1 cup cornmeal. After a few hours the clams will suck in the cornmeal and expel most of the sand in their shells.

Storing Clams. A live clam carries its own seawater habitat inside the shell. At refrigerator temperatures, it can stay moist and fresh for several days after you buy or gather it but is best stored no more than 24 hours. Place live clams in a shallow pan and cover them with a wet cloth. Do not place them on ice. Discard any shellfish that die—the shells will spring open—they are not safe to eat. It is all right to open them for serving and keep them in the refrigerator for two or three hours.

Shucked clams are best right away, but you can hold them one or two days in the coldest part of the refrigerator, in a tightly lidded container, covered in their own liquor.

Freezing Clams. Freeze clams raw. If they are not yet shucked, wash them first in cold water, scrubbing the shells with a stiff brush. Shuck the clams over a bowl to save all the liquid from inside the shell. Shucking, by the way, is tricky business. It is accomplished by inserting a dull, rounded knife between the edges of the shell on either side of the hinge, then twisting the knife to sever the muscles there. The inexperienced shucker should have someone demonstrate the method before giving it a try. Or the fishmonger will do it for you. Ask him to save the natural liquid for packing the clams.

Wash the shucked shellfish in lightly salted water—1/4 cup salt to 1 gallon water—then place them in small (3-cup size or less), rigid containers that have very tight-fitting lids. Pour in the natural liquid you have saved, adding brine—1 teaspoon salt to 1 cup water—to cover the meat completely. Fill the container no higher than 1/2 inch below the rim, seal it, label it, and put it in a freezer that is 0° F. or colder.

Clams keep well for three months at 0° F. Thaw them in the refrigerator, in their container, allowing eight hours per 1-pound package to defrost.

Cooked clams, well covered, keep three or four days in the refrigerator. Do not freeze cooked clams because they will become much too tough and rubbery.

CRAB

The blue crab from the Atlantic comes in third behind shrimp and lobster in crustacean popularity. The soft-shell crabs on the market are blues that have momentarily shed their hard carapaces in order to grow. Snow crab, the market name for various spider crabs, is another huge Atlantic crop. Along the Southern coasts, stone crabs are abundant and from the Pacific come Dungeness and Alaskan king crabs.

Buying Crabs. Live crabs are best but are sold only rarely outside those markets close to the water's edge, usually in late summer. Live crabs should brandish their claws when poked. Soft-shell crabs should be translucent and completely soft.

Fishmongers are most likely to carry precooked crab, much of which has been frozen. In the shell, you will find stone crab and Maryland blue claws; long spidery king crab legs; and whole Dungeness. Precooked crab meat, which is either lump meat from the back, flake meat from the rest of the body, or claw meat, can be found pasteurized and vacuum-packed, frozen or canned.

Storing Crabs. Brisk, healthy live animals can stay in the refrigerator as long as two days. If you plan to store them, ask your fishmonger to pack them in seaweed inside an insulated carton specially designed for the job. Crabs, like other live shellfish, must stay alive until the time you cook them, so check them often and store them as short a time as possible.

To freeze crabs, scrub the shells first, then boil or steam them according to your favorite recipe. When they are fully cooked, remove the meat from the shell and pack it into airtight rigid containers. Cover the meat with a light brine (4 teaspoons salt to 1 quart water), leaving 1/2 inch head space. Crab can be held at 0° F. for three to four months.

Meat that has been cooked, frozen, and thawed for sale should be eaten the day of purchase. Fresh-cooked crab can be stored a day or two in the refrigerator; vacuum-packed can be refrigerated up to a month. Store frozen crab no more than four months and thaw slowly in the refrigerator. Keep canned crab a maximum of six months.

CRAYFISH

Crayfish are fresh-water crustaceans, living by the millions in streams all over the country. Lobsterlike varieties big enough for the table are farmed in Louisiana and the Pacific Northwest. Crayfish should be purchased live, when they are active and look healthy.

To store crayfish, follow the directions for crabs, immediately preceding.

LOBSTERS

They are expensive, but so popular you can find live lobsters swimming in supermarket tanks in Keokuk, Iowa, and Mesa, Arizona. The familiar American lobsters trapped in the cold waters off Maine, Nova Scotia, and Newfoundland are at their peak in summer, but farms produce good lobsters year round.

The spiny lobster is a coarser, less sweet species that is nevertheless excellent eating. Its tail has the only usable meat and often lobster tails sold separately in restaurants and supermarkets come from this more plebeian species. "South African" and "rock lobster" are other market names for the spiny.

Buying Lobsters. Choose vigorous live lobsters and plan to cook them the same day. In addition to showing energetic leg movement, a healthy lobster will curl its tail under when it is picked up. Size has no bearing on quality.

You may find frozen, uncooked lobsters, whole or the tail only. Choose well-wrapped packages that are frozen solid and have no trace of odor.

Buy precooked lobster tails only if you are certain they have been prepared the same day. Even cooked, the tails should be curled—this means the lobster was alive when it went into the pot. Whole precooked tails are a great deal better than presplit, but note that spiny lobster tails, the ones usually in the freezer case, *should* be split because the shells are all but unbreachable otherwise.

Storing Lobsters. Follow directions for storing crabs (p. 234).

OYSTERS

Inside its gnarled rock of a shell, the shapeless gray body of the oyster lies passively puddled in its own brine. This unlikely hero inspires lilting prose and legendary gluttony and everyone seems to remember the first oyster he or she ever swallowed.

Arguments about which kind of oyster is best can never be settled. Some prefer the cold-water varieties, like Cape Cod, Kent Island, and Blue Point, while others find warm-water oysters like Indian River from the Gulf Coast or the thin-shelled species flown in from Hawaii decidedly better.

Buying Oysters. Rather than having a choice, you will probably find only one live oyster variety available, the one from the closest source. The oysters may be graded by size, and blue point, confusingly, is a name applied to a smallish specimen. Half-shell oysters are slightly larger.

A live oyster's shell is shut tight and unbroken. Choose oysters with a fresh scent that feel heavy for their size. You can buy oysters at any time of year but they may be thin and watery in summer months.

Shucked oysters should be in absolutely clear liquid.

Storing Oysters. To store or freeze oysters, follow all instructions for clams (p. 233).

MUSSELS

The mussel's musky, salty flavor certainly rivals clam's or oyster's but the mussel is not nearly so beloved in the United States as in Europe, where hundreds of tons of mussels are harvested and consumed each year.

Mussels are best in fall and winter. The nearly smooth, blue-black shell of the mussel should be whole and shut tight when you buy it. Choose individuals that are neither too light nor too heavy for their size. (Heavy mussels may be laden with sand.)

Store mussels as you would clams (see p. 233), with some additional precautions. The shell may gape open after abrupt temperature changes. To determine if the mussel is still alive, hold it between thumb and forefinger and try to slide the upper and lower shell across one another. A fresh mussel won't slide.

If scrubbing doesn't remove the shell's silky beard and crust of dirt, use a dull-bladed knife to scrape it clean before cooking. After cooking, discard any mussels with closed shells.

SCALLOPS

Scallops are mollusks, like clams and oysters, but the resemblance stops there. Clams and oysters stay quietly rooted to one spot for their whole lives. Scallops dart through the water by clapping the two halves of their fluted shell together and shooting out a jet of water. The scallop is so muscular, only part of the meat is tender enough to eat. And because the shell does not close completely to seal the animal in its own juices, the scallop cannot be stored live. Fishermen shuck all scallops before they arrive in markets.

Tiny bay scallops live close to the shore from the Carolinas, around Florida, and across the Gulf Coast. The cold shallows around Long Island produce an excellent variety. The more abundant, much larger sea scallop is dredged out of the deep waters off Maine.

Scallops should be white but translucent, plump, and smell fresh. Store them no more than one day in the coldest part of the refrigerator or on a tray of crushed ice. Poach them before freezing and pack them in their own stock. Tightly sealed, scallops keep well for three months at 0° F.

SEA URCHINS

The sea urchin's shell is a sphere of dangerously sharp spines. These shellfish are very rarely seen in markets, and then only in coastal towns. You can gather them by thickly gloved hand, but be sure to ask the local fish-and-game department in advance to determine which species is edible and where the water is unpolluted.

Using scissors, cut a round out of the bottom of the shell to extract first the viscera (discard these) and then the edible roe. The delicate roe, usually eaten raw, cannot be stored for more than a few hours. Keep it refrigerated in a container embedded in ice.

SHRIMP

Though dozens of varieties of shrimp are fished up and down every coast —pink, brown, white, royal red, rock sidestripe, and spot shrimp among them—markets label the catch only by size. Even the size specifications vary from place to place. Usually a count of up to 15 per pound is jumbo; 16 to 20 per pound is extra large, and so on.

Buying Shrimp. Fresh shrimp come to market with the heads removed, sometimes without legs and shells. The very freshest shrimp have firm, resilient flesh that is translucent. It is far more common, though, to find raw shrimp with opaque flesh, which has been frozen and thawed. Fresh shrimp have a mild scent: a whiff of ammonia is a sign of spoilage.

Frozen, uncooked shrimp should be individually coated with a glaze of ice and sealed in an airtight package.

Cooked, peeled shrimp are standard fare at the fish counter. They keep fairly well but, if you can, buy them the same day they are cooked.

Storing Shrimp. Raw shrimp can be refrigerated for several hours or left, wrapped, on a bed of crushed ice in the refrigerator for one to two days. Seal cooked shrimp in a plastic bag and store no more than three days in the coldest part of the refrigerator.

Shrimp can be frozen raw or cooked, but raw keep better flavor and texture. Do not freeze shrimp if they have been frozen before. To freeze them raw, clean the shrimp and remove the head. They can be shelled or not. If you do shell them, put the shells in a pot of water and simmer for 10 minutes to make a light broth that is perfect for cooking the shrimp when the time comes.

Rinse the prepared shrimp, in or out of the shell, in a light brine—1 teaspoon salt to 1 quart water—and drain. Pack the shellfish into a rigid container, leaving no headroom. Freeze the broth separately. Shrimp and broth will keep at 0° F. from four to six months.

To cook fresh shrimp for freezing, first wash, then simmer in salted water until they turn pink—3 to 5 minutes, depending on their size. Cool, then shell and devein the shrimp. Pack them in rigid containers without headroom and seal airtight. Cooked shrimp will keep well for one to two months at 0° F.

UNIVALVES

A univalve's shell has an opening that allows the animal to stretch out a foot and use it to move around or cling to a surface. The delicate abalone from the Pacific, Florida's conch, North Atlantic whelks, and the ubiquitous periwinkle all fall into this special category of shellfish. None of them are widely available fresh—because of its scarcity, fresh abalone cannot be shipped out of California, for instance—but they may be found near where they are caught. Frozen and canned univalves are sold as well. Use the basic standards of freshness to select these shellfish. If bought in the shell, they should be alive. Fresh meat should be moist and mild-smelling. Frozen meat should be solid, uniformly colored, and odorless.

Abalone, conch, whelks, and periwinkles are at least as perishable as other shellfish. Store univalves according to the suggestions for clams (p. 233).

Snails, *petit gris,* the univalves that become escargots, are the same critters eating your garden in summer. If you are very industrious, you can collect, purge, and fatten them at home before you simmer them into an appetizer. Canned snails for escargot are not as tasty but certainly more convenient.

STORAGE TIMETABLE FOR FISH AND SHELLFISH

The time limits listed on the table apply to fish and shellfish that have been handled according to the instructions in the text. Fish and shucked shellfish held for more than a few hours in the refrigerator should be wrapped and placed on ice. Frozen fish and shellfish should be kept at 0° F.

Type of Fish	Days in Refrigerator	Months in Freezer
FAT FISH[1]	1	3
LEAN FISH[2]	1	6
CAVIAR:		
sealed	(1–4 weeks)	N/R
opened	1–2	
Pasteurized	(1–2 weeks)	

Type of Fish	Days in Refrigerator	Months in Freezer
CLAMS:		
Live	2–3	—
Shucked	1–2	3
CRABS:		
Live	1–2	—
Cooked[3]	1–2	3
CRAYFISH:		
Live	1–2	—
Cooked[3]	1–2	2
EELS	1	3
FROGS' LEGS	1	3
LOBSTERS:		
Live	1–2	—
Cooked[3]	1–2	2
MUSSELS:		
Live	2–3	—
Shucked	1–2	3
OYSTERS:		
Live	2–3	—
Shucked	1–2	3
SCALLOPS:	1	3
SHRIMP:		
Raw	1–2	6
Cooked[3]	3–4	2
SQUID	1	3
UNIVALVES	1	3
COOKED FISH	3–4	3
SMOKED FISH		
Whole	7–14	N/R
Sliced	3–4	N/R

[1] FAT AND MODERATELY FAT FISH include:
Barracuda, Bluefish, Bonito, Buffalofish, Butter-
fish, Eel, Grunt, Herring, Kingfish, Mackerel,
Mahi-Mahi, Mullet, Muskie, Pompano, Porgy,
Sablefish, Salmon, Sardines, Sea Bass, Shad,
Sheepshead, Smelt, Striped Bass, Sturgeon,
Swordfish, Lake Trout, Rainbow Trout, Tuna,
Weakfish, Whitebait, Whitefish, Whiting, Yellow-
tail.

[2] LEAN FISH include: Bluegill, Burbot, Carp,
Catfish, Cod, Cusk, Crappie, Croaker, Dabs,
Fluke, Flounder, Grouper, Haddock, Halibut,
Lingcod, Monkfish, Ocean Perch, Octopus, Pick-
erel, Pike, Pollock, Red Snapper, Rockfish, Sand
Dab, Scrod, Shark, Skate, Sole, Snapper, Squid,
Tilefish.

NOTE: The above listings are meant to help you
decide how to freeze a particular fish. Some of
the fish labeled fat are only moderately fat or fat
only certain seasons of the year.

[3] If thawed, store no more than one day.

5

Grains and Grain Products

FLOUR

Anything from bananas to soybeans can be ground up into a dry powder and called flour. Usually, though, when we use the word we are talking about grain in general and wheat in particular.

Grains were the first crop cultivated by man. The seed, or kernel, is the edible part of grain. The plump seed, in its entirety, contains an impressive amount of proteins, vitamins, minerals, and carbohydrates—enough nutrients to support a long and healthy life with only a little help from other foods.

Grinding grain seeds into flour was an early and necessary bit of technology developed because people simply can't chew whole ones efficiently.

Flour ground by stones contains all the parts of the cereal seed except the husk, just as it comes from the plant. Outermost in the seed is the bran. The endosperm is the starchy interior. Tucked inside each seed is also the germ, the embryo that would sprout if the seed were planted. Each of these parts has varying amounts of protein, fats, and other elements. In some products, such as stone-ground cornmeal or whole wheat flour, you can actually see the color and texture of the different seed parts.

The modern method of processing flour, developed in the mid-nineteenth century, involves separating out the parts. The smooth, homogeneous white wheat flour, what is called simply "flour" on the grocery shelf, is made up of only the wheat's endosperm—the bran and germ have been extracted.

Buying Flour. Before you can know how to store a particular kind of flour you must have some idea of its composition. If it is a whole grain product it contains fat. The more refined flours, with the germ removed, contain little or no fat. Flours that contain fat do not keep well for long at room temperature. The fat becomes rancid in a matter of weeks.

• WHITE FLOUR. White flour is not as simple a product as it seems. It starts out as wheat. After it is milled, some natural elements are removed and some synthetic substitutes are added. The reasons for all this meddling begin with what we already know about flour—it won't last as long with the fatty germ and the bran left in. The bran and germ are sifted out, leaving behind white flour, a soft, uniform powder that not only stores well but also produces lighter baked goods than whole wheat flour can. Because so many of wheat's natural nutrients are sacrificed when the bran and germ are extracted, manufacturers put vitamins and minerals back into the flour. This is what is meant by "enriched." Laws in all but fifteen states require that flour be enriched. The list of ingredients on a typical package of enriched all-purpose flour reads: wheat flour, malted barley flour, niacin, iron, thiamine, mononitrate, and riboflavin. (The malted barley flour supplements the enzymes that help make successful yeast-raised products.)

If you are struck by the apparent inconsistency of first stripping the flour of its nutrients and enzymes, then mechanically replacing them, you are not the first person to raise this issue. Feelings among food manufacturers and nutritionists run high on this subject. It is not a new controversy either. A well-known Victorian home economist, Mrs. Isabella Beeton, addressed the issue in her 1861 *Book of Household Management:*

> . . . in fact, we may lay it down as a general rule, the whiter the bread, the less nourishment it contains. Majendie proved this by feeding a dog for forty days with white wheaten bread, at the end of which time he died; while another dog fed on brown bread made with flour mixed with bran, lived without any disturbance of his health.

This is interesting not because of the (dubious) fate of the dogs, but because a hundred years of protest have yet to dislodge white flour from our cupboards.

It is true that more varieties of whole grain flours are now available than ever before. White flour is no longer the only choice we have, but it still has an important function in most kitchens. White flour has an unparalleled ability to produce gluten, the critical element in the texture of baked goods. Very few whole grain flours will produce palatable baked goods when used on their own, and most whole wheat recipes call for the addition of some white flour. If we want moist, crumbly cakes and light, airy loaves, we must depend on white flour as a basic ingredient for baking.

White flour appears on the grocery shelf in quite a number of varieties. Here is a list of what to look for on the label:

Enriched means that thiamine, riboflavin, niacin, and iron have been added to the flour after milling. These synthetic nutrients do not replace all the vitamins and minerals that have been removed from white flour.

All-purpose flour is a mixture of hard, high-gluten wheat and soft wheat. The mixture is suitable for both bread and pastry making.

Unbleached means that no chemicals have been added to whiten or to age the flour artificially. Unbleached flour has a creamy, off-white color.

Cake flour or *pastry flour* is made from soft wheat. It will bake into light, fine-textured pastries.

Bread flour is entirely hard wheat, very high in gluten. Gluten is a protein that makes bread dough strong and elastic enough to trap the gases that yeast forms as the dough rises. Bread flour is not always readily available, but the large national flour processors do distribute it in consumer sizes.

Presifted all-purpose flour has been milled to a very fine texture. Whether or not to resift presifted flour is a decision best left to experience. Some recipes work better with the extra sifting and some are better without.

Self-rising flour has baking powder and soda added. It is used mostly in recipes for Southern cooking.

Bromated means that a maturing agent, potassium bromate, has been added to the flour. This chemical agent helps the flour to produce lighter baked goods.

Instant flour is white flour processed for use as a thickener. It will not work as a substitute for all-purpose flour.

Not all brands of all-purpose flour will be exactly alike, and cake flours will vary too. Even though you can't taste, touch, or smell the difference, flours can be drier, finer, softer, more glutenous—as different from one bag to the next as one tomato is to the next. All-purpose flour sold in the South, for example, usually has a higher proportion of soft wheats (presumably for better biscuits). You may not notice the variations in your baking, but some breadmakers can see a distinct difference between brands.

When buying white flour, the only way you can be sure the flour is fresh is to shop at a store with high turnover.

• WHOLE WHEAT FLOUR. Whole wheat flour, also called graham flour, contains all the parts of the kernel—bran, germ, and endosperm; nothing has been removed.

Depending on the brand, whole wheat flour will be coarser or finer grained. Stone-ground has more vitamins than steel-milled. Some whole wheat flour may contain potassium bromate. It will be labeled "bro-mated."

Whole wheat flour should have a sharp, fresh scent.

Cracked wheat is a coarse meal, cut rather than ground from the whole wheat kernel. You can use it cooked as a cereal or side dish. Cracked wheat added to homemade bread will make it nutty and crunchy.

Wheat berries are whole, husked wheat kernels. They can be cooked whole for breadmaking or for cereal; they can be ground into flour at home. For more information, see p. 253.

• CORNMEAL. You will most likely find two kinds of cornmeal on your grocer's shelf. One is ground from yellow corn and the other from white. For cooking, they are virtually alike, although the yellow has more vitamin A. The labels on both will say "enriched" and "degerminated," meaning that, as with white flour, the germ has been removed and the nutrients niacin, iron, thiamine, and riboflavin have been added.

Water-ground or *stone-ground* cornmeal still contains the fat-rich germ of the corn kernel. The texture can vary from powdery fine to very coarse. It has a fuller flavor than degerminated cornmeal.

Smell water- or stone-ground cornmeal and avoid packages with a stale, rancid odor.

Blue cornmeal from New Mexico is bluish grey and has a stronger, toastier flavor than yellow cornmeal. *Masa harina* is a finely ground corn flour used for tortillas and tamales.

• BARLEY FLOUR makes a sweet, light-textured loaf of bread. It must be combined with white flour because it has no gluten of its own. Barley flour is not usually whole grain.

• BRAN. Remember the outer layer of the wheat kernel that is removed to make white flour? Here it is, ready to add fiber and flavor to your cooking. It may be ground into flour or it may come totally unprocessed as "miller's bran." Bran contains a certain amount of fat.

• BUCKWHEAT FLOUR is famous for pancakes. You will find it either natural or toasted, the toasted type having a stronger flavor. Buckwheat is a grain completely distinct from wheat, so it has no gluten and should be combined with white flour for baking. It has a very high fat content.

• GLUTEN FLOUR is hard wheat flour stripped down to a high-protein, low-starch powder used as an additive in breadmaking.

• OAT BRAN is an especially healthy form of dietary fiber used in cooking and as a breakfast cereal.

• OAT FLOUR. Health food stores carry oat flour, which is ground, husked oat groats (kernels). The same product may also be called oatmeal. This flour is not to be confused with rolled oats (see "Instant Hot Cereals," p. 253), the traditional breakfast cereal.

• POTATO FLOUR. Sometimes called potato starch, this can be used in baking and as a thickener.

• RICE FLOUR is ground from either white or brown rice. The white variety has almost no fat; the brown has a considerable amount.

A variety of white rice flour called "sweet" rice flour can be used only as a thickener, so read the package label to be sure of what you are getting.

• RYE FLOUR. Rye is a nonwheat grain that will give you a heavy, pungent bread. The dark and light varieties are interchangeable in recipes, but the dark does have a stronger flavor. Rye flour is whole grain flour frequently combined with white flour in breadmaking and you may find the two packaged as a mix. Rye flour may be bromated.

• SEMOLINA. Used in pasta making and Italian puddings, this is a flour made from durum wheat, the hardest type of wheat grown, the highest in gluten. It is not a whole grain flour; only the starchy endosperm of the wheat kernel is used in semolina.

• SOY FLOUR is ground from whole soybeans. Sometimes called soy powder or soya, it packs a heavy nutritional punch when added to white flour for baking. Breads and cakes made with soy flour will be moist, fine-grained, and very high in iron, calcium, and protein. It has none of its own starch or gluten, so it must always be combined with another flour. (It does work by itself as a thickener, though.) Most soy flours have a very high fat content, but there is a defatted variety available.

• TRITICALE (tri-ti-cay'-lee) is a hybrid of hard wheat and rye grains. The high-protein flour is used in breadmaking in combination with white flour. You will probably find triticale only in specialty stores. It is a whole grain flour.

• AMARANTH. The tiny seeds of the amaranth plant contain 16 percent protein, compared to 12 to 14 percent in wheat. Rich in lysine, flour ground from the seeds has more *complete* protein than other grains and is high enough in gluten to produce light-textured pastries.

Called pigweed or redroot in the wild, certain varieties of amaranth have green leaves very like, but more delicate than, spinach. Difficult to find in markets, amaranth is a good candidate for the home garden.

• WHEAT GERM. The heart of the wheat kernel is the germ. It comes flaked or in a coarse meal, either raw or toasted. Wheat germ adds a pleasant, nutty flavor to baked goods, while boosting their protein and mineral content. It is the most oil-rich and therefore the most perishable part of the wheat. You will find it at the supermarket in vacuum-packed jars that must be refrigerated after opening. In health food stores, wheat germ comes packed in plastic bags kept in the refrigerator case. If the bags are not refrigerated, don't buy the wheat germ.

Storing Flour. Flour does not spoil in an obvious way. When it deteriorates, the changes are subtle but real. It can absorb moisture from the air and consequently give unpredictable results in baking. The fat in flour can alter and begin to have the off odor and off flavor characteristic of rancid fats. Over a period of time certain proteins in flour spontaneously degenerate and lose their nutritive value. Like every other food, flour has an optimum shelf life and suffers from long, improper storage.

White flour and other refined flours containing no germ, including degerminated cornmeal, gluten flour, potato flour, white rice flour, semolina, and defatted soy flour, can be stored at room temperature. It is important to transfer these flours from their original bags or boxes into airtight canisters. The only alternative is to seal the bag or box inside an airtight plastic bag. Keep the flours cool, dry, and dark and they store perfectly for six months to a year—six months if your pantry is warm and humid, a year if it is 72° F. and dry.

Whole grain flours that still contain the germ and bran, including barley flour, bran, buckwheat flour, water- or stone-ground cornmeal, brown rice flour, rye flour, soy flour, triticale, and wheat germ, are quite perishable at room temperature. Store these flours in the refrigerator or in the freezer in lightproof, tightly lidded containers, or sealed in airtight plastic bags. When this is not practical, take care to buy no more of the flour than you can use in a month. These flours keep no more than four weeks at room temperature. Wheat germ should always be kept in the refrigerator. In the refrigerator or freezer, whole wheat flour and triticale keep up to one year. Other whole grain flours keep two to three months in the refrigerator or freezer.

Before you use chilled whole grain flours for baking, measure out the amount you need and allow time for it to warm to room temperature. Put the storage container back in the refrigerator or freezer immediately so moisture does not condense on the inside. Don't sift them because you will sift out some of the tiny, nutritious particles that give whole grain flour its flavor.

STORAGE TIMETABLE FOR FLOURS

Type of Flour	Months in Pantry	Months in Refrigerator or Freezer
WHITE	6–12	12
WHOLE WHEAT	1	12
CORNMEAL, Degerminated, Masa Harina	6–12	12
CORNMEAL, Stone-Ground, Blue	1	2–3

Type of Flour	Months in Pantry	Months in Refrigerator or Freezer
BARLEY	1	2–3
BRAN	1	2–3
BUCKWHEAT	1	2–3
GLUTEN	6–12	12
OAT BRAN	1	2–3
OAT FLOUR	1	2–3
POTATO	6–12	12
RICE, WHITE	6–12	12
RICE, BROWN	1	2–3
RYE	1	2–3
SEMOLINA	6–12	12
SOY	1	2–3
SOY, Defatted	6–12	12
TRITICALE	1	12
WHEAT GERM	N/A	2–3

CEREALS AND WHOLE GRAINS

Bowls of cereal, hot or cold, brimming with milk, are taken for granted as quick, easy breakfast food. The grains that go into the bowls are actually very serious business. Of all the land farmed throughout the world, more than half is devoted to growing grains, principally rice, wheat, corn, barley, rye, and millet. As flour for bread, or cooked as whole grains in dishes such as couscous, these foods are the principal source of proteins and other nutrients for a large part of the world's population.

Buying and Storing Cereals. The majority of cereals at the supermarket —cornflakes, shredded wheat, and the like—are highly processed. Most are made from only the endosperm of the cereal grain and are cooked and dried. But health food stores carry a number of cereal grains as whole, husked kernels called groats. Groats consist of an outer layer of bran, starchy endosperm, and an inner core of germ. Whole kernels may be cracked or ground coarsely. Groats, both whole and cracked or ground, become porridge when they are boiled in salted water until soft.

All cereals keep best in airtight containers that keep out moisture, dust, and insects (remember this when you buy in bulk from a specialty store). At home, a tightly sealed plastic bag is sufficient protection.

• READY-TO-EAT CEREALS. Corn, wheat, and rice are the favorite grains for ready-to-eat cereals. The nutrients lost when the bran and germ of these grains are removed are replaced to some degree by minerals and artificial vitamins added during processing. There are a few types of ready-to-eat cereal that contain as little as 3 to 7 percent sugar. Many, many more have from 20 to 50 percent sugar. The extensive labeling on ready-to-eat cereals is not really helpful about exactly how much sugar is in the product, but *Consumer Reports* regularly publishes exact figures for 32 varieties in their *Buying Guide*. Ready-to-eat cereals almost always contain some preservatives, which are listed on the label. Also look for a "use by" date on the package.

Store these products in unopened boxes up to one year in a cool, dry place. After opening, refold the inner lining of the box to cover the cereal. Use it within three months or before "use by" date. Restore crispness to soggy cereal by spreading it in a baking pan and putting it in a 350° F. oven for 5 minutes.

• INSTANT HOT CEREALS are often whole grain products that have been processed and partially cooked so that they cook in a few minutes or need only be moistened with boiling water. Oats, rice, and wheat are the grains used in the most common instant hot cereals. Rolled oats are perhaps the most popular in this category. They are whole oat kernels that have been steamed and flattened into flakes.

Instant hot cereals keep up to one year in their original box, but a tightly lidded container is better.

• WHOLE GRAINS. Since whole grains are rich in natural oils, you should sniff before you buy to make sure the oils are not rancid.

Barley comes as groats, as pearls (called soup barley) that have had the bran removed, and as grits, which are coarsely ground pearl barley.

Buckwheat groats may be raw or toasted. The toasted groats are often called whole kasha, or simply kasha when they are ground into grits.

Hominy is a variety of corn sold cracked and with the bran removed. Hominy grits, coarsely ground kernels, are widely available. *Posole*, oversized white corn that has been parched and dried, is a Southwestern version.

Millet comes only as tiny yellow whole kernels.

Oats are available as whole kernels. The more common rolled oats are processed whole kernels.

Quinoa is a nutty, light-brown grain originally from Peru. It is high in protein and fiber and offers sturdier flavor and texture than rice, for which it can be substituted.

Rice is discussed on pp. 257–59.

Rye is sold as cracked whole kernels.

Wheat berries are wheat groats. They come whole and cracked. Cracked wheat berries that have been steamed and dried are called bulgur.

Teff is a whole grain so tiny it can be used as flour. Because of its size, teff contains a high ratio of bran and germ.

Store whole grain cereals—whole groats, cracked or ground—in tightly lidded containers or sealed plastic bags in the refrigerator. They do keep up to one month at room temperature but stay fresh in the refrigerator four to five months. Pearl barley and hominy are the exceptions; they keep up to one year on a cool, dry shelf.

STORAGE TIMETABLE FOR CEREALS
AND WHOLE GRAINS

Type	Time on Shelf	Time in Refrigerator
READY-TO-EAT:		
Unopened	12 months	——
Opened	3 months	——
INSTANT HOT CEREALS	12 months	——
WHOLE GRAIN CEREALS:		
Groats and Kernels	1 month	4–5 months
Pearl Barley	12 months	——
Posole	12 months	——
Hominy and Hominy Grits	12 months	——
Rolled Oats	12 months	——
Quinoa	1 month	4–5 months
Teff	1 month	4–5 months
COOKED CEREALS	——	2–3 days

PASTA

Pasta is a minor miracle that happens when you combine flour and water or flour and eggs. These simple mixtures, with the right techniques, can be made into toothsome noodles instead of library paste.

Buying Pasta. In the supermarket a distinction is usually made between prepackaged pasta and egg noodles. The pasta, which may also be labeled "macaroni product," is made from durum wheat flour, a very hard, glutenous flour called semolina, and water. Egg noodles are made from regular flour and eggs. Both products are dried.

Pasta comes in innumerable shapes and sizes. There are tiny star-shaped stellete; broad, flat lasagne; corkscrew rotini; farfalle that look like little bow ties; the very familiar spaghetti; and many more. The eating quality of the pasta depends entirely on the wheat flour from which it is made—hard flours absorb less water in cooking. Pasta imported from Italy is strikingly superior to domestic brands.

Some specialty food shops have freshly made pasta for sale. Depending on how long ago it was made, it still may be very limp or, as time passes, only a little pliable. In addition to white pasta, you may find green, made with spinach, or red, made with tomato or beets. Whole wheat or buckwheat flour produces a toasty brown noodle.

Cellophane noodles, also known as bean threads, are made from mung bean flour. Other Oriental noodles are made from rice flour; saifun may contain potato starch, cornstarch, or mung bean starch. Soba are buckwheat noodles.

There is no way you can look at dried pasta or noodles and know if they are fresh, although some brands have dating codes on the package. It is, however, very unlikely that you will run across a box of spaghetti or a bag of noodles that has been sitting on the shelf long enough to grow stale. The only thing to check is whether the product is crushed.

Storing Pasta. You can leave pasta and egg noodles in their own packages on a cool, dry shelf for a month or two. If you like to display pasta in canisters, that is adequate protection for a month or two as well.

For longer storage, it is best to keep pasta and egg noodles in airtight containers in the dark. This way, pasta keeps up to eighteen months and egg noodles up to six months. Though the change is subtle, proteins in aging pasta do deteriorate, so you shouldn't plan to store it endlessly.

Fresh-made pasta should be stored in a plastic bag in the refrigerator. Plan to use it within one or two days.

There are a couple of ways to store fresh-made pasta longer. Wrap the still moist pasta airtight in a plastic bag and freeze it for up to two months at 0° F. You can take it out of the freezer and drop it directly into boiling water to cook.

Homemade pasta can also be air dried. Spread the moist pasta in a single layer on a lightly floured cloth and turn it occasionally as it dries or drape it on a pasta rack to dry. Air drying takes from 10 minutes to 24 hours, depending on the kind of pasta, its thickness, and the humidity in the air. Continue drying until the pasta is brittle. Fresh pasta is much more fragile than the prepackaged kind, so handle it gently. Seal the completely dry pasta in a plastic bag or glass jar. Check the container after a half day. If there is any condensation on the inside, take the pasta out and dry it some more. Store dried homemade pasta in the refrigerator up to two months.

Store cooked pasta and noodles, coated in sauce or oil, in a covered container in the refrigerator up to four days.

STORAGE TIMETABLE FOR PASTA

Type	Time on Shelf	Time in Refrigerator	Months in Freezer
PASTA, DRIED:			
Commercial	18 months	——	——
Homemade	N/R	N/R	2 months
PASTA, FRESH	N/R	1–2	2 months
EGG NOODLES	6 months	——	——
COOKED PASTA AND EGG NOODLES	——	4 days	2 months[1]

[1] Only if frozen in sauce.

RICE

If you are ever washed ashore on a desert island, hope that a decent quantity of rice, in a nice clean, airtight container, floats up onto the beach right behind you. Rice outlasts almost any other kind of food you can name—it tastes good, holds on to its nutrients longer, and goes well with fresh-caught fish.

At least part of the reason rice has such remarkable longevity lies in its processing. As soon as rice grains are harvested they are cleaned and dried. Milling starts with further cleaning. First dust and chaff are winnowed out, then the husk is stripped from each grain. At this point, with the outer coating of bran intact, it is brown rice.

The next step in milling is to polish away the layers of bran. What is left is the familiar grains of white rice. Very often vitamins and minerals are added to the grains to make up for what has been lost in milling. The end product, white rice, is impervious to most kinds of decay.

Buying Rice. Rice is classified by the length of the grain. *Long-grained rice* cooks up dry and fluffy, with each grain separate. *Basmati,* both white and brown, is a flavorful variety of long-grained rice that tends to be slightly starchy. *Short-grained rice* looks plump, sometimes almost round; it is creamier when cooked and tends to stick together. Italian rice, *arborio,* and Japanese glutinous rice are both short-grained. *Medium-grained rice* may share the characteristics of both long- and short-grained types, depending on the particular variety.

As long as rice looks dry and clean there is little to be concerned about when you buy. There is a small chance that brown rice or preseasoned rice has been sitting on the grocer's shelf too long, but this is unlikely in a busy store with high turnover of goods. To be on the safe side, you can sniff the package for a telltale rancid odor.

The label on a package of rice tells you something about the way it has been processed:

• BROWN RICE is actually light brown in color. It still has the bran coating on each grain.

• WHITE RICE, also called polished rice, has had its bran removed. White rice is often enriched, meaning various nutrients have been added during processing.

• CONVERTED OR PARBOILED RICE is white rice that was treated with steam before the bran was removed. This partial cooking fuses into the grain nutrients that would otherwise be lost with the bran. Converted rice is usually cream-colored rather than stark white.

• INSTANT RICE has been fully cooked and dried, so you simply rehydrate it before serving. Instant rice is no match for regular rice. Its texture is not nearly so appetizing.

• PRESEASONED RICE may be regular, converted, or instant rice packaged with dried seasonings. This type of rice cannot be stored as effectively as plain rice because the seasonings become stale long before the rice.

• WILD RICE is the seed of an aquatic grass that grows wild in the Great Lakes region of the United States. The long green grains do resemble rice and are served like rice, but otherwise the two foods are not related.

Wild rice is very expensive, but it lasts as long as white rice. Look for packages with whole, unbroken grains. The color of the rice varies but does not necessarily reflect the quality. However, fancy grade wild rice is usually darker than lesser grades.

Storing Rice. White rice, regular and converted, and wild rice need very little special care. If you use a lot of rice you can leave it in its original container for several weeks at room temperature.

For longer storage or for storage during warm weather, transfer white and wild rice to an airtight canister to keep out dust and moisture (and insects). Both white and wild rice keep indefinitely at room temperature with virtually no loss of nutrients or eating quality.

Brown rice still contains natural oils that do eventually go rancid. You can store it up to a month at room temperature. For longer storage, keep brown rice in an airtight container in the refrigerator. It should last six months.

Preseasoned rice can be stored no more than six months at room temperature.

Cooked rice stores extremely well too. Properly reheated, it is indistinguishable from fresh-made rice. In the refrigerator, cooked rice keeps seven days as long as it is in a tightly covered container. Plastic wrap will do.

You can freeze cooked rice just as successfully. Sealed in airtight plastic bags, it keeps six to eight months.

To warm up the stored rice, put it in a saucepan with 2 tablespoons of liquid—water or bouillon, for instance—per cup of rice and simmer, covered, about 5 minutes. A block of frozen rice should be broken up into small pieces before reheating, and it may need a dash more liquid and a minute or two longer on the stove.

You can also make an enticing, crunchy patty with leftover rice, one that, not surprisingly, has a long storage life.

Start by thoroughly heating a greased heavy frying pan. Spread cooked rice in an even layer over the bottom of the pan and cook it, uncovered, over very low heat for about an hour. When the rice is perfectly dry and can be lifted out in one piece, set it out on a wire rack overnight. Break the pan-size chunk into 3-inch pieces and seal them in a plastic bag. The rice patties can be stored in a plastic bag, at room temperature, two to three months. To serve, deep-fry in oil for 10 seconds at 400° F. The rice will puff up and crisp. Drain the pieces on a paper towel and serve immediately. The rice patties can be eaten alone as a snack or as a base for Oriental dishes. Drop them into hot Oriental soups for a spectacular sizzle and a crisp texture.

STORAGE TIMETABLE FOR RICE

Type	Time on Shelf	Time in Refrigerator	Months in Freezer
UNCOOKED RICE:			
White, Converted, Wild, Instant	indef.	—	—
Brown	1 month	6 months	—
Preseasoned	6 months	—	—
COOKED RICE	—	7 days	6–8

BREAD

Whatever form bread takes—soft dark loaves of whole wheat, light crusty French, sweet heavy nut breads, or chewy bagels—it faces the same enemies. Air dries bread out. Molds, always present in the air and on kitchen surfaces, flourish on bread, growing blue-green in its folds and between slices. With time, bread becomes stale as its starch crystallizes, leaving the texture corky and the taste flat.

Buying Bread. Almost without exception, prepackaged breads you find in grocery stores are stamped with a "sell by" date. The bread should still be fresh on that date and stay fresh a couple of days longer. Bakery breads are either sold fresh-baked the same day out of the oven or are marked as "day old."

The feel of bread tells you whether or not it is perfectly fresh. Soft-crusted breads and rolls should feel spongy and a bit heavy for their size. Hard-crusted breads and rolls should feel light and the crust should be very firm and crisp. Soft tortillas and pita bread are pliable when they are fresh and should be under refrigeration.

Though all breads have lots of protein, vitamins, and minerals, there are, according to Consumers Union, a few ingredients that can make bread even better for you: milk, nonfat milk solids, eggs, whey, and soy products. If any of these are on the label, the bread is especially nutritious, and if there are whole grains, the bread is high in fiber, another benefit.

Along with baked goods, there are several unbaked bread doughs, biscuits, and dinner rolls available in supermarkets. Most are vacuum-packed in cardboard tubes and displayed in the refrigerator case. All brands have "use by" dates clearly stamped on the labels. Frozen bread doughs are also dated. Be absolutely sure these doughs are frozen solid as a rock. Both refrigerated and frozen doughs should be in tightly sealed packages with no signs of breaks or tears—the doughs are perishable and exposure to air ruins them very quickly.

The freezer case may also contain a few frozen quick breads—usually muffins, pancakes, and waffles. Again, they should be frozen hard and well wrapped.

Prepackaged bread crumbs, croutons, and bread stuffings are not usually dated because of their relatively long shelf life. These are useful when you are pressed for time but much more expensive than homemade.

Storing Bread. Different styles of bread have slightly different storage requirements:

• SOFT-CRUSTED BREADS. Prepackaged sliced bread, whole homemade or bakery loaves, quick breads, dinner rolls, pita, tortillas, sweet rolls, and raised doughnuts generally keep longer in the refrigerator than on the shelf. The wrapping is very important, because the air in the refrigerator can dry out the bread. Store the bread in tightly closed plastic bags and double-wrap if necessary.

Bread stored in the refrigerator feels firm and may be mistaken for stale. However, when the bread is again at room temperature, it feels moist and fresh once more.

You can also store any bread, still securely wrapped, at room temperature, and some lighter, drier loaves will keep better this way—you should experiment with the breads you use regularly. At room temperature, mold will move in more quickly. The very softest breads, such as Wonder Bread and Holsum Bread, contain preservatives that prevent mold almost forever, though they stale as other breads do.

Home-baked breads must be completely cooled before they are wrapped for storage; otherwise too much moisture will condense and collect on the inside of the bag, creating a perfect atmosphere for spoilage.

Sliced bread, quick breads, rolls, and raised doughnuts last from four to seven days in the refrigerator. Expect two to four days of shelf life at room temperature. Whole loaves can last from one to two weeks before they stale in the refrigerator. Breads made with eggs, butter, or oil stay moist and fresh longest.

A touch of mold on bread can be cut away and the rest of the bread is safe to eat, but very moldy bread should be discarded.

When yeast breads begin to dry they are by no means ready to discard. You can't restore them to their former moist, crumbly glory, but you can put them to use in a number of ways. Slightly stale bread still makes perfect toast. And any dried-up ends of bread can be turned into bread crumbs—rather than tossing out bits and pieces of stale bread, grind them up in the blender or food processor for a few seconds and tuck them away for future use. Make your bread crumbs slightly coarse and they will have a lot more taste. Crumbs you make from rich breads, especially home-baked loaves, are much more delicious than store-bought when used to coat fried foods, to make stuffings, to blend meat loaf, and to top casseroles. Store your accumulation of bread crumbs in a plastic bag or tightly lidded jar. They keep in the refrigerator for several months or in the freezer up to a year.

To freshen quick breads, wrap them in foil with a few drops of water sprinkled inside, then heat them in a 300° F. oven for 10 to 15 minutes.

For long-term storage, soft-crusted breads freeze very successfully. Fresh bread comes out of the freezer in almost exactly the same condition it went in—moist, soft, and fragrant. Freezing works every time, as long as you are meticulous about wrapping the bread. Always wrap it in a double thickness of foil or in two plastic bags.

Bakery yeast bread, sliced yeast bread, and yeast rolls keep up to four months in the freezer. Whole loaves of home-baked yeast bread can be stored at 0° F. from six to nine months.

Baked quick breads such as muffins, biscuits, corn bread, and fruit and nut breads, well wrapped, do very well in the freezer for up to three months.

Tortillas and pita bread can be frozen for up to four months.

Thaw soft-crusted breads at room temperature, still inside their wrapping. The condensation that forms on the inside of the package during defrosting eventually distributes itself evenly throughout the bread. You can also thaw bread one slice at a time in minutes at room temperature or put it right in the toaster.

• HARD-CRUSTED BREADS. Hard-crusted breads, such as French or Italian loaves, don't last nearly as long as soft-crusted varieties. If they are wrapped in plastic bags, the very dry crust draws moisture from the soft interior of the bread and soon becomes soggy. If they are left in the open, the center dries and hardens in a day. To keep the crust crisp and the center chewy, store the bread or rolls loosely wrapped in paper, at room temperature, preferably in a drawer or bread box. Hard-crusted breads keep fresh at the most two days.

A dried-out loaf can be turned into crumbs or croutons. To recrisp a whole loaf, run it quickly under cold water, then warm in a moderate oven for 5 to 10 minutes.

Hard-crusted breads freeze well but must be handled differently from their softer counterparts. Wrap them loosely in freezer paper, allowing for a little air circulation. They will keep up to three months at 0° F. Thaw the bread, still in its wrapping, at room temperature. Before serving, heat the bread in a 400° F. oven for 5 minutes to restore crispness.

• BREAD DOUGHS. Bread doughs can be frozen and there are even recipes for doughs designed to be frozen. But, once frozen, the dough becomes unpredictable. It is better to bake the bread and freeze the loaf.

Quick bread doughs are more at home in the freezer but still give uneven results. You can freeze precut biscuit dough for two weeks and bake without thawing. Muffin dough can be frozen right in the muffin tin and wrapped for later use. Store muffin dough up to two months in the freezer and thaw an hour before baking. Fruit and nut bread doughs should not be frozen—they are just not reliable afterward.

Store-bought doughs packed in vacuum cans must be refrigerated and used before the date on the label. Do not freeze them—the package will burst. Commercial frozen bread doughs should be stored at 0° F. and used before the date on the label. If you cannot use the dough in time, bake it and freeze the bread.

• BATTER BREADS. You may not think of pancakes, waffles, and fritters in terms of storage, but there are mornings when you whip up more batter than you need for breakfast and occasions when you want to make, say, dessert crepes ahead of time. And of course there are frozen, ready-made waffles and the like that may find their way into your freezer.

The first approach for homemade batter breads is to save the batter. Freezing, even for one day, is best because the milk and raw eggs in the mixture are vulnerable to dangerous bacterial growth. Pour the batter into a clean rigid container, seal it airtight, and store at 0° F. for no more than one week. Thaw it in the refrigerator—it will probably take overnight—and thin it with milk if it is too thick.

For convenience, you can also store home-cooked crepes, pancakes, and waffles. Stack them up with squares of waxed paper between and slip them into a plastic bag. Seal the bag tightly and either refrigerate up to one day or freeze up to two months. There is no need for thawing; just pop the waffles or pancakes into a hot oven for 2 or 3 minutes and serve.

Keep commercially frozen waffles and pancakes at 0° F. up to three months.

STORAGE TIMETABLE FOR BREAD

Type	Days on Shelf	Time in Refrigerator	Months in Freezer
SOFT-CRUSTED BREADS:			
Sliced yeast bread, tortillas, pita	2–4	4–7 days	4

Type	Days on Shelf	Time in Refrigerator	Months in Freezer
Quick bread, rolls, sweet rolls, raised doughnuts	2–4	4–7 days	3
Yeast bread: whole loaves	5–7	7–14 days	6–9
HARD-CRUSTED BREADS	2	N/R	3
DOUGH[1]			
Bread[2]	—	—	2
Biscuit[2]	—	—	1/2
Muffin	—	—	2
BATTER BREADS:			
Batter	—	N/R	1/4
Crepes, Pancakes, Waffles	—	1	2–3

[1] When possible, bake breads before freezing.
[2] Store commercial doughs according to package directions.

CRACKERS AND CRISP BREADS

The whole point of crackers is to be dry. There is not much pleasure to be had from sinking your teeth into a soggy cracker. And because crackers and crisp breads are dry, they won't support any nasty molds. You must take measures to keep these foods dry and therefore fresh. You can't depend on simple cardboard boxes to do the job.

Buying Crackers and Crisp Breads. There seems to be no end to the variety of crackers and crisp breads on the market. Square, salty soda crackers; long tubular bread sticks; matzoth ten inches across; crackers flavored like bacon or pizza; dark grainy rye crisp; sweet, whole wheat graham crackers; twisty pretzels—these are a mere fraction of the types made with wheat. There are also rice crackers and taco shells, made with corn.

An important contributor to shelf life in your kitchen is the way the crackers are packaged by the manufacturer. They are all sealed airtight in plastic or waxed paper to begin with, but the question is, will the package continue to protect even after it has been opened?

What you want, if you can get it, is some form of double wrapping. Some crackers come sealed in stacks, with three or four stacks packed into a box. There are crackers boxed in partitioned trays that can be unsealed one section at a time. Whatever form it takes, the purpose of double wrapping is to allow you to use a few crackers at a time and keep the rest of your purchase secure against air and moisture.

Storing Crackers and Crisp Breads. Crackers and crisp breads get stale when exposed to too much air and too much moisture. Two things can happen—they lose their crispness and the fats used to bake them turn rancid.

While these products are still sealed in their original package, you can keep them on a cool, dry shelf up to about six months before they start to lose quality.

As soon as crackers and crisp breads are open you should be ready to transfer them to a protective container. Any glass, plastic, or metal canister with a snug-fitting lid can keep loose crackers fresh at room temperature for a month or so. A plastic bag will do as well but doesn't protect against crushing. A container with a rubber gasket in the lid that seals out air can keep crackers and crisp breads even longer—three to four months on a cool, dry, dark shelf.

Warm, humid weather cuts these storage times down by more than half. In the summertime you can put crackers in a plastic bag in the refrigerator and they will stay fresh for the duration—three to four months. When you take the bag out to help yourself to the crackers, reseal it and put it back in the refrigerator quickly so there is no time for condensation to form on the inside of the bag.

It is convenient to store different kinds of crackers and crisp breads together in the same canister or bag but watch out for two potential problems. Don't mix very fresh products with ones that have been stored for a while. Otherwise, in a month or two you will be stuck with the job of picking out the stale crackers one at a time. And don't mix highly seasoned crackers with blander ones because the flavors tend to cross over.

Rancid crackers and crisp breads have a stale, mildly unpleasant odor and should be promptly thrown out. Crackers that have simply lost their crispness can be saved in a number of ways.

Spread soggy crackers or crisp breads out on a cookie sheet, put them in the oven at 300° F. for about 5 minutes and they will crisp right up. Cool them on a rack and seal them in a canister or plastic bag. They should stay crisp for a week to ten days.

Another approach to uncrisp crackers is to use them for cooking. Crush the crackers or crisp bread into fine crumbs (use a blender, food processor, or put them between two pieces of waxed paper and mash them with a rolling pin). Cracker crumbs are the equivalent of bread crumbs for any recipe—in meat loaf, as a topping for casseroles, and as a coating for fried fish or chicken, for instance—and add an interesting flavor of their own, especially when they have been made from different kinds of crackers. You can store cracker crumbs in a plastic bag in the freezer and they keep up to one year.

STORAGE TIMETABLE FOR
CRACKERS AND CRISP BREADS

Type	Months on Shelf	Months in Refrigerator
UNOPENED PACKAGES	6	—
OPENED PACKAGES	1	3–4

NOTE: Crackers and crisp breads can be frozen for up to six months.

CAKES, COOKIES, PIES, AND PASTRIES

As long as you are going to eat the devil's food cake with all its forbidden sugar, butter, and calories, it might as well taste as good as it possibly can. Most baked desserts stale fairly quickly, just like bread. You must protect them from drying air and, if there are any still uneaten after one or two days, freeze them to keep them fresh.

Buying Baked Desserts. The baked desserts in the supermarket fall into three main buying categories. First there are certain fresh-baked cakes, pies, and cookies, usually from a local bakery, that come in boxes stamped with the last date for sale. The shelf life of these products extends up to one week after the "sell by" date—maybe longer for cookies. Some do and some do not contain preservatives.

Second, there are prepackaged snack cakes and cookies, usually nationally advertised brands, with a much longer shelf life. Most of these contain some preservatives and are meant to last at least two to three months inside their packaging. A few of these also carry "sell by" dates.

Frozen baked desserts are the third category. They keep a number of months in the supermarket's freezer and rarely carry dating codes.

Desserts you buy direct from a bakery should have been baked within the previous forty-eight hours. If not, they should be marked and sold at a reduced price or otherwise dated.

The best baked desserts are made with real eggs, real butter, real flavorings such as natural chocolate, and real fruits. A look at the list of ingredients on the label can tell you how rich and how real the food is.

When you buy these products off the shelf, check the dating code, of course, and check for evidence of crushing. Frozen baked goods should be stored in the supermarket case below the frost line. Don't buy packages that are heavily frosted, that are sticky on the outside, or that have frozen liquid showing inside the wrappings: all are signs the food has thawed and refrozen. Shake the box and don't buy frozen baked desserts that rattle with broken bits and pieces.

Pies and pastries with cream or custard fillings, such as éclairs, cream puffs, napoleons, and whipped cream cakes, must be refrigerated or frozen when you buy them to be safe to eat.

If you are inclined to buy refrigerated cookie or pie dough, look for the dating code always stamped on the label and pick the freshest.

Storing Baked Desserts. You can store *unopened* commercially baked desserts up to the "sell by" date. If there is no date, you can probably count on storing *unopened*, prepackaged cookies, cakes, fruit pies, and ready-to-eat pie shells from three to four weeks (in this category are the products from Hostess, Nabisco, Pepperidge Farm, and similar bakers). Plan to keep commercially frozen baked goods in your home freezer no more than six months. After these foods are opened or defrosted, the storage recommendations for fresh-made goods apply.

Packaged mixes for making cakes, brownies, cookies, and pie dough can be stored up to one year on a cool, dry shelf. If your pantry is warmer than 72° F. for any length of time, use these mixes within six to nine months.

The following recommendations are for baked desserts that are homemade or bought fresh from a bakery. They also apply to cakes, cookies, and pies that have reached their "sell by" date.

• CAKES. Keep cakes, frosted or unfrosted, well wrapped in plastic or foil or seal them inside airtight containers. There are lidded plastic boxes made especially for storing layer cakes. Cakes become noticeably dry and stale after about two days at room temperature or three days in the refrigerator.

Cake with whipped cream filling must be stored in the refrigerator and will stay fresh no more than two days.

For longer storage you can freeze angel food, butter, chiffon, sponge, and pound cakes. With proper handling, they come out of the freezer as fresh as they went in. To freeze, wrap unfrosted cakes tightly in heavy-duty aluminum foil or freezer paper, using the drugstore wrap (see pp. 180–81).

Frosted cakes are a bit trickier. Uncooked, powdered sugar frostings freeze best, but other kinds work too. The problem is wrapping the cake so the frosting stays more or less intact. What you can do is freeze the cake unwrapped until the frosting is hard, then seal it in freezer paper or heavy-duty aluminum foil. When the cake is homemade, consider freezing the unfrosted cake and the frosting separately, then putting them together after they have thawed.

You can keep unfrosted angel food, chiffon, and sponge cakes up to two months at 0° F. Cakes with denser texture, such as devil's food, yellow, pound, and carrot cake, can stay frozen four to six months. Frosted cakes, cupcakes, and frosting itself should be frozen no more than two months.

Thaw cakes at room temperature, still in their wrapping, so that condensed moisture will be reabsorbed evenly. A full-size frosted cake takes about two hours to thaw; a single layer of unfrosted cake thaws in an hour; a cupcake takes about half an hour.

Do not try to freeze cakes with whipped cream or other soft fillings, because they end up soggy.

You can freeze fruitcake and plum pudding up to one year. Refrigerated, they keep even longer covered in cheesecloth inside plastic or foil wrap. The cake or pudding stays moist if you unwrap it every few months, drizzle it with rum or whiskey, and rewrap it. This works for up to two years.

• COOKIES. Too much air can make crisp cookies soggy and moist cookies stale, so once again an airtight container is a requirement. In a tightly lidded jar or sealed plastic bag you can expect cookies to stay fresh for one to two weeks at room temperature, depending on the type of cookie and the humidity in the kitchen.

Just about every kind of cookie freezes successfully. A rigid container protects cookies from breaking: gently layer the cookies inside the box or jar, using waxed paper as a cushion between layers. You can use an empty coffee can or similar recycled container but you should line it with a plastic bag that seals airtight around the cookies.

Another method for storing homemade cookies is to freeze the dough. Dough for refrigerator cookies and drop cookies can be formed into a cylinder and sealed in heavy-duty foil for freezing. To bake, simply slice rounds off the cylinder and place on a cookie sheet. You can freeze bar cookie dough in a greased pan overwrapped with plastic or heavy-duty foil. For rolled cookies, roll the dough, cut out shapes, and store them in a rigid container as you would baked cookies.

Baked cookies and cookie dough both keep from nine to twelve months at 0° F.

Keep commercially prepared cookie dough that is vacuum-packed in tubes refrigerated and use it before the date on the label. Do not freeze it.

Commercially frozen cookie dough can be stored in your freezer up to a year.

• PIES. Store pies, loosely covered in foil or plastic, in the refrigerator. Fruit pies and pecan-type pies keep four to five days; custard and chiffon pies keep two to three days; cheesecakes keep seven to ten days.

Before serving leftover fruit pie you can bring it back to its former glory by slipping it into a moderate oven for 5 to 10 minutes or into a micro-wave 30 to 60 seconds.

Leftover pies freeze just as well as other baked desserts. Freeze a fully baked fruit or chiffon pie unwrapped, sitting level on the shelf. After it is frozen, put a paper plate upside down over the pie and seal it airtight in heavy-duty aluminum foil.

Baked fruit pies keep up to six months in the freezer. Fully prepared chiffon pies can stay frozen up to one month. Pumpkin pie keeps up to two months in the freezer.

Do not freeze cream, custard, or meringue pies, because they are wa-tery after thawing. You can freeze cheesecakes up to one month, but the texture will not be as smooth afterward.

To thaw fruit or pumpkin pies, leave them at room temperature, in their wrappings, for about 30 minutes, then warm the whole pie in the oven at 350° F. for another thirty minutes or warm individual slices in the microwave for about one minute. Thaw chiffon pies at room tem-perature for about 1 hour before serving.

When you make pies that you plan to freeze, use plenty of shortening in the crust and freeze them unbaked. Freeze them first then invert a paper plate over each pie and wrap it airtight in heavy-duty aluminum foil. Do not cut steam vents in the crusts until you are ready to bake the pies. There is no need to thaw the pie before baking; simply add 10 to 15 minutes to the usual baking time.

You can freeze pie shells either baked or unbaked. Fit the rolled dough or crumb crust into a pie pan (there are inexpensive, disposable foil pie pans at the supermarket) for freezing. Bake it or not, then wrap it very securely in heavy-duty foil or seal it in a plastic bag. Baked pie shells keep four to six months, but they are fragile and must be wrapped and stored so they don't break apart. Unbaked pie shells can be frozen up to two months. Crumb crusts and commercially frozen pie shells can be stored two months, as well.

You can thaw baked pie shells, unwrapped, at room temperature or heat them for 5 minutes in a 325° F. oven. An unbaked shell can be thawed at room temperature or bake it from the frozen state 8 to 10 minutes at 475° F.

• PASTRIES. Elaborate constructions of creams, custards, thin layers of buttery cake, sweetened fruits, flaky crusts, and, of course, chocolate are the high art of the baker. They are rich and expensive and deserve to be eaten when they are very fresh. If you must store pastries, keep them well wrapped in the refrigerator no more than two days. Freezing them at home is unpredictable and not recommended.

STORAGE TIMETABLE FOR CAKES, COOKIES, PIES, AND PASTRIES[1]

Type	Time on Shelf	Time in Refrigerator	Months in Freezer
CAKES:			
Angel, Chiffon, Sponge, and All Frosted Cakes	2 days	3 days	2
Whipped Cream	——	2 days	N/R
Fruitcake	1 week	2 years	12
Others	2 days	3 days	4–6
COOKIES	1–2 weeks	1–2 weeks	9–12
COOKIE DOUGH	——	——	9–12
PIES:			
Fruit	——	4–5 days	6
Pecan Type	——	4–5 days	6
Custard and Chiffon	——	2–3 days	N/R
Pumpkin	——	4–5 days	2
Cheesecake	——	7–10 days	1
PIE SHELLS:			
Baked	——	2–3 days	4–6
Unbaked	——	1 day	2

[1] These times are for fresh-made baked desserts.

6

Staples

COOKING STAPLES

Among the raw materials of cooking are those ingredients that resemble the trappings of a chemistry lab more than they resemble food. There is, for instance, baking powder to leaven dough, cornstarch to thicken sauces, and cream of tartar to acidify egg whites for whipping.

Some cooking staples you use every day and some almost never. The boxes, bags, and bottles of them accumulate on your shelves and do require attention—certain ingredients lose their effect after a while, others cake up, some evaporate. Few things in the pantry—just salt and sugar, really—are immune to time.

Buying and Storing Cooking Staples. There is usually not a lot to choose from when you buy staples. The ingredients and the package sizes are fairly standard. One thing you can look for is special packaging designed to help you store the product—an example is the snap-on plastic lids common on cans of baking powder.

• ARROWROOT. A fine white powder ground from the roots of various tropical plants, arrowroot thickens sauces, glazes, and custards. It can be stored in its own box on a cool, dry shelf up to a year. It is important to keep moisture out of the box, so if your kitchen is humid you should seal the box inside a plastic bag for extra protection. If you have trouble with insects in your pantry, store arrowroot in a tightly lidded container.

• BAKING POWDER. Baking powder is a combination of an acid, an alkali, and a starch that keeps the other ingredients stable and dry. The powder reacts with liquid by foaming and the resulting bubbles can aerate and raise dough.

Almost all baking powder now on the market is double-acting, meaning it has one acid that bubbles at room temperature and another acid that only reacts at oven temperatures. Unless a recipe specifies otherwise, this is the type to use. Tartrate, or single-acting, baking powder creates all its potential gas bubbles within two minutes of contacting liquids.

Don't expose baking powder to steam, humid air, wet spoons, or any other moisture. Store it in a tightly lidded container for no more than six months.

Even bone-dry baking powder eventually loses its potency. To test its strength, measure 1 teaspoon powder into 1/3 cup hot water. The mixture should fizz and bubble furiously. If it doesn't, throw the baking powder out.

• BAKING SODA. This plain white powder is a simple alkali, a major ingredient in baking powder. It is also called bicarbonate of soda.

As long as it is kept dry and fairly well covered, baking soda can be stored up to eighteen months.

There are many uses for extra quantities of baking soda. It is a very effective deodorizer—a half box left wide open in the refrigerator absorbs odors for a month. Baking soda in warm water is the best cleaning solution for the inside of the refrigerator, as well. Baking soda is a soothing additive to the bath and can substitute for toothpaste in an emergency.

• BOUILLON CUBES AND CRYSTALS. Made from meat or vegetable extracts along with a host of additives, bouillon cubes and crystals must be stored in a tightly closed container. Too much moisture makes them lumpy and sticky. They will keep up to a year at room temperature.

• CORNSTARCH. This silky white powder is a thickener for sauces, puddings, and fruit-pie fillings. Look for a dating code on the box. There is often a "best if purchased by" date that allows for a reasonable storage time at home.

Store cornstarch as suggested for arrowroot (p. 275).

• CREAM OF TARTAR. When the sediment that collects in wine casks is refined, it becomes the tiny white crystals we know as cream of tartar. It is called for in some sugary recipes, such as frosting and candy, because it helps to produce a creamy texture. And cream of tartar is the acid that, combined with baking soda, makes single-acting baking powder. (To make the equivalent of 1 teaspoon double-acting baking powder, mix 1 level teaspoon cream of tartar with a rounded 1/2 teaspoon baking soda. Use only fresh ingredients and don't try to store the mixture.)

Cream of tartar usually comes in a can with a snug lid that is well suited for storage. Keep the can on a cool, dry shelf up to a year.

• EXTRACTS. The intense flavors of extracts are usually produced by dissolving the essential oils of certain foods in alcohol. Some common extracts are vanilla, maple, orange, and peppermint. There are also imitation extracts made from chemicals that taste similar to their natural counterparts.

Buy only small amounts of extract at a time, because they do not store well. In tightly capped bottles, extracts keep at room temperature for three to four months and a bit longer in the refrigerator. Extracts won't spoil, but the flavors change during storage.

• GELATIN. Gelatin is a dry, powdered protein made from animal byproducts. In addition to unflavored gelatin for making aspics and jellied soups, there are flavored, sweetened gelatins meant for salads and desserts.

Keep flavored gelatins cool and dry and they should last up to a year at room temperature. You can store unflavored gelatin on the shelf as long as three years.

Prepared gelatins should be refrigerated. Put them in a rigid container with a tight-fitting cover. They keep their quality about one week but get progressively more rubbery during storage.

In order to freeze prepared gelatin successfully, you must make it with only three fourths the normal amount of liquid. Gelatin salads blended with cream, cottage cheese, cream cheese, and mayonnaise freeze well, and any added fruit, meat, or vegetables keep their texture. Freeze the gelatin hard, then wrap it tightly in plastic. Keep it frozen no more than two weeks. Unwrap and set out at room temperature about an hour before serving.

• SALT. Table salt usually contains some additives—dextrose or sodium bicarbonate for instance—that help keep it from caking up. If your kitchen is so humid that table salt tends to get lumpy, put a few grains of rice in the box or shaker to absorb extra moisture. There is no time limit on storing table salt.

The coarse crystals of kosher salt (also combined with anticaking agents) and the even chunkier rock salt used in ice cream makers both keep indefinitely.

Sea salt tastes better than ordinary table salt but it costs more. English sea salt from Malden is usually preferred over the Mediterranean brands. Sea salt, like its relatives, is all but indestructible.

Seasoned salts, like onion salt or celery salt, are not so stable. They can be kept, tightly capped, at room temperature for no more than one year for best quality.

• TAPIOCA. You can find tapioca in the form of little pellets called pearls —these are used almost exclusively for making pudding. Granulated tapioca is a thickener for sauces, puddings, and fruit-pie fillings.

Store tapioca as suggested for arrowroot (see p. 275).

• VINEGAR. Vinegar is the very sour liquid fermented from a distilled alcohol, usually apple cider or wine. White vinegar comes from grain alcohol; malt vinegar from barley; and Oriental sweet-sour vinegar from sake.

Fine vinegars are allowed to ferment naturally, while lesser versions are treated with heat and chemicals and consequently taste harsh. A host of vinegars come bottled with herbs, peppercorns, fruit, and even flower petals. It is usually wiser and less expensive to add these flavorings to taste in your own kitchen.

Vinegar, tightly capped, keeps up to one year at room temperature, or until sediment appears at the bottom of the bottle. Stored vinegar some-times forms a harmless, cloudy substance called "mother." Unless the vinegar smells unpleasant, you can continue to use it: strain out the mother by pouring the vinegar through several layers of cheesecloth and into a sterilized bottle.

You can make vinegar at home if you have the two essential ingredients: fine wine and a good "mother." Many inexpensive wines have either been pasteurized or treated with preservatives and cannot produce vinegar. As for the "mother," best get it from a practicing vinegar maker. Otherwise make your own by first pouring 1/4 cup unpasteurized vinegar into a wide-mouthed crockery jar. Let it sit for 30 minutes, then pour in a bottle of well-chosen wine. Cover the jar with cheesecloth and leave it on a cool, dark shelf for several weeks. A thin disk of "mother" should have formed in the brew. Gently place the "mother" on top of the liquid, recover with cheesecloth, and wait another few weeks. Save the "mother" for future batches and seal your vinegar in a sterilized glass container. A bad smell at any stage in the process means that uninvited bacteria have spoiled the wine. Should this happen, throw everything out and start over.

• YEAST. There are two kinds of yeast on the market. *Dry yeast* is granu-lated and usually comes in small foil envelopes. *Compressed* or *fresh yeast* is in the form of small moist cakes. Both types of yeast have "use by" dates stamped on the label.

Dry yeast in foil envelopes keeps for months on the shelf, up to the expiration date. If your dry yeast is in a jar, close the cap tightly after opening and store the jar in the refrigerator until the expiration date.

Fresh yeast must be refrigerated and lasts no more than two weeks. You can, however, freeze fresh yeast, tightly wrapped, for up to six months at 0° F.

Fresh cakes of yeast should not have a sour odor or mottled coloring. You can test either type of yeast to be sure that it is still active. Mix a small amount of the yeast with an equal amount of sugar in warm (105° F. to 115° F. for granular; 95° F. for fresh) water. Within 5 minutes, active yeast will become bubbly and begin to expand.

STORAGE TIMETABLE
FOR COOKING STAPLES

Type	Time on Shelf
ARROWROOT	12 months
BAKING POWDER	6 months
BAKING SODA	18 months
BOUILLON CUBES AND CRYSTALS	12 months
CORNSTARCH	12 months
CREAM OF TARTAR	12 months
EXTRACTS	3–4 months
GELATIN, DRY, BOXED:[1]	
Flavored	12 months
Unflavored	3 years
SALT	indefinitely
TAPIOCA	1 year
VINEGAR	1 year
YEAST:	
Dry[2]	up to date on label
Fresh[3]	N/A

[1] Refrigerate prepared gelatin 5–7 days; freeze it up to 2 weeks.
[2] Refrigerate opened jars.
[3] Refrigerate up to date on label.

FATS AND OILS

Fats and oils are very similar to one another in their chemical composition —they are both mixtures of saturated and unsaturated molecules called fatty acids—but there is a practical distinction. Fats are solid at room temperature and oils are liquid. The higher the proportion of saturated fatty acids in the mixture the more solid the fat.

Buying and Storing Fats and Oils. The different types of fats and oils all work in deep frying, baking, and sautéing but won't necessarily give exactly the same results. If you want to avoid unhappy surprises, use the type of fat or oil specified in a recipe.

What to remember when shopping for oils: darker oils usually have more flavor but less storage life than paler varieties; cold-pressed oils are smoother-tasting; read the label to spot preservatives you may not want; when possible, buy oils in dark or opaque containers; buy no more than you can use within a few months.

Exposure to air, heat, and light eventually changes the flavor of fats and oils. Store them tightly sealed, in the dark, on a cool shelf or, when necessary, in the refrigerator.

• ANIMAL FATS. Chicken fat, called *schmaltz*, and suet, the fat from beef and sheep kidneys, may come into your kitchen as by-products of other cuts of meat or may be bought separately. This is also true of highly prized goose and duck fat (see p. 221).

Unrendered animal fats should be stored, covered, in the refrigerator no more than one month or sealed airtight in a plastic bag and frozen at 0° F. no more than three months.

For best results, these fats should be rendered for storage. Cut the solid fat into cubes and melt it very slowly in a heavy pan with a small amount of water. When the fat is entirely liquid, strain it through cheesecloth. You can line a sieve with the cheesecloth and pour the warm fat through it into a wide-mouthed container. Cover the container and refrigerate up to six months.

Any bacon drippings you save should also be strained through cheesecloth. Another method is to pour the drippings into a clean glass jar and wait for the sediment to sink to the bottom. When the drippings have cooled and solidified, you can skim the clear fat off the top and discard the cloudy brown dregs. Store the clear fat, well covered, in the refrigerator up to six months.

Commercially rendered animal fats last up to a year on a cool shelf.

• AVOCADO OIL. This delicate, buttery oil has remarkable properties: it is lower in calories than other oils, it is monounsaturated, and has a very high smoke point. Use it for both salad dressing and sautéing. Keep it cool, dry, tightly capped, and in the dark, for up to six months after opening.

• CORN OIL. This light, almost tasteless, all-purpose oil is good for deep frying because of a high smoke point. Use corn oil within a year of purchase, or within six months of opening the bottle.

• GRAPESEED OIL. Pale, mild grapeseed oil is a favorite in France and Italy. Its high smoke point makes it suitable for fondues. Capped, kept cool and in the dark, it keeps up to six months after opening.

• LARD. Lard is pork fat that, when properly rendered, is delicately flavored and long-lasting. Lard is frequently sold purified and processed and will keep up to a year on a cool shelf. *Leaf lard,* from around the pork kidneys, is considered the best fat, even better than butter, for making pastry. Render and store it as you would the animal fats described above.

• OLIVE OIL. *Virgin* olive oil is deeply colored and richly flavored. It is graded, from highest to lowest, Extra, Fine, and Lampante. The best olive oil is hand-pressed (or cold-pressed) extra-virgin. If no grade is specified, the oil is between Fine and Lampante. The lower grades are more acidic. *Pure* olive oil is a more refined product, paler and blander than *virgin.* Domestically packaged olive oils are less expensive but generally inferior to imported. Furthermore, they do not carry these grades on their labels.

If it is kept dark and well sealed, olive oil may last up to a year on the shelf, but it is wiser to buy no more than four to six months' supply. If its original package cannot be tightly sealed after opening, pour the oil into a container with a snug lid. Move olive oil to the refrigerator if the pantry gets hotter than 72° F. to 80° F.

• PEANUT OIL. Another very light oil, standard peanut oil has little or no taste and a high smoke point. Specialty shops often carry less refined peanut oils with more distinct flavor.

Store it, tightly capped, on a cool, dark shelf no more than one year unopened, or six months after opening.

• SAFFLOWER OIL. Safflower oil is often the first choice for those on low-cholesterol diets. It is very light, virtually tasteless, and has a very high smoke point. Store as you would peanut oil.

• SESAME OIL. Domestic sesame oil has a pleasant, nutty taste. It makes a good salad dressing and adds a touch of flavor when it is used to sauté fish, poultry, or vegetables. Because of its low smoke point, it is best to mix sesame oil with a lighter oil for frying. The darker, imported varieties are much heavier in flavor. A few drops go a long way.

Sesame oil can turn rancid relatively quickly. Seal tight and store on a cool, dark shelf no more than two months.

• SUNFLOWER OIL. A very light, almost flavorless oil, sunflower oil is usually 100 percent pure and stores as well as peanut oil.

• SOYBEAN OIL. This light-colored oil has a heavier flavor than peanut or safflower oil but can be stored the same way. It is low in saturated fats and has a high smoke point but is not recommended for deep frying because it foams.

• VEGETABLE OIL. A mixture of different vegetable oils designed to be all-purpose, the blend varies from brand to brand and may contain additives and less desirable oils such as coconut and cottonseed. Store tightly sealed on a cool, dark shelf up to one year.

• VEGETABLE SHORTENING. Shortening is solid. Though most often made from pure vegetable oils that have been hydrogenated, it may be a combination of animal and vegetable fats. It is appropriate for deep frying and sautéing. For baking, you can substitute shortening for butter, volume for volume.

After opening, tightly lidded cans of vegetable shortening can be stored up to one year on a cool shelf.

• WALNUT OIL. A darker oil, redolent of walnuts, this is too expensive to use as anything but a salad oil. But as a salad oil it is supremely good. Keep the bottle firmly closed, keep it cool, and use the oil within two months.

NOTE: It is not really necessary, but you can preserve oils a bit longer and a bit better in the refrigerator. They may thicken and turn cloudy, but they usually clear after they climb back up to room temperature and no harm is done.

Reusing Fats and Oils. Both fats and oils used for frying can be saved and used again if you are careful with your cooking and storage methods. Heat the fat or oil slowly up to the desired temperature. Never heat them to the point of smoking. At the smoke point, fats and oils decompose and form off-tasting, possibly unhealthy chemicals.

While you are frying, skim out small particles of food before they burn. When you are finished cooking, allow oil to cool completely, then strain through cheesecloth or a coffee filter into a clean container. Cover the container and refrigerate. Follow the same procedure for fats, but strain while they are still warm enough to be completely liquid.

You can clarify the fat or oil further by adding another step. Before you cool it, add a couple of slices of raw potato to the hot fat or oil, cook the slices until they are brown, and discard them. The potato absorbs flavors left over from the food that has been fried.

Fats and oils may be reused several times, but don't keep them longer than one month after the first use. Discard them if they foam during heating, turn too dark or too light, or acquire an odd odor.

STORAGE TIMETABLE FOR FATS AND OILS

Type	Time on Shelf	Time in Refrigerator
ANIMAL FATS (lard; suet; chicken, duck, and goose fat; bacon drippings):		
Unrendered[1]	——	1 month
Home rendered	——	6 months
Commercially rendered	12 months	——
OILS:		
Avocado, Corn, Grapeseed, Peanut, Safflower, Soybean, and Vegetable		
Unopened	12 months	——
Opened	6 months	——
Olive	4–12 months	6–12 months
Walnut and Sesame	2 months	4 months
VEGETABLE SHORTENING	12 months	——

[1] Freeze up to 3 months.

SUGAR AND OTHER SWEETENERS

Fructose is the sugar in fruit; lactose is the sugar in milk; levulose is a sugar in honey. Table sugar—granulated sugar—is sucrose, a highly refined product made from sugar beets or sugar cane. It is so refined that it is nearly 100 percent pure and virtually indestructible. Powdered sugar and brown sugar are simple variations on granulated sugar and share its interminable shelf life.

Liquid sweeteners are not quite so long-lived as dry sugars. Honey, maple syrup, corn syrup, and molasses may mold and crystallize after a while. These syrups are chemically not as simple as table sugar and therefore lose flavor and otherwise break down over a long period of time.

Buying and Storing Sugar and Other Sweeteners. About the only thing you have to be concerned about when you buy sugars is that the package or bottle is clean and dry.

• GRANULATED SUGAR. There are several forms of white, refined table sugar on the market. The standard is *Fine. Ultrafine* or *Superfine* is white sugar that dissolves very readily and is used in baking, for frostings, and for mixing drinks. Fine sugar may be packaged as compressed cubes.

Granulated sugar never spoils. The only thing that can happen to it during storage is caking—the sugar gets damp, even from humid air, then hardens into a lump when it dries.

Small lumps can be mashed with a fork. If you have a considerable amount of hardened sugar, use whatever tools come to hand. Put it in a bag and pound it with a hammer or put the chunks into the blender to break them up.

Caked sugar will dissolve in water, making a syrup you can use for sweetening beverages, freezing or poaching fruit, and caramelizing for flan.

You can prevent caking by keeping sugar in an airtight container. A canister with a rubber gasket in the lid works best. Store granulated sugar at room temperature.

• POWDERED and CONFECTIONERS' SUGAR. Both names refer to the same product: refined white granulated sugar that has been pulverized. For commercial bakers, there is a range of textures from coarse to ultra fine. Consumer packages may or may not specify the grind, but they usually contain either very fine (6X) or ultra fine (10X). Manufacturers usually add a small amount of cornstarch to prevent caking.

Powdered sugar is as durable as granulated sugar as long as it is kept dry. It will absorb the least amount of moisture in the air and begin to harden, so you have to keep powdered sugar *really* dry.

The best technique is to store it in an airtight plastic bag. When you seal the bag, force out extra air and bind the top very tightly with a twist tie. If you live in a humid climate, you can store the bag inside an airtight jar for more protection.

Once hardened, powdered sugar is difficult to save. You can try forcing it through a sieve. This works if the sugar is not too far gone. Grind it in the blender or food processor or, as a last resort, dissolve it in water as described above.

If you are in the middle of a recipe when you discover your powdered sugar is like a pound of granite, make a substitute by processing granulated sugar in the blender with a pinch of cornstarch.

• RAW SUGAR. It is illegal to sell raw sugar in the United States because it has so many impurities. Anything labeled raw sugar—for instance, turbinado sugar—has undoubtedly been processed in some way—usually washed and steamed to make it clean. Turbinado sugar should be stored like brown sugar.

• BROWN SUGAR. Brown sugar consists of fine crystals coated with highly refined, colored, molasses-flavored syrup. Light brown sugar has less molasses flavor; dark brown sugar has more.

Granulated brown sugar and liquid brown sugar are also available. They don't have the same quality as regular brown sugar, but they do not dry out and harden as readily.

Brown sugar gets hard when it gets dry. Storing it in a thick, airtight, moistureproof plastic bag helps to keep it moist and free-flowing and it can be stored indefinitely at room temperature.

There are a couple of ways to revive hard brown sugar. Put it into a screw-top jar and cover it loosely with plastic wrap. Wet a paper towel, damp but not dripping, and place it on top of the plastic wrap. Close the jar very tightly and leave it at room temperature. In about twelve hours the sugar will have absorbed the moisture from the towel and softened considerably. Or use an apple slice in place of the damp towel.

If you need to measure out the brown sugar right away, put it on a piece of aluminum foil on a cookie sheet in the oven. Heat it at 250° F. until it is soft enough to spoon. The brown sugar will harden again as it cools.

Hardened brown sugar can be put to use as pancake syrup. Mix it with water and bring it to a boil. Flavor the syrup to taste with fruit jelly or orange juice.

• CORN SYRUP. This is a liquid sugar refined from cornstarch. Light corn syrup is bland and not as sweet as sugar. The darker syrup, which tastes similar to molasses, has refiners' sugar (a by-product of granulated sugar) added. Both light and dark corn syrup often contain some flavoring or preservatives. Both kinds are used in baking and they are useful in candy making because they do not form crystals when heated.

Corn syrups often have a "best if sold by" dating code on the label.

Store corn syrup in its original bottle, tightly capped, in a cool, dry place. Unopened bottles keep about six months from the date on the label. After opening, keep the corn syrup four to six months. Be on the lookout for mold or bubbles caused by fermentation on the surface of the syrup. These are signs of spoilage and, if they are present, throw the syrup out. Always take care to wipe drips off the outside of the bottle before you put it back on the shelf.

Since corn syrup is one of the few sweeteners with a limited shelf life, you may be faced with the problem of using it up in a hurry.

Light corn syrup, though not as sweet, can take the place of granulated sugar, teaspoon for teaspoon, in many cases. Don't use it as a replacement in baking or candy making, but you can use it to sweeten fruit, drinks, and hot cereal, for example.

Dark corn syrup has enough flavor to be used over pancakes or as a substitute for molasses in baked beans and spice cakes.

Either corn syrup works as a substitute for honey in any recipe, but they are not as sweet.

• MOLASSES. Molasses consists of the plant juices pressed from sugar cane. The syrup is purified and concentrated by boiling. At the first stage of refining sugar, light molasses is produced; at the second stage, dark molasses; and the final by-product is blackstrap molasses. Unsulfured molasses is not a by-product but a finer grade of syrup, manufactured for its own sake. All of these dark, viscous syrups have a heavy, rich flavor.

An unopened bottle of molasses can be stored up to one year on a cool, dry shelf. After opening, you can store it another twelve months on the shelf. Wipe the outside of the bottle clean, especially the lip, after each use.

You can serve a little leftover molasses as syrup for pancakes or stirred into hot cereal.

• HONEY. Honey is much sweeter than table sugar, and easier to digest. The precise composition of honey depends entirely on what kind of nectar the bees were eating when they made the honey. Bees feeding on orange blossoms produce a milder honey than bees fed on buckwheat, for instance. Not just the flavor but also the degree of sweetness varies.

The rather pale, thin variety most often found at supermarkets has sometimes been diluted with other liquid sweeteners. And unless it is labeled "raw," liquid honey has been heated and filtered to prolong its shelf life. Comb honey is honey stored in the natural, waxy substance the bees made for that very purpose, and some raw honey has bits of comb floating in it. Eat the honey, chewy comb and all.

Granulated or spun honey is honey with some moisture removed to make it thicker and easier to spread. It looks cloudy rather than clear.

You can store liquid honey in its original jar up to a year in a cool, dry, dark place. Be sure the cap is tightly closed. When honey is exposed to too much air and moisture it quickly darkens and crystallizes. It can also absorb odors from other foods in the pantry.

Honey won't mold but after a while it will harden. You can thin honey that has begun to crystallize by placing it in a pan of hot water from the tap.

Comb honey deteriorates more quickly. Plan to use it up within six months or refrigerate it. If you keep it chilled, you will have to warm the jar in hot water each time you want to use it.

• MAPLE SYRUP. Maple syrup is a little bit sweeter than table sugar and has a particularly winning taste. Pure maple syrup tends to be expensive. Many pancake syrups are corn syrup with either natural or artificial maple flavoring.

An unopened bottle of pure maple syrup can be stored on a cool, dry shelf one or two years. The syrup may darken, but that is not a sign of trouble.

Once opened, the syrup should be refrigerated. The cold syrup does not pour easily, so you should leave it at room temperature an hour or so before serving. It lasts about a year in the refrigerator.

A haze of mold may form on the syrup, but you can sterilize and recan it easily. Pour the moldy syrup into an enameled pan and bring it to a boil over high heat. Take it off the heat and skim the surface until the syrup is clean and clear. Have pint-size home-canning jars ready—sterilize jars and lids in boiling water for 15 minutes and keep them hot. Pour the hot syrup into the hot jars, leaving ½ inch head space. Allow the jars to cool and then transfer them to the refrigerator or to the freezer, where they can be stored indefinitely.

Maple-flavored syrups should be stored as for corn syrup.

• SUGAR SUBSTITUTES. These are chemicals, sold as powder, liquid, or tablets, that simulate the sweetness of natural sugar. Sugar substitutes can be stored indefinitely at room temperature, in their original wrapping. Aspartame brand-named NutraSweet, cannot be stored in hot liquids for more than a couple of hours without losing sweetness and chemically decomposing.

STORAGE TIMETABLE FOR
SUGARS AND OTHER SWEETENERS

Type	Time on Shelf
SUGAR:	
Granulated, Powdered, Brown	Stores indefinitely
CORN SYRUP:	
Unopened	Up to date on label
Opened	4–6 months
MOLASSES	12–24 months
HONEY	12 months
MAPLE SYRUP:	
Unopened	2 years
Opened[1]	
SUGAR SUBSTITUTES	Store indefinitely

[1] Up to one year in the refrigerator or indefinitely in the freezer.

HERBS AND SPICES

Herbs are leaves and soft stems of plants. They are most intense, most fragrant when fresh but can be used dried as well. Spices come from other parts of a plant—bark, roots, or seeds, usually—and are always dried.

Herbs and spices do not amount to much on their own. But when applied with precision these seasonings put color and contour into bland foods and bring artful surprises to overfamiliar ones.

Buying Herbs and Spices. Most supermarkets carry all the herbs and spices you need for everyday cooking and even for French, Italian, and other European cuisines. You may have to search out ethnic and specialty markets or even shop by mail to get seasonings for cooking Chinese, Japanese, Indian, Thai, Korean, Caribbean, Latin American, and other exotic foods. A few fresh herbs are sold regularly at retail but, for a consistent supply, consider growing your own.

- ALLSPICE. Dried berries from the West Indian allspice tree, they taste like nutmeg, cinnamon, and cloves combined, with a whiff of juniper berry.
- ANNATO. Earth-red seeds used for both color and flavor in Latin American cooking, they may also be labeled *achiote*.
- ANISE. Seeds with a sweet, licorice taste, often added to cookies, cakes, and breads. In the Middle East, an infusion of anise seeds is used to brew flavored tea. Star anise, from China, is a different, star-shaped seed, but with a similar flavor.
- BASIL. A savory herb, perfect partner for tomatoes, seafood, eggs, eggplant, and all meats. Available fresh in the summer, cut, or as a broad-leaved potted plant. To preserve fresh basil, grind the leaves with just enough oil to make a thick paste and freeze in small quantities for salad

dressing or pesto sauce. Dried basil is a dull but often necessary substitute for fresh.

• BAY. An herb sold most often as slender, whole, aromatic leaves. Imported bay leaves are much milder than the more common California variety, while bay leaves fr   Mexico are milder still.

• BOUQUET GARNI. A bunc   of fresh or dried herbs, often bay, thyme, basil, and parsley. In classic French cooking, the mixture is tied with string or wrapped with cheesecloth and simmered in soups, stews, and other long-cooking dishes.

• CAPERS. A tart addition to sauces, capers are tiny flower buds packed in salt or vinegar. The best are *nonpareilles* from the South of France.

• CARAWAY. Dark, acrid seeds, often baked into rye bread, but also complements to cabbage, sauerkraut, potatoes, and pork.

• CARDAMOM. Dried seeds. Usually expensive, cardamom has a sweet, powerful, almost fermented aroma. It is popular in Scandinavian desserts and a common ingredient in curry. Cardamom may come whole in green or straw-colored pods; the seeds inside should be shiny and moist-looking. Extract the seeds or use the whole pod—one pod can flavor a large pot of coffee.

• CAYENNE. A very spicy powder ground from various hot peppers grown in the Cayenne region of Africa.

• CELERY SEED. Seeds from a plant, smallage, similar but not identical to the familiar vegetable. The seeds do strongly suggest the flavor of celery. *Celery salt* is regular salt mixed with ground celery seed.

• CHERVIL. A delicately flavored herb tasting like parsley with undertones of pepper and licorice. Add it to dishes just before serving, as its flavor fades as it cooks.

• CHILI POWDER. A mixture of ground peppers, cumin, garlic, oregano, and other spices, invented in the American Southwest.

• CHIVES. Members of the onion family, chives are often sold potted, as living plants. Trim the fine, narrow leaves close to the base of the plant and add them just before serving as their flavor turns harsh after even a short spell of storage.

• CINNAMON. Virtually all packaged cinnamon sold in the United States is actually cassia. Both cassia and true cinnamon are dried bark, and they share the same sweet, warming taste. Cassia is darker and stronger with a slight bitter undertone. True cinnamon, usually from Ceylon or the Malabar coast, is soft, pale, and mellow in comparison. Sticks of cassia are rolled from both edges so that two scrolls meet in the middle; a true cinnamon stick is rolled into a single tight scroll.

• CLOVES. Dried flower buds, very pungent. The best come from Zanzibar and Madagascar. Often partnered with cinnamon and nutmeg in sweets, cloves perform well in meat dishes as well. A fresh whole clove is oily when crushed.

• CORIANDER. A dried seed or leaf both sweet and pungent. Fresh coriander leaves, called cilantro in Latin American cooking, and Chinese parsley in Oriental cooking, are powerful. The dried seed, common flavoring for hot dogs, cannot be substituted for the fresh herb.

• CUMIN. Hot, bitter seeds, sold whole and ground. Cumin is an essential ingredient in chili powder and curry, and it is at home with cheese and sauerkraut concoctions.

• CURRY POWDER. A mixture of turmeric, coriander, black and red peppers, cumin, ginger, cinnamon, and other spices. If you like curry that makes your tongue burn and your eyes water, try blending your own powder fresh from whole spices.

• DILL. Also called dillweed, it is the flavoring in dill pickles but is also a very elegant addition to fish and fresh vegetables. Dill is sold as fresh, feathery leaves and as dried seeds or leaves.

• FENNEL. These dried, licorice-scented seeds are the customary spice for mild Italian sausage and go well with fish. Fresh fennel is discussed on p. 43.

• FENUGREEK. Small red, bitter, celery-like seeds that are one component of curry powder.

• FILÉ. Ground leaves of the sassafras tree, used in gumbo.

• GARLIC. Dried flakes, powder, and garlic salt are on the grocer's spice shelf; chopped garlic in oil and garlic paste in tubes may also be found; but fresh garlic is altogether more satisfactory (see pp. 57–58).

• GINGER. Dried, ground ginger root is a sharp spice often used in cookies and cakes; it is not a substitute for fresh ginger in recipes (for information on fresh ginger, see p. 25). Preserved ginger, boiled and packed in syrup, can be chopped and added to desserts or sweet vegetables like yams but, again, cannot replace fresh.

• JUNIPER BERRIES. The bittersweet fruit of the juniper tree, these blue-black berries are a favorite with game and the basic flavor in gin. They are soft when fresh.

• MACE. The seed covering of the nutmeg, mace tastes nearly identical.

• MARJORAM. A sweet, mild herb, closely related to oregano, sold as dried, crushed leaves. Marjoram has a particular affinity for lamb but suits all meats.

• MINT. A cool-tasting herb, sold both dried and fresh. Spearmint is the most common variety, though there are countless others.

• MUSTARD. Mustard powder is what is left when oil is pressed from whole mustard seeds. To blend your own prepared mustard, combine 3 parts mustard powder with 2 parts liquid—use wine, cream, beer, vinegar, or water, for instance—add herbs, garlic, sugar, or other seasonings, and let the mixture rest for about 10 minutes. This homemade mustard can be stored no more than a hour or two. (See also Prepared Mustard, p. 298.)

Whole mustard seeds are used for pickling.

• NUTMEG. A hard seed about the size of a large marble, nutmeg has a sweet flavor particularly suited to desserts, but it is capable of quietly improving beef, cheese, and vegetables. It is sold whole or ground but tastes far better freshly grated—a simple matter with the classic, inexpensive nutmeg grater.

• ONION. For convenience only, onions are processed into dried flakes, powder, and salt. Read about fresh onions on pp. 59–61.

• OREGANO. The crushed leaves are used in hearty dishes, particularly Italian and Mexican sauces. Uniquely, oregano is stronger dried than fresh.

• PAPRIKA. Mild ground red chili pepper. The Hungarian variety, bright and fiery, is far superior to domestic paprikas.

• PARSLEY. The herb most often used fresh. The common curly variety makes a hardy green garnish, but seek out flat-leaved Italian parsley for the best cooking. Avoid dried parsley.

• PEPPER, BLACK. The condiment on everyone's table. Black peppercorns are dried berries, sold whole, crushed, or ground fine. Pepper ground in your own mill is fresher, sharper, and altogether tastier than preprocessed; and whole peppercorns can be stored three to four years.

• PEPPER, RED. Crushed or ground red chili pepper, a bit milder than cayenne.

• PEPPER, WHITE. The bleached inner core of black pepper berries, white pepper is slightly milder but sometimes preferred for its relative invisibility.

• PEPPERCORNS, GREEN. These are the soft, unripe berries of the black pepper plant, a tangy flavoring for meats and sauces.

• PEPPERCORNS, PINK. These rose-colored pepper berries are slightly sweet.

• PICKLING SPICE. A blend of peppercorns, mustard seed, bay, dill seed, peppers, ginger, cloves, mace, or cinnamon. Each packager puts together a different mix.

• POPPY SEEDS. Nutty-tasting seeds used in baking, the best come from Holland. Dark, shiny seeds are fresh; dull seeds may be rancid.

• POULTRY SEASONING. Primarily ground thyme and sage in combination with other herbs, depending on the manufacturer's recipe.

• ROSEMARY. An intense savory herb, its dried, needlelike leaves come whole or crushed. Always used cautiously, rosemary goes with lamb, game, duck, and rabbit.

• SAFFRON. Dried, threadlike flower parts, saffron seasons rice, fish, and pastries, and colors the food a rich orange yellow. It is so expensive because 75,000 hand-picked blooms of the autumn crocus yield only 1 pound of saffron.

• SAGE. A pungent herb, sold dried and crushed, sage is often used to flavor sausage and poultry stuffing.

• SAVORY. A peppery herb, sold dried and crushed, summer savory is preferred.

• SESAME SEEDS. Sweet, nutty seeds used frequently in baking. Toasted seeds are a delicious high-protein addition to green salad.

• TARRAGON. A luxurious, bittersweet herb, it is most often available crushed and dried—the best is French. If you have fresh, try it with eggs, veal, or chicken.

• THYME. A pleasant, outdoorsy-scented herb, sold as tiny, whole leaves and powdered.

• TURMERIC. Sold finely ground, acrid turmeric is the earthy yellow ingredient in curry powder.

The *dried* herbs and spices sold in most grocery stores are generally of good quality. They are packed in tins or screw-top jars and are therefore well protected. The drawback to buying supermarket seasonings is that you must buy each kind in a relatively large quantity. It can take years to use up 1/2 ounce of tarragon, which, in volume, is about 1/2 cup of crushed leaves. After one year, tarragon and other herbs will have lost almost all their punch as flavorings.

In some stores you may run across smaller quantities of herbs and spices packed in cellophane bags. The trouble there is that cellophane does not protect the seasonings from the air and they may be stale before you buy them. Preferable are herbs and spices in small plastic bags, though these are still not entirely airtight.

There is no perfect solution to the dilemma. If you live in a large urban area you may be able to find a specialty store that sells herbs and spices in bulk, so that you can buy only what you need. Of course, they must be fresh when you buy them, so shop at stores with high turnover and good storage methods of their own.

You can grow your own herbs.

You can share with friends.

You can resign yourself to emptying the spice rack into the trash can once a year and starting over. Or you can simply settle for very much less than perfect herbs and spices that have passed beyond their satisfactory shelf life.

The situation is somewhat relieved when you buy dried herbs and spices whole rather than ground. Ground seasonings do not stay fresh and pungent even half as long as whole.

You can grind your own herbs and spices as you need them. In addition to the usual kitchen graters and grinders, there are simple hand grinders and small electric grinders designed just for herbs and spices. All these devices must be thoroughly cleaned between uses so the various strong flavors don't mix.

When you buy *fresh* herbs, look for a healthy green color and a clean scent. Don't buy limp, yellowing herbs. Parsley is the fresh herb easiest to find, but many stores carry fresh dill, mint, basil, and coriander. Chives and sweet basil are often sold in the produce department as whole, live plants in pots.

Storing Herbs and Spices. Dried herbs and spices lose the strength and character of their essential oils when exposed to heat, light, and air. They can be stored at room temperature, but keep them away from the stove, dishwasher, refrigerator, toaster, and any other appliance that gives off heat.

Certain spices suffer more than others from high heat. During the summer months you should consider keeping cayenne, paprika, chili powder, and red pepper in the refrigerator. Poppy seeds and sesame seeds are rich in oil—they get rancid within several months unless they are stored properly. They keep best in the refrigerator.

The bottles and tins most herbs and spices come in are perfect storage containers, but the tins have a slight edge because they are lightproof. Cap the containers tightly to keep out air and moisture.

If you buy seasonings in bulk, transfer them to small glass jars with snug screw-top lids. Make sure the jar is immaculately clean—boil it for a few minutes if necessary—because herbs and spices readily absorb left-over odors.

Light is not the worst enemy of your seasonings, but some precautions are called for. Don't store herbs and spices in direct sunlight, for instance. When you have an expensive seasoning, such as saffron, store it in a drawer or in a lightproof container, to prolong its shelf life.

Whole dried herbs and spices can be stored for about one year. Ground and powdered seasonings last no more than six months before they stale.

If, after a year of storage, your dried herbs and spices still have a strong, distinctive aroma, there is no reason not to go ahead and use them. To get the most flavor from aging dried herbs, rub them briefly in your hands, sauté them in butter, or simmer them in liquid before you add them to your recipes.

Freezing dried herbs and spices definitely prolongs their storage life but is not a very practical use of limited freezer space. Furthermore, when they go in and out of the freezer, moisture forms inside the container, a condition that damages flavor.

Wash and dry fresh herbs and put them in an airtight jar, one with a rubber gasket in the lid, and refrigerate it. Parsley can keep up to two weeks in the jar; other herbs last a week to ten days.

For three to four weeks of storage, put fresh herbs in a jar of water, as you would cut flowers. Cover the herbs and jar with a plastic bag and refrigerate. Every five days, change the water and discard wilted and rotting sprigs.

Drying Fresh and Homegrown Herbs. It is very simple to preserve fresh herbs by drying them yourself.

Most homegrown herbs should be harvested before the plant begins to bloom, but herbs in the mint family—mint, dill, and oregano, for example —are best harvested when in flower.

Wash the stems and leaves in a sinkful of cold water and gently shake off the excess moisture. Cut off any yellow, damaged, or wilted leaves.

Tie small bunches of herbs together at the stem end with regular string. Put the bunches, stem ends up, inside a brown paper bag that has several 1/2-inch holes cut in the sides. Tie a string around the top of the paper bag to close it. Hang the bag somewhere that has fairly even temperatures and good air circulation. In five to ten days the leaves will be dry enough to crumble easily. Roll the bag gently back and forth to loosen the leaves from the stems.

Separate out the dried leaves and put them into an airtight glass container. For the first few days, check the container to see if there is any condensation on the inside. If there is any moisture in the jar, dry the leaves further before storing them.

Keep the dried herbs cool and sealed airtight. They keep six months to one year.

Freezing Fresh and Homegrown Herbs. Frozen herbs are more potent and closer to fresh than dried herbs. There are two ways to freeze fresh herbs.

Blanch a handful of whole stems for 5 to 10 seconds in boiling water. Dry them thoroughly on a paper towel, seal airtight in heavy-duty aluminum foil, and freeze. To use the herbs, take out the package, chop off what you need, rewrap the stems, and put them back in the freezer.

Another method is to wash, trim, and chop herbs as for drying. Leave them to drain on a paper towel an hour or so until they are quite dry, so they won't stick together when frozen. Freeze the chopped herbs in small plastic freezer bags or in tightly lidded glass jars. To use, remove the amount you need, reseal the container, and keep it in the freezer.

Fresh herbs keep two to four months at 0° F.

STORAGE TIMETABLE
FOR HERBS AND SPICES

Type	Months on Shelf	Weeks in Refrigerator	Months in Freezer
DRIED HERBS AND SPICES:			
Whole	12	—	—
Ground	6	—	—
FRESH HERBS[1]	—	1–4	2–4

[1] See text for storage techniques.

CONDIMENTS, SAUCES, AND DRESSINGS

The American palate is accustomed to a legion of prepared sauces and dressings. It is not a hot dog unless it is slathered with mustard. A hamburger demands ketchup and a turkey sandwich without mayonnaise is a dry and pitiful thing.

Buying and Storing Condiments, Sauces, and Dressings. These foods almost invariably come in bottles that allow you to look at what is inside. Although it is very rare to find a bottled sauce that has been sitting in the market too long, you can easily tell if you have found one. The contents will be duller in color than its shelfmates'.

Presuming they are fresh when you buy them, these foods should last a long time, though they will undergo some slight changes in character. As noted before, they tend to darken, and they also very slowly lose the bright flavor they had when fresh. Whether the storage time is spent in an unopened bottle on the shelf or in an opened container in the refrigerator, condiments, sauces, and dressings should be used within one year of purchase.

After opening, a number of these products—mayonnaise, mustard, and ketchup in particular—develop dark brown or black crusts around the rim of the bottle. This crust is a result of oxidation and does not mean the food is spoiled. You can wipe off the crust with a damp paper towel and continue to store the product to its full storage time limit.

- MAYONNAISE AND VARIATIONS. Commercial mayonnaise is made from vegetable oil, egg, and lemon juice or vinegar. Most also contain seasonings.

Substitutes for real mayonnaise have the same color and consistency but are made with water and cooked starch. Some have chemical additives. They must be labeled salad dressing or imitation mayonnaise to avoid confusion with the genuine article.

Both mayonnaise and its substitutes have dates stamped on the label indicating the last suggested date of sale, which allows for reasonable storage time at home after purchase.

Store unopened jars of mayonnaise and mayonnaise substitute at room temperature. Once opened, they must be refrigerated. You can safely store them as long as one year after opening if you keep the jars chilled and tightly lidded, but the total storage time, unopened or opened, should not extend more than six months beyond the last date of sale on the label.

• KETCHUP AND VARIATIONS. Tomatoes are the soul of red ketchup. It also contains sugar (enough to alarm the health-conscious), vinegar, salt, and seasonings.

Chili sauce is a spicier, slightly thinner version of ketchup. Most barbecue sauces also have a tomato base, but a tangier flavor than ketchup.

Unopened bottles of ketchup, chili sauce, and barbecue sauce can be stored up to one year on a cool, dry, dark shelf. You can store opened bottles, tightly lidded, on the shelf up to about one month. For longer storage, keep them in the refrigerator.

• PREPARED MUSTARDS. Prepared mustards are made from ground mustard seeds blended into a liquid such as wine, beer, vinegar, or plain water. Added seasoning makes some mustards hot and some mild. The color of commercial mustards ranges from Crayola yellow to russet.

Mustard can be stored unopened on the shelf for up to one year. After opening, mustard can be refrigerated up to one year. An opened jar keeps only a month at room temperature. In particular, fancy mustards spiced with horseradish, garlic, or herbs gradually lose their zest after opening, so buy no more than a six months' supply.

• SEASONING SAUCES. The dark brown sauces, such as Worcestershire and steak sauces, are made from proprietary recipes containing numerous vegetables and seasonings. Soy sauce, also called tamari, is a product of fermented soybeans and wheat. American-made soy sauce may contain sugar.

Pepper sauces (the best known being Tabasco) are very hot purees of red chilis, vinegar, and numerous seasonings. They are bright red when fresh.

These sauces all last up to a year at room temperature, whether they are sealed or open. If, however, your pepper sauce turns brown, throw it out.

• SALAD DRESSINGS. Bottled salad dressings made of oil, vinegar, seasonings, sugar, and additives are sold in an astounding variety of flavors. Most of these are shelf stable, but a few brands, containing no preservatives, need chilling and are sold only from the grocer's refrigerator case.

Store unopened bottles of salad dressing no more than one year on a cool, dry, dark shelf. Of course, salad dressings that require refrigeration (this information is on the label) must be stored in the refrigerator even unopened. Keep them no more than six months.

Another version of prepared salad dressings is dry and packaged in foil envelopes. The contents are to be added to fresh ingredients, such as oil and vinegar or mayonnaise. Dry seasonings for salad dressings keep up to one year on a cool, dry shelf.

If the temperature of your kitchen is frequently above 72° F., all condiments will deteriorate more quickly stored on the shelf. Reduce their overall storage time accordingly.

COFFEE

Coffee is much less a food than it is a partner in life. It is a most welcome companion on bleary-eyed mornings; coffee warms and stimulates every kind of social occasion during the day, then rounds off the evening as the essential *digestif* of an elegant dinner.

It is a pity that most people misunderstand and abuse this friendly brew. Ground coffee stales within days at room temperature. Coffee drinkers think nothing of leaving a 2-pound can of ground coffee on the shelf and dipping out ever more tasteless spoonfuls of the ruined granules as the weeks wear on.

Like freeze-dried stew, canned ground coffee is compact, easy to prepare, and, unopened, will store for long periods of time. But there is another less fortunate similarity: after the first pot of coffee is brewed (this one is often delicious), subsequent pots taste as dull as K rations. Using canned ground coffee, let alone instant, is an expedient that makes sense only in cases of absolute necessity.

There are times when the efficiency and economy of canned ground and even instant coffee make them a good choice. But the oily essence of the coffee bean, the part with all the flavor, is very volatile. It will diminish or escape altogether very quickly after the whole bean is ground. Because fresh ground coffee also gives off gases that would cause a closed can to explode, some time must pass before it is vacuum-sealed. During that time some flavor is necessarily lost.

If you truly love the flavor of coffee, if you are spending days cooking up the perfect dinner for friends, then it is more than worth the extra effort to buy fresh roasted beans or freshly ground coffee, chosen with the same care you bring to other fresh ingredients in your meals.

Buying Coffee. The first choice you make in buying fresh roasted coffee is the most important one: where to buy it. Just because a merchant artfully displays coffee and has two dozen or more kinds of beans doesn't mean that he is an expert or even that the beans are better than what you get out of a can.

The ideal retailer is one who roasts beans right on the premises and sells enough of them to ensure that nothing sits on the shelf for very long. Second best is a retailer who buys direct from a local roaster. Wherever you find fresh roasted coffee beans sold in bulk, whether it is a specialty shop, delicatessen, or supermarket, don't be afraid to ask the clerk where and when the beans were roasted. Whole roasted beans begin to stale within a week, so make sure the beans you buy have been sitting around for no more than a couple of days.

The knowledgeable retailer stores coffee beans in covered containers, protected from air, moisture, and foreign odors.

When you have decided where to buy the beans, you still must choose which beans you want. Coffee beans are identified by place of origin; for instance, Kona is a district in Hawaii, Java is an island in Indonesia, Mocha (to bend the rules a bit) is an ancient port that once shipped the coffee grown in Yemen and the coffee bears the name of the port.

In addition to the point of origin, the label on the beans may identify the character of the roast. Regular, American, and Medium Roast beans have been roasted until they are a medium brown color. Viennese or Light French Roast beans are slightly darker and have a trace of oil visible on the surface. Italian, French, Dark, and Espresso Roast beans are dark brown and gleaming with oil. These are only a few of the more common names. And many retailers use their own terminology.

In fact, the label gives only the barest hint of how the coffee will taste. You may get stuck with a bad batch of a coffee with the highest reputation. You may be so baffled by a retailer's peculiar labeling system that intelligent choices are out of the question. Since what matters most is the coffee's flavor, ask the merchant to brew a cup. If that is not possible, taste whichever coffees he has fresh in the pot and buy what you like best. If the store is not able to give you a taste test, then buy just a tiny amount of a few different types and try them at home.

To complicate matters further, most coffee sellers put together blends of different beans. Mocha-Java is a classic blend almost everyone has heard of—the distinctive qualities of each bean combine to make coffee that is smoother and richer than either bean alone can produce. Almost every serious specialty shop will have its own private blend of coffees for sale. These are often a very reliable buy—usually fresh and carefully controlled.

Having decided on which beans to buy, you still face another set of choices: how and where to grind the coffee and how much of it to buy.

The retailer can grind the coffee as coarse or fine as required by your coffee maker. Percolators use a coarse grind; drip pots use medium; vacuum pots use fine; pots that have paper filters need a very fine grind; espresso is made with coffee ground to a powder.

There is a good reason to grind your coffee at home, however. Once the beans are ground they begin to stale within hours. There are dozens of coffee grinders on the market and many of them are inexpensive. Your blender, spice mill, or food processor may be able to grind coffee; read the manufacturer's directions. Most grinders are easy to clean—just wipe them off with a damp cloth—so don't imagine you will have to wash some huge, clanking contraption every time you want a fresh cup of coffee. Whatever you use, you will need to experiment to arrive at the right grind for your coffeepot.

As for how much coffee to buy, start by calculating how much coffee you prepare in a week. Let's say it's 6 cups per day or 42 cups per week. One pound of coffee yields just about 40 cups. One pound of preground will keep satisfactorily for one week if refrigerated, so that is the amount to buy. If you brew 2 cups a day, buy 1/3 pound. Since beans are sold in bulk, you can buy whatever amount fits your planning. Whole beans keep longer (see below), so you can buy more at one time.

If you are quite content with canned ground coffee, you will get the best brews by using up the opened can within one week, so buy quantities accordingly.

The most determined coffee lovers can take shopping one step further and absolutely guarantee that the coffee they drink is the freshest possible. Specialty shops sell green, unroasted coffee beans that you can roast at home as needed.

Storing Coffee. Roasted beans and ground coffee stale surprisingly fast. Aging coffee has a flat, papery flavor you can avoid by buying in small quantities and storing it carefully.

• CANNED GROUND COFFEE. The average can of coffee has been on the grocer's shelf for one to three months and 95 percent of all canned coffee is bought within a year of processing. Since the shelf life of an unopened can of coffee is two years, you can expect to keep it unopened at home for at least a year without any loss in quality.

As soon as the can is opened, however, taste goes downhill fast. The only way to slow up the decline is to chill the coffee. Left at room temperature, it will stale almost immediately.

Keep open cans of ground coffee in airtight containers (the plastic-lidded cans are ideal) in the refrigerator or freezer. The coffee should keep well for a week to ten days.

When you take the coffee out to measure it into the pot, whisk it right back into the refrigerator; otherwise moisture will condense on the inside of the container and waste more flavor.

• INSTANT COFFEE. An unopened jar of instant coffee powder will store well on the kitchen shelf for about twelve months. Freeze-dried coffee is not quite as long-lived, so store it no more than six months before using it. Once opened, instant and freeze-dried both should keep another two or three months. However, the more humid the air the more moisture instant coffee will absorb; eventually it will clump together and even grow mold. It helps to keep the container cool, both when it is sealed and after it is opened. You can refrigerate the jar during the humid summer weather.

If your instant coffee does cake and harden, pour very hot water into the jar to dissolve it. Refrigerate the concentrated liquid, which will keep for up to one week. To make drinkable coffee, spoon the concentrate into a cup and dilute it with boiling water.

• WHOLE ROASTED COFFEE BEANS. Whole coffee beans begin to lose flavor after one week at room temperature. After three weeks they will produce a noticeable stale flavor when brewed. Unless stored in an airtight container, the beans can take on flavors from other foods in the vicinity—garlic, cheese, apples, or whatever.

You can extend their freshness considerably by storing them in the freezer. In a tightly lidded jar, the beans will keep three to four months at 0° F. Take the jar out of the freezer only long enough to measure out the beans, then put it right back in; otherwise moisture condenses on the inside of the container and dampens the beans.

• FRESH GROUND COFFEE BEANS. Transfer the coffee from its paper bag into a tightly lidded metal or glass container. Ground coffee keeps one week in the refrigerator and two weeks in the freezer.

• GREEN COFFEE BEANS. Before coffee beans are roasted they are quite stable. You can store them in an airtight container, in a cool, dry place for as long as one year.

• DECAFFEINATED COFFEE. You can buy decaffeinated coffee in any of the forms listed above. It will store exactly the same length of time as untreated coffee.

• FLAVORED COFFEES. There are a number of these coffees on the market—chocolate, rum, and almond, to name a few. You can buy flavored beans and flavored ground or instant coffees. The extra ingredients can easily acquire a rancid taste, so cut storage times in half for these products. You can spice up coffee at home simply by adding dry flavorings such as cocoa, cinnamon, cloves, allspice, cardamom, and orange peel to the grounds before brewing, or liquid ingredients such as hot milk, rum, or brandy into each cup as it is poured.

• BREWED COFFEE. Brewed coffee is pretty hard to hang onto. If you leave it over heat, within a couple of hours it will taste burned or bitter. At those times when it is inconvenient to make fresh pots of coffee, use a vacuum bottle to store the hot brew. In addition to the familiar lunchbox-style vacuum bottles that are good for traveling, you can buy insulated pitchers, handsome enough for the table, that keep coffee hot for eight to ten hours.

There is only one way to reheat coffee that restores its fresh-brewed taste. You can warm leftover coffee in a microwave and even after twenty-four hours get good results.

Keep brewed coffee you want to use as a cooking ingredient in a tightly lidded jar for no more than one day in the refrigerator; longer than that and it soon turns harsh-tasting.

STORAGE TIMETABLE FOR COFFEE

Type	Time on Shelf	Time in Refrigerator	Time in Freezer
COFFEE, CANNED GROUND:			
Unopened	1 year	—	—
Opened	—	7–10 days	7–10 days
COFFEE, INSTANT:			
Unopened	12 months	—	—
Opened	2–3 months	—	—
COFFEE, FREEZE-DRIED:			
Unopened	6 months	—	—
Opened	2–3 months	—	—
COFFEE, WHOLE BEANS	1–3 weeks	—	3–4 months
COFFEE, FRESH GROUND	—	7 days	14 days
COFFEE, GREEN BEANS	1 year	—	—
COFFEE, BREWED	—	1 day	—

TEA

A steaming cup of tea can soothe the spirit and smooth the brow. It is as sociable a drink as coffee, yet there is a peaceful quality to tea suitable for introspection and days at home in bed with the flu.

If you use tea only occasionally, it can hold its own on the shelf for quite some time, always ready for an emergency. If tea is your everyday drink, you can buy and brew it in enough varieties to satisfy the most dedicated connoisseur.

Buying Tea. There are three basic types of tea: green, black, and oolong. *Green tea* is not fermented. It is picked and immediately treated with heat to deactivate enzymes in the leaf. It makes a very light, yellowish-green brew. *Black tea* is fully fermented. The natural enzymes in the leaf are left alone to darken the tea and modify its flavor. It makes a red-brown brew with a lot of body. *Oolong tea* is partially fermented and brews up an amber liquid.

Black tea is graded by size. There are grades for whole leaves and for pieces. From the largest to the smallest, the whole-leaf grades are: *souchong, pekoe,* and *orange pekoe.* Smaller than these are the "broken" grades—broken pekoe and so on—down to fannings and dust. These grades in no way relate to the quality or flavor of the tea. Teas are separated by grade because small leaves, pieces, and particularly dust infuse quickly; large leaves slowly. Like different grinds of coffee, tea must be brewed to suit its size.

Green tea has an entirely different grading system, based on the leaf's age and shape, characteristics that do influence the flavor of the tea. The system varies from country to country. The highest grade of China tea is *gunpowder*, which is young leaves rolled into tight pellets. *Fine Young Hyson* is the highest grade of India's green tea; *Extra Choicest* is the best green tea from Japan.

Oolong has yet another set of grades, of which *Choice* is the best.

Other guides to the quality of the tea you are buying are the origin and the blend. Excellent black teas, for instance, come from Darjeeling and Assam, districts in India, and from Ceylon (Sri Lanka). Common blends are "English Breakfast," a strong mixture of India and Ceylon black teas, and "Prince of Wales," made of Chinese teas. Teas may also be scented with aromatic spices and perfumes such as cloves, fennel, and jasmine.

Specialty shops sell fine teas loose, either in bulk or in small tins. Buying bulk tea has two advantages. First, a large chest of tea stays fresh longer than tea in small quantities—less air reaches the leaves as they sit. Second, you can test bulk tea for freshness by taking a pinch and smashing it between your fingers. If the tea is dry and crumbly it is past its prime.

Supermarkets carry loose tea in tins, tea bags, instant tea powder, and tea to use in electric coffee makers. If you like the flavor of tea, loose tea is the best buy, and if it is vacuum-packed, so much the better. The finest teas usually don't go into tea bags, and instant or processed teas taste very little like the real thing. Instant iced tea mix has a particularly offensive chemical flavor.

Storing Tea. The enemies of tea are air and moisture. Keep every type of tea, including tea bags, in airtight containers on a cool dry shelf. Do not store tea in the refrigerator, because it quickly absorbs condensation. Tea also readily absorbs odors, so carefully clean your tea canisters between one buy and another. Glass jars are okay, but it is better to store tea away from light.

Go on the assumption that the tea you buy at retail is already one year old. Plan to keep it no more than six to twelve months. Green teas, except gunpowder, are the poorest keepers, and black teas the best.

Regularly brewed tea cannot be saved and reheated successfully, as is, but you can keep brewed tea bright and fresh-tasting for many hours if you use the right technique. Put a carefully measured ⅔ cup loose tea into a large teapot that has been warmed in a rinse of hot water. Bring 1 quart cold water to a full, rolling boil, then immediately pour it over the tea. Cover the teapot and allow the tea to brew for exactly 5 minutes. Immediately pour the tea into another container, using a sieve to strain out all the leaves. You can leave this concentrated tea at room temperature for four to six hours. To make a fresh, hot cup of tea, put 2 tablespoons of the concentrate into a cup and add boiling water to fill. This recipe is enough for 20 cups.

You can get surprisingly good results from "second-potting" green and oolong teas. Save the wet leaves from the first brew and you can use them again, even after a few hours, to make a brand-new pot of tea.

A good way to save any brewed tea is to transform it into iced tea. Cold tea keeps its smooth flavor for two to three days in the refrigerator. The tea may turn cloudy when it is chilled, but you can clear it up by adding just a little boiling water.

Buying and Storing Herbal Teas. Herbal teas may appear alone—you will find genteel camomile, perhaps, or bracing mint—or in blends of every conceivable flower, leaf, root, and fruit. Keep herbal teas dry, cool, and in the dark and try to use them in six to nine months.

STORAGE TIMETABLE FOR TEA

Type	Time on Shelf	Time in Refrigerator
TEA LEAVES, LOOSE AND BAGGED	6–12 months	——
HERBAL TEA	6–9 months	——
TEA, INSTANT	6–12 months	——
TEA, BREWED	4–6 hours	2–3 days

JUICES

You could look at juice as simply a very sensible way to store fruits and vegetables. Juice contains very nearly all the flavor and nutrients of the food it comes from, and it takes up a lot less space. You can subject juice to freezing and canning and still come out with a food that seems very close to fresh.

Buying Juices. A few juices go directly from the fruit into the bottle, but many are processed in some way before packaging. You will see lots of juice labeled "reconstituted" or "from concentrate." Both phrases mean the same thing. A concentrate is pure juice with water removed. The concentrate is easy to store and to ship. Reconstituted juice in bottles and cans is concentrate diluted with water back to normal strength.

Many juices—frozen orange, grapefruit, and apple, for instance—are sold directly to the consumer as concentrates.

Fruit and vegetable juices may be sweetened or unsweetened, and may have preservatives added. The package label lists all ingredients.

• APPLE JUICE AND APPLE CIDER. There is no absolute or legally established definition that distinguishes apple juice from sweet apple cider. Consequently the same product might be labeled one or the other depending on the manufacturer and on local custom.

Traditionally, natural sweet cider is the juice, just as it comes pressed from the apples. It is cloudy—almost opaque—and develops a dark golden-brown color because of fine particles of apple suspended in the liquid. If no preservatives are added, sweet cider begins to ferment within a few days: microorganisms in the cider convert sugar into alcohol to produce mildly intoxicating hard cider, safe to drink.

Natural sweet cider may be treated in a number of different ways before it is bottled. Filtering it removes apple pulp and leaves a clear amber liquid. Pasteurizing destroys enough microorganisms to prevent fermentation but it also oversimplifies the flavor of the juice.

It is not unusual to find apple juice that is pasteurized but unfiltered, or filtered but labeled cider. The one thing the label tells you for sure is whether the product is 100 percent pure—most are.

Very few apple juices are sweetened or have other additives, but if they do the label spells it out.

- CRANBERRY JUICE. Commercial cranberry juice is almost always made from concentrate and bottled with sweeteners.
- GRAPE JUICE. Unsweetened grape juice is available bottled in purple and white varieties, usually made from concentrate. Frozen grape juice concentrate may or may not have sugar added.
- GRAPEFRUIT JUICE. Grapefruit juice is packaged in the same way as orange juice.
- LEMONADE. Just about anything can be called lemonade. Frozen lemonade is most often lemon juice, sugar, water, and lemon oil, but canned, bottled, and granulated lemonade may have little or no lemon juice at all.
- LEMON JUICE. Bottled lemon juice is usually reconstituted and has lemon oil and a preservative added. Pure lemon juice, frozen in a plastic container, is also available. It keeps two months after thawing.
- NECTARS. Nectar is fruit juice that has a certain amount of fruit pulp suspended in the liquid. The most commonly found nectars are made from apricots, peaches, and pears and contain sugar or other sweeteners.
- ORANGE JUICE. Frozen concentrate is the most popular form of processed orange juice, but you won't have any trouble finding regular-strength orange juice: canned, it is displayed on grocers' shelves; juice in bottles and paperboard cartons is refrigerated; and fresh juice may be bottled right there in the produce department (only as good as the orange it comes from and only at its peak if squeezed before your very eyes).

Commercial orange juice is a mixture of the juices of many different types of oranges, tart and sweet, combined for consistent flavor.

Unless clearly labeled otherwise, orange juice is pure and unsweetened.

- PINEAPPLE JUICE. Pineapple juice is sold bottled, canned, and as a frozen concentrate. It is usually pure and unsweetened.
- PRUNE JUICE. Unsweetened prune juice is available bottled and canned.
- FRUIT DRINKS. A beverage that contains only a percentage of juice

must be labeled "drink." A fruit drink is usually juice diluted with lots of water and heavily sweetened.

• TOMATO JUICE. Tomato juice and other vegetable juices come in cans and bottles. Almost without exception, they are pure juice from concentrate, sometimes seasoned with salt.

Storing Juices. Juices tend to take on distinctly off flavors when they are stored carelessly, and warm temperatures have an especially bad effect.

• FROZEN JUICE. A small can of frozen concentrate can defrost very quickly between store and kitchen, so have the grocer bag all your cold and frozen foods together, and then get them home quickly. Partially defrosted juice is safe to refreeze but a lot of flavor is lost.

You can keep frozen juice eight to twelve months at 0° F.

Once it is reconstituted, keep the juice in the refrigerator, covered. The bright taste lasts only two or three days but can be stored as long as ten days before it actually begins to spoil.

JUICE IN BOTTLES AND CANS. You can store most bottled and canned juices up to one year on a cool, dry shelf, but citrus juices need more care. Keep bottled orange and grapefruit juice chilled. Unopened, bottled juice keeps up to a year in the refrigerator. Canned citrus juice takes on a metallic flavor when stored in warm temperatures and, in any case, should be held no longer than about six months at room temperature.

Store opened cans and bottles of citrus and other juices in the refrigerator a week to ten days. Canned juice will taste better if you transfer it to a glass container after opening.

Paperboard Cartons. Juice boxes, rectangular aseptic cartons on the supermarket shelf, are good four to six months at room temperature—they usually bear a "use by" date. The paper containers in the grocer's refrigerator case must be kept chilled. Unopened, they store well up to three weeks. Once opened, juice in cartons can be refrigerated a week to ten days.

Only a couple of days after opening, most juice begins to taste a bit stale. You can revive any juice with a little aerating. The simplest way to do this is to pour juice back and forth from the glass to the container or whip it with a whisk until it begins to foam. You can aerate a quart of juice all at once by putting it in the blender for a few seconds of high-speed agitation. These methods work especially well with orange juice. A splash of soda water in a glass of juice can also refresh the taste.

STORAGE TIMETABLE FOR JUICES

Type	Time on Shelf	Time in Refrigerator	Months in Freezer
FROZEN JUICE:	——	——	8–12
Reconstituted	——	7–10 days	——
BOTTLED JUICE:			
Unopened (except citrus)	1 year	——	——
Unopened (citrus)	——	1 year	——
Opened	——	7–10 days	——
CANNED JUICE:			
Unopened (except citrus)	1 year	——	——
Unopened (citrus)	6 months	——	——
Opened	——	7–10 days	——
JUICE IN CARTONS:			
Standard	——	3 weeks	8–12
Juice boxes	4–6 months	——	8–12
Opened	——	7–10 days	——

SOFT DRINKS

Even soft drinks don't stay fresh forever. Bottles and cans of sweet, effervescent beverages eventually begin to taste stale and, in some cases, develop a sediment at the bottom of the container. The flavor of powdered soft drink mixes fades if stored too long, and they will cake up if exposed to humid air.

Buying and Storing Soft Drinks. You have little to be concerned about—at least in terms of storage—when you buy most soft drinks. They move in and out of stores so quickly, chances are very slim that you will buy any that are too old.

The bottled soft drinks labeled "contains no preservatives" are another matter. They should be stored in a refrigerator case and be bought from stores that you know have high turnover. Refrigerate them at home for a maximum of two to three months. Given the right conditions, these drinks begin to ferment and turn into an unpalatable wine. Preservative-free soft drinks in cans are fine stored at room temperature, but use them within six months for best flavor.

Nearly all soft drinks in bottles and cans do contain preservatives. They can be stored up to a year or more on a cool, dry, dark shelf, but after several months the flavor deteriorates noticeably. Heat and light will break down the flavor even more quickly. They will always be safe to drink.

Not too many years ago, the metal caps on soft drink bottles had cork linings. Over time, the caps would rust and the cork would dry out. To keep contaminants out and the fizz in, the plastic-lined caps now used were introduced.

Even bottled sparkling water will begin to taste musty if it is stored more than one or two years.

After opening, carbonated drinks begin to go flat. If the container is resealed very tightly, the drink may stay fizzy for several weeks. There are a number of devices on the market that reseal soft drink bottles and cans quite effectively. The more often the cap is removed the fewer days the carbonation will last.

Powdered soft drink mixes in unopened packages can be stored on a cool, dry shelf up to two years. Once opened, they begin to absorb any available moisture from the air and, even in well-sealed containers, last no more than several months. You can keep them on the shelf or refrigerate them. Store the prepared drink, covered, in the refrigerator seven to ten days.

7

Snacks and Sweets

CHOCOLATE

An addiction to chocolate is so common and so deeply felt that enthusiasts have tried to analyze it. Does chocolate contain substances that soothe the careworn and rock the weary to sleep? Maybe so but, the way chocolate tastes, there hardly needs to be any more reason than that to crave it.

Buying Chocolate. Chocolate comes from the almond-sized bean of the tropical cacao plant. After cacao beans are fermented, dried, roasted, and hulled, the fragrant meat of the beans is ground between stones or steel blades. During grinding the beans' natural fat, cocoa butter, is separated out, leaving a dark paste called chocolate liquor. Whether they are 10-pound blocks for professional bakers or 4-ounce candy bars, chocolate products are all some form of this basic liquor.

 • UNSWEETENED CHOCOLATE. Also called bitter or baking chocolate, unsweetened chocolate is chocolate liquor hardened into cakes. It usually still contains about 50 percent cocoa butter and occasionally has vanilla or vanillin as a flavoring. Flavorings are listed on the label.

Recipes that specify "1 ounce" or "1 square" of chocolate usually mean unsweetened, often packaged as eight 1-ounce squares.

The "premelted" liquid chocolate, packaged in foil envelopes meant to replace one square of unsweetened chocolate and spare the cook the trouble of melting, is not true chocolate but a combination of cocoa powder and vegetable oils. It does not taste the same as solid chocolate.

 • SWEETENED CHOCOLATE: SWEET, SEMISWEET, AND BITTERSWEET. Sweetened chocolate is made up of between 35 and 50 percent chocolate liquor, mixed with cocoa butter and sugar. Some flavoring, particularly vanilla, may be added, as indicated on the label. Sweetened chocolate may be packaged in 1-ounce squares to be used in recipes for frosting and cake fillings, or packaged as candy.

These chocolates may taste cloying, waxy, bitingly bitter, or perfectly balanced between smooth and sharp: there are scores of manufacturers with unique recipes. You need only your own preferences as a guide, because all of these products are interchangeable in cooking when sweetened chocolate is specified.

- CHOCOLATE CHIPS. Usually chocolate chips are semisweet chocolate. They can be used as chips or melted for recipes calling simply for sweetened chocolate.
- MILK CHOCOLATE. A favorite candy, milk chocolate is 10 percent chocolate liquor, 16 percent milkfat and milk solids, and the rest is sugar, cocoa butter, and flavorings. Milk chocolate *cannot* be substituted for other chocolates in recipes.
- COCOA. Cocoa is dry, powdered, unsweetened chocolate from which some of the cocoa butter has been removed. If it is labeled "breakfast cocoa," it is rich cocoa with the highest fat content, at least 22 percent. Other types have less. Dutch cocoa has been treated to reduce its acidity; it is darker and has its own distinct taste.

Cocoa can replace unsweetened chocolate in many recipes: 3 tablespoons cocoa and 1 tablespoon butter equal 1 square.

Instant cocoa contains sugar. Hot cocoa mix contains sugar and dry milk solids. Instant cocoa and cocoa mix make hot and cold beverages but are not suitable as substitutes for regular cocoa in recipes.

- COMMERCIAL COATING CHOCOLATE. Not generally available in retail stores, commercial coating chocolate (also called couverture) is used in candy making and baking—cookies may be dipped in it, or cakes iced with a thin layer. Some commercial coating chocolate is a pure form of chocolate liquor and cocoa butter; others are compound chocolate, a mixture of cocoa and vegetable fat. Compound chocolate is not as intensely flavored but is very easy to use in intricate recipes for decorating cakes or making candy. True commercial coating chocolate must be tempered—heated and cooled to precise temperatures—or it will discolor after cooking. The deep brown of tempered chocolate should be absolutely uniform and richly glossy. Stored improperly, it can lose its stability and start to show gray streaks.

Chocolate lovers purchase commercial coating chocolate from bakery suppliers, specialty shops, and by mail order.

- CHOCOLATE SYRUP. Chocolate syrup is made of cocoa, sweeteners, and a number of flavorings and preservatives. It cannot be substituted for other chocolate in cooking.

• CHOCOLATE SUBSTITUTES. The word "imitation" usually applies to chocolate made from cocoa powder mixed with vegetable oils and other natural ingredients. On the other hand, some chips and candies bear no relation to the cacao bean at all. They will be labeled as "chocolate-flavored" or "artificially flavored."

When the surface of any solid chocolate has a grayish sheen, called "bloom," you know that it has been kept too warm or cooled and warmed during storage. Unless you plan to use it in precise, delicate candy and coating recipes, the bloom can be overlooked. The chocolate is certainly safe to eat and only the most exacting palate will notice the slightest difference in taste.

Storing Chocolate. Chocolate is sensitive to temperature. Above 78° F., not an unusual temperature in the summer, chocolate begins to melt and the cocoa butter separates and rises to the surface. When this happens, the chocolate develops a grayish-white sheen. If sweetened or milk chocolate is refrigerated, then allowed to warm, its sugar dissolves and rises to the surface, causing a similar dull white film. So, if possible, store chocolate in a cool, dry place. It should be wrapped so that it can't absorb odors from other foods. Moist air can alter the texture of the chocolate, so in warm, humid weather storing it in the refrigerator or freezer is your only choice.

On the shelf at 70° F. or in the refrigerator, dark chocolates will keep up to one year, milk chocolates up to six months, filled chocolate candy up to three months. All chocolates keep for one year in the freezer.

It is difficult to imagine but if you do have chocolate on the shelf that is nearing the end of its storage life, try using it up as shavings. Refrigerate the chocolate—sweetened, sweet, and milk chocolates work best but you can use even unsweetened—then grate with a fine grater or use a vegetable peeler to make curls for garnish.

Cocoa must be kept in a tightly lidded container so it won't absorb moisture and odors. Cocoa and instant cocoa keep up to eighteen months at room temperature. Cocoa mix containing dry milk solids keeps up to a year.

You can store unopened liquid chocolate up to eighteen months. Chocolate syrup can be stored on the shelf until it is opened. After opening, store it in the refrigerator. Its total shelf life, for best quality, opened or unopened, is one year.

CANDY

Concoctions made mostly of sugar and corn syrup with a parade of flavorings—fruits, nuts, vanilla, mint, chocolate, cream, and butter—candies inhabit the dreams of children and make nightmares for nutritionists and dentists.

All kinds of candy—hard candy, chewy ones, and creamy fondants—store and freeze well. Only a few precautions are necessary to keep candy fresh.

Buying Candy. Read the label on candy you buy in the supermarket. Some are made with a preponderance of artificial ingredients and some are not.

Other than knowing what it is you will be eating, you only have to check the candy for texture. Give the soft candies like caramels and jelly beans a little squeeze to make sure they have not hardened. Look at hard candies for signs of stickiness. They should be glossy and their wrappers should be perfectly clean, not sugary.

You can buy fresh-made candy from a reliable specialty shop and be assured that it will store just as well as prepackaged varieties.

Storing Candy. Commercially made candies should keep up to a year and homemade candies up to about a month, if they are stored properly. The only exception is divinity, which can be stored no more than a couple of days. Keep all candies in airtight containers in a cool, dry spot. Store chocolates according to the recommendations on p. 319.

Store hard, brittle candies separate from softer types. Coated candies, such as chocolate-covered bonbons, should be individually wrapped or they should be stored in a single layer.

Temperatures above about 72° F. begin to have a bad effect on most candies, so transfer them to the refrigerator or freezer in hot, humid weather. All candies keep up to a year in the freezer, as long as they are sealed in an airtight container.

Packaged fudge mixes can be stored up to a year on a cool, dry shelf.

SNACK FOODS

It is an unusually austere household with no stock of popcorn, potato chips, and other such nonessentials. As long as you are going to indulge in these crisp-and-salties, there is no point in eating them limp and stale.

Buying Snack Foods. Many snack foods come in cellophane bags, the rest in boxes. The most obvious problem to watch for is crushing. Peer into the bag, on the lookout for those sliver-sized chips that are too small for digging into dips, or shake the box and listen for their telltale rustling.

• CHIPS. Potato chips, corn chips, and cheese puffs almost invariably have a date stamped on the label. On the freshest chips, the date is about one or two months away. Try to buy chips well in advance of this date on the label, and never after.

• POPCORN. Buy plain, unpopped popcorn in a jar with a screw-top lid when you have the choice. Unpopped corn packaged in its own oil, has a dating code on the label. This type does not keep nearly as long as plain popcorn.

Popped popcorn is sold in cellophane bags that are dated. Buy it as you would chips.

Storing Snack Foods. Both chips and popcorn need careful packaging.

• CHIPS. Chips must be protected from air because they readily absorb moisture and become soggy and because air quickly turns the oil in the chips rancid. An unopened bag can be stored on the shelf until the date stamped on the label, no more than two months.

Once opened, transfer chips to an airtight jar or plastic bag. They should stay fresh one to two weeks. Ziploc bags work particularly well for chips, not only because they are easy to open and close, but they also are stiff enough to prevent crushing. In humid weather, store the jar or plastic bag in the refrigerator. There, chips easily keep up to two weeks.

Chips that are limp but still smell fresh can be made crisp again. Spread them on a cookie sheet and put them in a warm oven for a few minutes, then put them back in an airtight container. Chips that taste or smell stale have rancid oil. It won't hurt you to eat them, but they do not taste good and never will.

• POPCORN. Plain unpopped popcorn doesn't absorb moisture, it loses it. What makes the popcorn pop is the water inside the kernel expanding from heat. Popcorn that dries out won't pop. Keep it in a tightly lidded container to hold in the moisture. You can store plain popcorn a year or two at room temperature.

You will know when the popcorn begins to dry by the number of kernels in a panful that won't pop. To recondition it, put about 3 cups popcorn in a 1-quart glass jar with a snug lid. Pour 1 tablespoon water into the jar, put on the lid, and give the jar a few good shakes every five or ten minutes. When the popcorn has absorbed all the water, put the jar, still capped, on the shelf for two or three days. After that, the popcorn should pop perfectly.

Keep the kind of popcorn packaged with oil and salt on the shelf for no more than three months or until the date on the label. Store popped popcorn as you would chips.

STORAGE TIMETABLE FOR SWEETS AND SNACKS

Type	Time on Shelf	Time in Refrigerator	Time in Freezer
CHOCOLATE:			
Plain	1 year[1]	1 year	1 year
Milk	6 months[1]	6 months	1 year
Filled Candy	3 months[1]	3 months	1 year
Liquid	18 months	——	——
Syrup, Unopened	1 year	——	——
Opened	——	1 year	——
COCOA:			
Regular and Instant	18 months	——	——
Instant Mix	1 year	——	——
CANDY:			
Commercial	1 year	——	1 year
Homemade	1 month[2]	——	1 year
SNACKS:			
Chips, Unopened	2 months	——	——
Opened	1–2 weeks	2 weeks	——

Type	Time on Shelf	Time in Refrigerator	Time in Freezer
Popcorn, Unpopped Unopened	1–2 years[3]	—	—
Popped	2–3 months	—	—
Opened Popped	1–2 weeks	2 weeks	—

[1] No warmer than 70° F.
[2] Except divinity, which can be stored only 2–3 days.
[3] Store preseasoned popcorn up to date on label.

8

Kitchen Systems

WHAT MAKES FOOD SPOIL

Food spoilage and decay are not mistakes. They are the result of processes as natural as growth.

Enzymes

It is in the very nature of things that food decomposes. All plant and animal tissues contain organic compounds called enzymes which act, within each cell, as catalysts for the multitude of creations and transformations that keep a plant or animal alive. Enzymes carry on in fruits, vegetables, and grains even after harvest, and in meat after slaughter, sending signals that eventually cause the plant and animal tissues to decompose. Unchecked, enzymes break down cell structure and chemical components such as fats and carbohydrates. What begins as enormous changes in texture, color, and flavor ultimately ends in complete decay.

Certain proteins in foods also have, during storage, an innate tendency to transform into molecules that have little nutritional value. Flour, for example, that has been stored more than six months supplies less usable protein than flour that is fresh.

Environment

Rough treatment obviously spoils food before its time: a bruised peach or banana develops a bitter brown spot, resulting from enzyme reaction and from microorganisms that breach broken tissues; a cracked shell opens the egg to bacteria; a smashed potato chip is useless with clam dip.

But the seemingly passive qualities of air and light can also ruin food. Fats react with oxygen and turn rancid. Dry air can dehydrate foods: an unprotected carrot becomes limp, withered, and loses its vitamin A when moisture inside its cells evaporates. Foods that *should* be dry can absorb too much moisture from the air. Salt left out in humid air, for instance, soon clumps into a sticky mass.

A fresh potato stored in the light develops a green pigment on its skin that contains a toxic chemical. Sunlight and fluorescent light can break down one of the amino acids in fresh milk, giving it a noticeable off flavor.

Microorganisms

By far the most destructive spoilers are microorganisms—bacteria, molds, and yeasts—that invade and literally consume food before you can. Bacteria, yeasts, and molds are pervasive. They are in the air, water, and soil. All food is at some point exposed to them.

In order to use food, microorganisms must first break it down into digestible components. As they proliferate, microorganisms convert natural proteins, fats, and carbohydrates into different, simpler substances. These substances very often have foul odors and odd flavors that spoil the food for human consumption.

Microorganisms may also manufacture toxic by-products that cause illness. When they leave a trail of repulsive tastes and smells, the microorganisms and their poisons are not likely to be—and should not be—eaten. But certain bacteria produce toxins that leave no evidence that they have contaminated your food. For information on food poisoning and its control, see pp. 361–75.

• MOLDS. Molds like cool, damp places. Colonies of mold can usually be recognized by their furry appearance. They may be green, pink, blue, white, or black. They often grow on bread, fresh fruits and vegetables (especially damaged ones), cheese, and meat. Just about any food left in the refrigerator long enough will end up supporting a colorful patch of mold.

Particular strains of mold may be harnessed to produce desirable effects: the blue-green veins in Roquefort are edible mold, for example. In contrast, a few common molds manufacture poisons called mycotoxins that can be dangerous.

- YEASTS. Yeasts prefer food high in sugar. They convert the sugar into alcohol and bubbles of carbon dioxide. This tendency is useful for winemakers who want the alcohol, and for bakers who want the bubbles to raise dough. Yeasts are less welcome in fruit juice and jelly.

- BACTERIA. There are thousands of kinds of bacteria competing for our food. Most multiply in warm, moist food that is neither too acid nor too alkaline. Some prosper in cold, others in high heat; there are bacteria adapted to almost any combination of conditions.

Certain bacteria break down proteins in food, creating malodorous chemicals like ammonia, hydrogen sulfide, and the aptly named putrescine in the process. The slime on spoiled meat is a product of bacteria, as are the curds in spoiled milk. Not all bacteria destroy food. For instance, it is a species of benign bacteria that makes yogurt out of milk.

Techniques for storing food must control one or all of the factors that spoil food. The next section describes in detail what those techniques are and how they work.

WHAT KEEPS FOOD FRESH

There is no method of keeping food forever. A refrigerator will hold a peach at perfect ripeness for one or two days, while canning the peach preserves its flavor and color for many months. But nothing stops spoilage altogether.

FOOD IN THE REFRIGERATOR

Chilling food is a universal and effective way to extend its storage life.

Both enzymes and microorganisms slow down in cold temperatures. A ripe plum at room temperature turns watery, dark, and bitter within a few days. At 32° F. the same plum will still be firm and sweet after a month.

How to Store Food in the Refrigerator. Refrigeration has some tremendous advantages over other storage methods: it doesn't change the look, taste, or feel of your food, nor does it subtract any nutrients beyond those inevitably lost over time.

On the other hand, your refrigerator has some limitations that you must work around. A refrigerator cannot thwart enzymes, bacteria, yeasts, and molds indefinitely.

Another shortcoming in your home refrigerator is that the air inside is fairly dry and, like air outside the refrigerator, it is laden with microorganisms. Further, because it is in a closed box, the air picks up and holds odors.

Almost everything you put in your refrigerator needs packaging to protect it from the air. When the free water in fruits and vegetables evaporates into dry air, vitamins C and A go with it. Foods, like fresh celery or asparagus, that need humidity to stay crisp keep longer when refrigerated in plastic bags. (Storing vegetables whole, unpeeled, and uncut also prevents excessive evaporation.)

Eggs, milk, butter, and many other foods absorb any odors at large and should be kept in closed containers for protection. Conversely, strong-smelling foods such as sliced onions and cantaloupe should be wrapped very securely or even double-wrapped to trap their aromas. It helps to use an odor absorber, either baking soda or a charcoal filter designed for refrigerators.

Secure packaging is also a bar to most of those microorganisms so eager to consume the food before you do. In addition, there are a couple of steps you can take to keep down their population. Since raw meat, fish, and poultry leak juices loaded with bacteria, store them in a separate meat drawer or wrapped so drippings can't contaminate other foods. The walls, shelves, and bins in the refrigerator should be cleaned regularly with a solution of water and baking soda. Spills, crumbs, and scraps are fertile ground for yeasts, molds, and bacteria, so wipe them up immediately.

For some foods, the dry air of your home refrigerator is a benefit. Dried fruits, pasta, crackers, cereal, whole grain flour, and nuts all keep longer in the refrigerator than they do on the shelf because they don't absorb extra moisture. Furthermore, the cold slows their chemical deterioration to a near standstill. These foods still need some kind of packaging—plastic bags or lidded containers—to keep odors and molds away.

To get the most out of that box of cold air that you are paying night and day to run, always leave space around each container so air can flow freely. Pack hot foods loosely in the smallest possible amounts before refrigerating so that they can chill quickly. Densely packed food in large containers can take many hours to chill thoroughly, time enough for dangerous bacterial growth.

Other Ways to Chill Foods. An old-fashioned root cellar can hold fresh produce—root vegetables, winter squash, and apples—in large quantities all winter, but most modern houses have warm dry basements that are not too helpful for food storage. To duplicate the conditions of a root cellar, with moist air that remains constant between 32° F. and 40° F., you must do some building, either indoors or out. For more information, get "Storing Vegetables and Fruits in Basements, Cellars, Outbuildings and Pits" from the Superintendent of Documents, U. S. Government Printing Office, Washington, D.C. 20402.

Some foods need temperatures warmer than your refrigerator but cooler than your pantry. Fresh potatoes and winter squash, for example, keep best at a temperature of around 50° F. to 55° F. An unheated porch, garage, or attic may be the right spot for storing quantities of these foods for longer than a week or two.

FROZEN FOODS

Freezing is simple; it preserves food for a long time and it does relatively little damage in the process. Food that has been frozen is much closer to fresh, in appearance, taste, and texture, than food that has been preserved by canning, drying, or pickling.

What harm freezing does is a consequence of sharp-edged ice crystals that rupture cell walls, softening foods' tissues and releasing their natural juices. Delicate fruits and vegetables are more vulnerable to this kind of damage than meats and baked goods, but all food suffers some deterioration during freezing.

The faster ice crystals form the smaller, and therefore the less harm they do. Commercial methods of freezing are so efficient, the water in food crystallizes in less than thirty minutes. The same process in a home freezer takes much longer and so causes more damage. Of course, the lower the temperature of the freezer the faster freezing takes place.

Freezing food puts a stop to the growth of microorganisms: some are destroyed and the rest are completely incapacitated at such low temperatures. Enzyme activity in frozen meat, poultry, fish, and fruit is inconsequential, but enzymes in most frozen vegetables continue the aging process, so they must be destroyed by blanching—a brief immersion in boiling water or steam.

Proteins, fats, carbohydrates, vitamins, and minerals survive freezing nicely. Of course, food heated, trimmed, or chopped before freezing will lose something. Blanching vegetables, for example, greatly reduces their vitamins C, B_1, and B_2. But once in the freezer, nutrients are stable. Commercially frozen vegetables, because they are processed so soon after harvest, may actually contain more vitamins than fresh vegetables that have been refrigerated for weeks.

How to Freeze Food. Within the sections on each type of food you will find specific suggestions about preparing that food for the freezer. Carrots require blanching, for instance, and apricots freeze best packed in sugar syrup with ascorbic acid. Once the food is prepared, follow these three steps to freezing food successfully: seal the food in airtight pack-

ages; label the package so you can find and use the food within its recommended storage time; freeze the food quickly.

Sealing Food Airtight. The merits of different types of containers and
wraps are discussed on pp. 349–54. Whichever kind you choose for freezing food, it must be vaporproof. Rigid containers made of metal are
suitable, but only those glass and plastic containers designed for the
freezer and so labeled should be used. Rigid containers should have very
tight-fitting lids. Heavy-duty aluminum foil is good for freezing and so is
plastic-coated freezer paper. Plastic bags and plastic wraps are effective
only if their labels specify that they can be used in the freezer.

Don't use waxed paper, wax-coated freezer paper, butcher paper, brittle plastic containers, bread wrappers, Styrofoam containers, cottage
cheese cartons, or the plastic film used by supermarkets. None of these is
sufficiently airtight and some crack at low temperatures.

When you pack foods in rigid containers, choose the right size for the
job. A 1-quart jar storing 2 cups of soup will also contain 2 cups of air. Air
on the inside of the package can be just as damaging as air from the
outside.

A certain amount of air space is necessary, however, when you are
freezing liquids, fruits, and vegetables. These foods expand as they freeze
and need a little room to do it in. Fill a 1-pint jar no higher than 1/2 inch
below the rim. Leave 1 inch of head space in a quart-size container filled
with liquid; 1/2 inch is enough for semiliquids, plain fruits, and vegetables.

Once a rigid container is packed, close the lid as securely as you can.
Any time you have doubts about the fit of the lid, seal it all around with
freezer tape.

To protect foods packed in freezer bags, first place the food in the bag
(not too full), then lay it on a hard surface and force out all the air left in
the bag. Seal Ziploc bags by pressing the lock-top together from end to
end. More flexible plastic freezer bags should be twisted very tightly at
the top. Fold the twist over before you seal it shut with a twist tie or
rubber band. Close heat-sealed plastic bags according to manufacturer's
directions.

Wraps—heavy-duty aluminum foil, freezer paper, and plastic film—are
most appropriate for meats. Fold meats into the wrap, using one of the
techniques described on p. 180: the butcher wrap or the drugstore wrap.
The wrap should fit very snugly around the food and you should press out
any extra air before you seal the package. Use only freezer tape for
sealing. Other tapes lose their hold at freezer temperatures.

Labeling the Foods You Freeze. Every package of food you freeze needs a
label with a description of the contents, the date frozen, and a "use by"
date.

The description of the food should include what it is, how much there is, and how you have prepared it. For instance; "1 lb. ground beef patties" or "1 cup whole eggs, sugar added."

Consult the storage timetables in this book to decide on the maximum storage time you want for the food.

Here is a sample of a complete label:

June 1, 1985
1 1/2 lb. boned chicken breasts
Use by January 1, 1986

Freezing Food Quickly. To keep food as close to fresh as possible:
• Food should be no warmer than room temperature when it goes into the freezer. Refrigerate hot foods before you freeze them.
• Package food in the smallest quantities that are practical for your purposes. Small packages freeze faster than large ones.
• Place packages to be frozen directly on freezer surfaces, not on other frozen foods. Leave some space around each new package to allow air circulation.
• Freeze no more than 3 pounds of food per cubic foot of freezer capacity in a twenty-four-hour period.
• When freezing a lot of food at one time, lower the temperature control on your freezer for twenty-four hours.

How to Store Food in the Freezer. Above 0° F., enzyme activity and other spontaneous chemical changes continue, however slowly, to spoil food. Food that keeps well for one year at 0° F. deteriorates within six months at 10° F. Fluctuating temperatures that allow food to partially thaw and then to refreeze are very destructive, particularly to fruits and vegetables.

Check the temperature of your home freezer. In this book, the recommended maximum storage life of frozen foods is based on a temperature of 0° F. If your freezer is warmer than that you must reduce the storage times accordingly.

Frozen food is not immune to other kinds of spoilage. Air contact dehydrates frozen food and makes fats rancid. As frozen food dries out, it develops freezer burn, which appears as discolored, papery patches— freezer burn ruins the taste and texture of foods. Packaging must be airtight.

Careful thawing is as important as proper freezing methods. Some vitamins inevitably drip away as food defrosts, but quick thawing reduces these losses. You can cook many foods directly from the frozen state. Most commercially frozen foods have instructions on the label that tell you how. Home-frozen vegetables can go right into simmering water. Home-frozen meat, fish, and poultry require some adjustments in cooking time but can be thawed as they cook. Certain plastic bags, made to be heat-sealed, are strong enough to go from freezer to boiling water and can cook or reheat an array of foods from fresh fish to chili.

Slow thawing, accomplished in the refrigerator, provides time for food to reabsorb juices and is suitable for every kind of frozen food.

Because freezing tears into a food's cell structure, thawed, uncooked food is more susceptible to bacteria and so should be cooked or served within twenty-four hours of thawing.

Cooking for the Freezer. Almost every entree in your repertoire can be adapted for long-term storage in the freezer. Dishes that include broth, sauce, or gravy will keep particularly well: soup, stew, spaghetti sauce, casseroles, creamed chicken, hash, pot roast, chili, macaroni and cheese, and baked beans for instance.

The guidelines below will help you alter your recipes to minimize the bad effects of freezing.

When you prepare a large quantity of food, some to serve now and some to freeze, separate the two portions at an appropriate time so that the meal meant for the freezer can be cooked according to these suggestions.

• Undercook vegetables, beans, pasta, and rice. When the frozen entree is reheated, these ingredients will become just tender rather than overcooked.

• Use less garlic, pepper, and cloves than usual. Freezing intensifies the flavor of these spices.

• Use more onion and herbs, as these tend to lose flavor during freezing.

• Use little or no salt. Salt inhibits freezing and can easily be added later.

• Artificial salt, sweeteners, and flavorings do not hold up well in the freezer.

• Leave off toppings, such as cheese and bread crumbs, because they get soggy in the freezer. Add them later when reheating the dish.

• Use new potatoes and cut them into small pieces. Mature potatoes will fall apart after freezing. It is even better to add potatoes during reheating—even new potatoes lose their pleasing texture in the freezer.

• Use only regular or converted rice. Quick-cooking rice becomes very mushy after freezing.

• Leave hard-cooked egg whites out of any entree to be frozen. They are like rubber after freezing.

• Cover sliced cooked meat with sauce or gravy to extend storage life.

• Make sauces and gravies thicker than usual so they will be less likely to separate.

Once the food is ready, you must package it and freeze it rapidly:

• Cool the food quickly by setting the cooking pan in the refrigerator or in a shallow sinkful of cold water. Stir the food to cool it even faster.

• Store the food in the smallest possible meal-size packages so it can freeze quickly. Small containers also make it possible to thaw exactly the amount you need and no more.

• To make your own compact, airtight freezer container, line a casserole dish or loaf pan with heavy-duty foil. Pack the entree into the liner and seal the foil tightly, using the drugstore wrap (see p. 180). Place the pan in the freezer until the food is frozen solid. You can now remove the wrapped food to store in the freezer and return the dish to your cupboard. (To reheat, place the frozen food, still in its liner, back in the dish and warm it in the oven.)

Equally convenient are plastic bags designed for heat sealing. They are extremely airtight, take up little space in the freezer. The sturdiest bags can be boiled for quick reheating.

• Fill rigid containers no higher than 1/2 inch below the rim to allow for expansion during freezing. Many plastic or ceramic containers can go directly from the freezer to the microwave, but be sure the container seals airtight to prevent freezer burn.

Combination dishes keep well up to eight weeks at 0° F. Some loss of quality may be noticeable after three months, and after six months the entree is likely to be bland and spongy. Fried, smoked, and cured foods deteriorate even sooner—within two months.

Thawing isn't necessary for reheating frozen precooked entrees. Slip the food out of its wrappings and right into the pan or oven. Always use moderate heat, to keep the food from scorching.

In general, a 2-quart frozen casserole will take about 1 1/2 hours to reheat in a 350° F. oven. The process will take less time in a double boiler but will require more vigilance to prevent burning. Dishes that have lots of liquid, like soups and stews, can be warmed up over direct heat in a saucepan, but it is a good idea to stir them frequently. A microwave oven will reheat your frozen entrees most efficiently as long as you rotate the dish or stir the contents once or twice. Consult your microwave manual for appropriate times and oven settings.

Reheating won't always go off without a hitch—flour-based gravies and sauces usually separate during reheating. They taste smooth but look odd. The only way to deal with this is to thaw the sauce, then stir it vigorously or process for a few seconds in a blender. When the sauce is frozen in combination with other foods, you have to thaw the food, then separate out the meat or vegetables before you can stir the sauce. You have to decide whether it is worth the trouble.

Sauces made with milk or cream tend to curdle but they will still taste fine. You get the best results from sauces made with cheese. They thaw very smoothly but must be warmed over very low heat or they get rubbery.

After freezing, your sauce may be too thick, in which case you can thin it by adding a little milk or water as it reheats. If the sauce is too thin, thicken it by adding a mixture of 1 tablespoon flour (or 1/2 tablespoon cornstarch) and 2 tablespoons water per cup of sauce.

Pasta and rice sometimes absorb too much liquid during preparation. Check the pan or oven once the entree is thawed; if the food seems too dry for any reason, add a little milk or stock.

If you want to thaw the frozen food before reheating it, defrost it in the refrigerator, never at room temperature.

It is safe to use raw ingredients—meat, fish, poultry, vegetables—that have been frozen to make combination dishes for the freezer. Cooked foods, on the other hand, should not be refrozen once they have thawed.

Sandwiches in the Freezer. When you prepare sandwiches for the freezer you need to start with very fresh bread. Fine-textured and egg-enriched breads work best because they don't stale as quickly. Start by coating each slice of bread all the way to the edge with butter or margarine to keep the filling from soaking in. If you plan to use a very moist filling such as tuna salad, you can freeze the buttered slices of bread first for even more protection for the bread. Peanut butter sandwiches don't need the butter coating, but be sure to spread the peanut butter on both slices of bread to keep the jelly from soaking through.

Most any kind of sandwich filling can be frozen. You can use sliced meat and poultry; meat, poultry, and fish salads; luncheon meats; liverwurst; cream cheese; and peanut butter. A touch of mustard, ketchup, or pickle relish should come through the freezing fine too. You can't freeze hard-boiled eggs, lettuce, or tomatoes in sandwiches.

Mayonnaise tends to thin and separate when you freeze it alone. The best dressing for salad sandwich fillings is a mixture of mayonnaise and sour cream or, for the smoothest dressing, mayonnaise and cream cheese.

Double-wrap prepared sandwiches in heavy-duty foil or in freezer paper. Another possibility is to freeze fillings in small portions wrapped in foil, plastic, or packed in small airtight containers. You can then pack fresh bread separately and make up the sandwich at lunchtime.

Most sandwiches taken out of the freezer at breakfast time will be thawed by noon. Sandwiches with heavy breads and thick fillings can be moved from the freezer to the refrigerator the night before they are needed so they will thaw in time for lunch.

Slipping a frozen sandwich into a school lunch bag is a real timesaver during the morning rush. It has another advantage as well. The thawing sandwich poses very little threat of food poisoning because it spends so little time in the dangerous temperatures that encourage microbial growth.

CANNED FOODS

Canning works because, in the process, food is heated hot enough and long enough to destroy all enzymes and virtually all microorganisms it may carry. The sterilized food is sealed inside sterile, airtight cans or jars so that no contaminants can reach the food until the container is opened.

Canning drastically alters food—the food has been cooked and no longer resembles fresh, raw food. The color changes; the flavor and texture change. And commercial canners usually add salt or sugar.

The heat destroys some of the food's nutrients that would not be lost if the same food were cooked fresh, but many more nutrients are preserved than disappear, and the losses may be no more serious than losses from poor handling of fresh food.

Home Canning. Home canning is an intricate and demanding task. Unless correct procedures are faithfully followed, home-canned food can easily spoil and can even harbor dangerous poisons.

Safety depends on processing food at the right temperature for the right amount of time. Different types of food call for different treatments depending on their acidity and density. Almost all vegetables must be canned under pressure, which requires special equipment.

Commercial canners heat food to temperatures high enough to kill bacteria and their spores ten times over. The chances of a microorganism surviving the process is one in one billion. This overkill is not possible at home.

Of course many people do can at home successfully. It is not impossible, but extreme care must be taken to do it according to reliable instructions. You can further reduce the risks by learning how to identify signs of trouble in stored home-canned food (see p. 369).

How to Store Canned Food. Properly canned food is *safe* to eat as long as the container is intact.

But that is not the whole story. Canned food is not in suspended animation. Inside the container, slowly but measurably, chemical reactions continue to occur. In addition to physical deterioration in the food, any liquid inside the can leaches nutrients out of the solids. A can of crab meat, packed in brine, loses about 10 percent of its protein, about 50 percent of its niacin and thiamine, and around 75 percent of its riboflavin in a few months when stored at room temperature. Canned fruit juice loses about 25 percent of its vitamin C within a year. The changes are not all invisible. Canned asparagus not only loses vitamins, it fades to pale yellow after about one year. Canned food is distinctly less appealing after long storage.

The way canned food is stored makes a lot of difference in how long it keeps well. Stored at 67° F., canned food stays at its peak of quality twice as long as it would stored at 85° F. When storage instructions specify a "cool, dry shelf," 67° F. is closer to the mark than the 72° F. of the average kitchen.

You can plan on storing most canned food for a year if your kitchen stays cool. Asparagus, beets, citrus fruits, green beans, fruit juices, pickles, peppers, sauerkraut, mixed fruits, fruit juices, and all tomato products should be stored no more than six months. If your kitchen is warmer than 75° F., even for a short time in the summer, cut these storage times in half.

A "dry shelf" is necessary because cans rust when they get wet. A can may rust all the way through, exposing the food inside to contamination.

Light can accelerate certain chemical reactions, so food packed in jars is best kept in the dark.

Once a can or jar of food has been opened, it is perishable. The food should be stored in the refrigerator in a covered container for no more than three to five days.

DRIED FOOD

Drying preserves food because, without sufficient water, microorganisms can't function and therefore cannot spoil the food. Enzymes are relatively unaffected by drying, so in many cases food to be dried is treated with moist heat or chemicals that inhibit enzymes.

It is impossible to generalize about how dried food compares to fresh. While it is obvious that a dried apricot is darker, more leathery, more intensely flavored, and in fact altogether different from a fresh apricot, other dried foods like rice, wheat flour, oats, and breakfast cereals are for all practical purposes recognizable only in the dried form. Instant coffee, dried herbs, and seasonings such as onion flakes are dried foods widely accepted as substitutes for fresh. You may not buy dried milk or dried eggs, as is, but they are probably in your kitchen as ingredients in baking mixes, pudding mixes, and the like.

The process of drying does not in itself cause major vitamin losses in foods. But once the food is dried, exposure to air and moisture can reduce vitamins C, A, and E.

Home Drying. You can sun-dry fruits if you live where there are many days of above 90° F. temperatures and low humidity. A reliable oven that can hover between 120° F. and 140° F. is all you need for drying many fruits, vegetables, and meats. There are home dehydrators available, small appliances that make the job of drying food even simpler.

There are few dangers associated with home drying, but getting good results requires precise timing and temperatures and good storage techniques. *How to Dry Foods*, by Deanna DeLong (Tucson: HP Books, 1979), is a comprehensive reference book on the subject.

How to Store Dried Food. Keeping out air and moisture is the most important factor in storing dried foods. All dried foods should be stored in airtight plastic bags or airtight rigid containers.

Dried foods cannot be stored indefinitely. A certain amount of natural deterioration goes on, so that eventually they lose their character. High temperatures accelerate the natural decay, so store dried food in the coolest place you can. Under 60° F. is best, but you can safely store dried food in a 72° F. pantry for a short time. If you have the room, you can refrigerate dried foods to extend their storage life. Dried meat, fish, and poultry need refrigeration to keep the fats from becoming rancid. Freezing dried fruits, meats, and vegetables is not a good alternative, because it causes undesirable changes in texture.

To store home-dried food, you must first be sure it is thoroughly dry or it can mold during storage. Every time you open a container of dried food, moist air rushes in. Put home-dried food in small packages so that when you use it you will not be exposing your whole supply to the air.

Keep dried fruit and vegetables at room temperature no more than one to two months. In the refrigerator, they last six to twelve months. Dried chili peppers keep longer at room temperature, up to six months. Store dried herbs and seasonings at room temperature no more than twelve months. White flour and ready-to-eat cereals keep up to a year at room temperature. Store whole grain flours in the refrigerator up to six months.

OTHER WAYS TO PRESERVE FOOD

Adding chemicals to foods can make life difficult for invading microorganisms. Pickled foods are treated with acid, usually vinegar. Cured meats contain salt and other chemicals that not only retard microorganisms but also add flavor. Smoke from a wood fire contains antimicrobial substances that prolong the storage life of smoked meats and fish. In jams and jellies, the high concentration of sugar is most inhospitable to bacteria and yeasts.

In general, pickling, curing, smoking, and jelly making add only a little storage life, but treated foods are usually canned and therefore very stable. Cured meats may be canned or vacuum-packed for added shelf life, but most still require refrigeration.

All the means of preserving discussed so far—refrigerating, freezing, canning, drying, pickling, salting, smoking, and jelly making—can be done in the home. The food industry has many more techniques at its disposal. Take refrigeration: warehousers control the amount of moisture in their huge cold storage rooms—cabbages, for example, are stored for months at around 32° F. and 90 percent humidity. The composition of the air, as well, is adjusted to accommodate crops like apples, which keep longer in an atmosphere of about 3 percent oxygen (air normally contains 20 percent).

Commercial vacuum packs, such as the thick plastic film on bacon, is a sophisticated version of the wrapping you do at home. Vacuum packs store food in a nearly oxygen-free environment. Without oxygen, most microorganisms cannot survive and fats do not become rancid as readily. (The bacteria that cause botulism become active only when there is no oxygen. Bacon and other processed meats are treated with sodium nitrate and other chemicals that prevent the bacteria from producing deadly toxin.)

While you might use sugar, salt, or vinegar for preserving in your own kitchen, industry has at hand a vast dispensary of additives to extend foods' storage time.

Antioxidants, for instance, are a class of additives that prevent oxidation of fats and vitamins. Oxidation occurs when certain molecules react with oxygen and change form, darkening food, tainting its flavor, or reducing its nutritive value. Common antioxidants are BHT and BHA.

Another class of additives prevent spoilage by microorganisms. No one knows precisely how all these chemicals interfere with microbial growth, but they are effective. Sodium nitrate and sodium nitrite, as mentioned before, put a stop to the microbe that causes botulism. Other preservatives you are likely to see on food labels are benzoic acid, sulfur dioxide, and propionic acid. Sorbic acid added to cheese retards mold.

Anticaking agents absorb moisture without themselves becoming wet. They are added to salt and dry mixes to keep the food dry and free flowing during storage. Aluminum calcium silicate and magnesium silicate are common anticaking agents.

There are hundreds and hundreds of chemical additives that go into processed food but the majority of them do not affect storage life. Thirteen hundred flavorings are approved for use by the Food and Drug Administration. Artificial nutrients are routinely added to processed foods as are large quantities of sugar and salt.

Questions about the safety and desirability of chemical additives are outside the scope of this book. Commercial methods of treating food, including the use of many additives, can be justified to an extent since these methods preserve food that might otherwise go to waste. These methods also make available a variety of foods that would be impossible to distribute using any other means. However, if you choose to avoid additives, particularly the ones meant to artificially color, flavor, and alter the texture of food, the choice should have very little effect on storage practices in your kitchen. Cured meats without nitrates or nitrites must be stored frozen and breads without preservatives last longer in the freezer. On the other hand, soup, tuna, and applesauce canned without additives last as long on the shelf. Frozen meats and vegetables need no additives to keep well. White flour, sugar, and baking powder easily keep for a year on the shelf, just as long as commercial cake mixes that contain artificial ingredients.

A so far uncommon commercial method of preservation involves sterilizing food by bombarding it with radiation. Irradiation kills bacteria, molds, yeasts, viruses, and insects, but it also disrupts foods' natural chemistry, causing, in some cases, off flavors and odd textures. It does *not* make food radioactive. A relatively expensive process, irradiation is now in limited use in the United States almost exclusively with spices that go into prepared foods such as bottled sauces and frozen pizzas. Irradiation is a process that alarms many consumers because its effects are to some extent unknown. The usual alternative to irradiation is to fumigate spices with ethelyne oxide, a questionably safe product itself. Spices are so often carriers of dangerous microbes—salmonella, shigella, and clostridium, for example—that they must be sterilized one way or another. These naturally occurring bacteria are by no means the lesser evil.

YOUR KITCHEN IS A WAREHOUSE

After you have roasted the meat, served dinner, washed the dishes, switched off the light over the sink, and gone to bed, your kitchen is still at work. Night and day the cupboards, refrigerator, and freezer in your kitchen store your food supply. Whether you use all the food you buy or throw some of it away depends largely on how well they do the job.

Of all the things to consider when you are organizing your kitchen, the storage life of your foods is probably not the first but it should not be the last.

SHELVES, CUPBOARDS, AND DRAWERS

All you need is a thermometer to help you find the best places in your kitchen to store food. Food that would stay at its peak of quality for a year at 70° F. may begin to deteriorate after six to nine months at 78° F. The cupboards that are the coolest are the ones you should use for food.

These cupboards must be at a distance from stoves, dishwashers, dryers, and other sources of heat. Direct sunlight can heat up a small area of your kitchen briefly every day. Heat shortens the storage life of all food, including fruit left out to ripen, and light is damaging to dried herbs, garlic, potatoes, and flour among others. When you are searching for the right shelves for food, leave the thermometer in each place for one or two days to see how much the temperature rises and falls.

You want to avoid too much heat, too much light, and also too much moisture. Do not store food under the sink or near any other open plumbing. Water dripping or condensing on cans may rust them through; water on packages of dry foods can penetrate and spoil the contents.

Once you have located the places in your kitchen that are best for storing food, you will want to make the most out of the space available. Here are some ideas on how to use all the room you have:

• Mount shallow shelves on the inside of closet and cupboard doors. These shelves can hold herbs and spices, pasta, cereal, and other light-weight foods.

• Take pots, pans, and utensils out of cupboards and drawers that are better suited to food storage and put them on display. Hang pots from racks or pegboards. Gather utensils into baskets or jars on counter tops.

• Use a plastic lazy susan (Rubbermaid manufactures them in different sizes) in deep cupboards so items that would otherwise be out of sight in the back can be swiveled to the front. Rubbermaid also makes roll-out shelves that can be mounted in deep wood or metal cupboards.

• Assess the height of the shelves in your cupboards. They are often taller than you need. If you have vertical space going to waste, put up extra shelves no taller than you need to hold canned goods or other short containers. You can also find portable shelf organizers, racks shaped like miniature tables, that instantly add an extra layer of shelving.

• Use the space between counter tops and hanging cabinets. Put up racks or pegboards on the wall to store utensils. You can also buy drawers and shelving units that can be mounted on the underside of the cabinets.

• Take seldom-used equipment like ice cream freezers, fondue pots, and picnic coolers out of the kitchen and store them in less valuable space.

REFRIGERATOR/FREEZERS

Your combination refrigerator and freezer is at the heart of your food storage system. An energy-efficient refrigerator/freezer that maintains correct temperatures saves your food and saves you money.

Buying a Refrigerator/Freezer

When you buy a new refrigerator/freezer you have several decisions to make. You must choose the right size, the right style, and the right price for your situation.

Before you go shopping list all the things you like and all the things you don't like about your old one. Judge each model against the list to find the one that suits your storage habits.

When you look at the price, also look at the energy guide attached to every new refrigerator/freezer. The energy guide gives an estimate of the annual cost of operating the unit. The money you can save in electric bills may offset a higher price tag.

Choosing the right size is more than a matter of knowing how big a box will fit in your kitchen. There is no point in buying more capacity than you need and paying to cool the wasted space. Plan on 8 cubic feet of refrigerator capacity plus 1 more cubic foot per person. A family of 2 needs about 10 cubic feet; a family of 4 needs 12. The freezer compartment should measure 3 cubic feet for 2 people, plus 1 cubic foot for each additional family member. If you do a lot of entertaining you may want a little more space.

Simply measuring the inside of the compartments won't give you the exact capacity. Bins, racks, ice makers, and other appurtenances subtract from the total. You should consult a buying guide that rates the different brands and models to get an accurate assessment.

The size and shape of your kitchen will to some extent dictate the style of refrigerator you choose. There must, of course, be enough room to open and close the doors. These are the four most common types of refrigerators on the market:

• *Single-door* refrigerators are the least expensive, but the freezer compartment is enclosed in the same cabinet with the fresh food compartment and is not cold enough for long-term food storage.

• *Two-door, top freezer* refrigerators have separate doors, one to the freezer above and one to the fresh food compartment below. This configuration is particularly energy efficient.

• *Two-door, bottom freezer* units have separate doors to the freezer below and the fresh food compartment above. The height of the raised fresh food compartment makes it easier to load and unload.

• *Two-door, side-by-side* models have vertically mounted doors for each compartment. These have greater freezer capacity than the other two door models, but the arrangement of space can be limiting. For instance, a side-by-side freezer is too narrow for a cookie sheet (used for tray freezing) or a very large turkey.

• *Three-door, side-by-side* refrigerators divide the vertical freezer space into two parts, with two separate doors. The top section is meant for ice cubes and other everyday items. The bottom section, opened less frequently, stays colder and consequently keeps food longer.

Refrigerator/freezers come with a whole array of features, only a few of which directly affect the storage life of your foods:

• *Bins, drawers,* and *crispers* are fairly standard. The ones designed for vegetables hold in moisture and keep out drying air flow. They do keep vegetables fresh longer. Meat drawers may have their own temperature controls, allowing you to store fresh meat between 28° F. and 35° F. This deep chilling can add several days to the storage time of meat, poultry, and fish. Roll-out baskets in a bottom-mounted freezer compartment help organize packages and bring them quickly into view so you don't stand there with the door gaping while you hunt for the frozen peas.

• *Shelves.* Adjustable shelving is worthwhile because you can arrange the refrigerator to hold exactly what you need to store from day to day and on special occasions.

Plastic shelves are easy to clean and prevent drips from coursing down all the way to the bottom of the box, but you can't see through them, so food may get lost. Wire racks allow more air circulation and make the whole compartment easy to see but don't stop drips. Glass is transparent, easy to clean, and dripproof but can break.

• *Egg Racks.* These are a waste of space since eggs should be stored covered. If you can't avoid them, make sure they are removable.

• *Dual Controls.* When you have separate thermostats for the freezer and the refrigerator compartments, you can set them to maintain the optimum temperatures in changing circumstances. It is sometimes impossible to adjust single-control models to stay at 0° F. in the freezer and 37° F. in the refrigerator all at the same time.

Using Your Refrigerator/Freezer

A refrigerator/freezer does not require a lot of attention, but there are things you can do to help it operate more effectively and more cheaply.

Buy two of the inexpensive thermometers designed for refrigerators and freezers and put one in each compartment. When you know exactly what the temperature is, you can adjust the thermostat to lower it (to save food) or raise it (to save electricity).

Leave the doors closed as much as possible to keep warm air out and cold air in.

Keeping the freezer full helps it to work more efficiently. If the freezer is not frost free, defrost it when the ice on the walls is more than 1/4 inch thick. (Directions for defrosting are on p. 349.)

Wash the refrigerator surfaces regularly with a solution of water and baking soda to clean and deodorize the inside of the cabinet.

Occasionally check the rubber gaskets around each door. Close the doors on a piece of paper. If the paper slips out easily, the gasket is loose and leaking and should be replaced. Keep the gaskets clean.

At least once a year vacuum underneath and behind the refrigerator to remove dust from the operating parts.

Also read "How to Store Food in the Refrigerator" (p. 330), "How to Freeze Food" (p. 332), and "How to Store Food in the Freezer" (p. 334).

FREEZERS

Having a freezer separate from your refrigerator has a number of advantages. The temperature in a separate freezer unit can be significantly colder than in a freezer that is part of a refrigerator. These lower temperatures make it possible to freeze foods that taste better and store longer. Because of the unit's relatively large capacity, you can buy and freeze more food.

The purchase price and cost of operating a separate freezer unit, on the other hand, may well outstrip any savings you gain from freezing inexpensive foods.

Buying a Freezer

Once you have decided to buy a freezer, your selection hinges on the size, the type, and the cost.

In deciding on the size, allow 3 cubic feet of freezer capacity for each member of the family, allow another 2 cubic feet per person for special purposes, and add 3 to 5 more cubic feet if you plan to buy wholesale meat cuts, freeze homegrown produce, or regularly prepare main dishes for freezing. Don't buy a freezer that is too large for your needs, because they run much less efficiently when they are not full.

There are two styles of freezers, chest and upright, each with advantages and disadvantages to consider.

Chest freezers are shaped like a trunk and they open at the top, which gives them their first advantage. Since cold air is heavier than warm, any time you lift the lid of a chest freezer, the cold air stays inside instead of flowing out. Chest freezers have no shelves or coils inside the cabinet, making it a simple matter to store bulky or odd-shaped packages. The air at the bottom of the cabinet does not circulate, which reduces freezer burn on the foods. Chest freezers are less expensive to buy and to operate than uprights.

Arguments against a chest freezer are that it takes up a lot of floor space; that food must be stacked instead of shelved; that packages are below waist level and therefore difficult to find and to reach. Chest freezers are difficult to defrost manually and they are not manufactured with an automatic defrost feature.

Upright freezers, which look like refrigerators, have shelves that make it easy to organize and to locate packages, and coils in the shelves help to quick-freeze foods. Uprights occupy less floor space than chests and are less tricky to defrost. Many upright freezer models defrost automatically. Compared to chests, uprights are more expensive to run and a certain amount of space is wasted above each shelf. Cold air escapes rapidly every time the door is opened.

All in all, a chest freezer is a good choice for a basement or an out-of-the-way location, and a good choice when you want a freezer for long-term storage and for foods bought in bulk. An upright is more convenient for everyday use.

When you have decided on the style of freezer you want, you should compare not only the price of different models but also the operating expense. Each freezer sold is tagged with an energy label that estimates its annual operating cost and compares that cost to other models.

Using Your Freezer

Place your freezer where the air temperature is moderate, never going below freezing or above 110° F. Locate it away from ovens, laundry equipment, furnaces, and other heat sources.

An upright freezer should have at least 4 inches headroom; all freezers should have 3 inches of clearance on all sides, including the back side, to allow for air circulation and easy cleaning. Consider raising a chest freezer a few inches up on blocks so you can more easily drain water off when you are defrosting.

At least once a year, vacuum around and under the base of the freezer and, if it is exposed, around the motor housing.

• DEFROSTING YOUR FREEZER. If your freezer defrosts manually, defrost it regularly. More than 1/4 inch of frost on the freezer wall decreases the efficiency of your freezer, reduces storage space, and raises the temperature inside the cabinet.

Before complete defrosting, set the thermostat at its coldest for about an hour to get the food as cold as possible before turning off the unit. Remove all the food and insulate it with blankets or newspapers to slow thawing. You can pack your picnic cooler with frozen foods too.

Scrape frost off the surfaces of the freezer with a plastic, rubber, or wooden instrument that can't scratch or puncture. Set pans of hot water in the freezer to speed up the melting.

Wash walls and shelves with a solution of warm water and baking soda to clean and deodorize. After defrosting and cleaning, thoroughly dry all surfaces. Any moisture left turns into new ice when the freezer goes back on. Towel dry all packages as you return them to the cabinet or they will ice up and stick together.

When the food is replaced, put the thermostat at its lowest setting and turn on the freezer. Leave the door closed as long as possible. When the temperature inside is where you want it, return the thermostat to normal.

Also read "How to Freeze Food" (p. 332) and "How to Store Food in the Freezer" (p. 334).

STORAGE CONTAINERS AND WRAPS

Even if you have done everything else right—bought food at its freshest and kept it at exactly the right temperature for exactly the right time— your storage system won't work unless you have sturdy, reliable containers and wraps. The storage life of fresh meats, fresh vegetables, fresh fruits, all the foods you cook and save, even flour and sugar and more, depend upon the packaging you use to protect them.

Packaging must keep air, excess moisture, and microorganisms out of your food and keep odors and desirable moisture in. To do this, the material must be vaporproof (meaning air cannot pass through it) and moistureproof.

Rigid Containers

Rigid containers with their own tight-fitting lids are the most useful for storage. A lid is preferable to a wrap or other covering on the shelf and in the refrigerator, but mandatory in the freezer.

The lids that give you the best seal are plastic lids with flanges that you press down over the rim of the container and any lid with a rubber or plastic gasket. Tupperware, for one, makes lids which, when pressed firmly, expel air from the container and also lock tight.

You should have rigid containers in enough different sizes so that you will always have the right one for the job at hand. A container that is too large allows too much air space around the food inside and takes up too much valuable storage space.

• One- and two-cup-capacity containers are good for freezing fruits and vegetables, and for freezing and refrigerating leftovers, serving-size entrees, sauces, gravies, and nuts. On the shelf, they are good for dried fruit, nuts, tea, candy, cornstarch, powdered sugar, and other foods that need protection from humidity.

• Quart containers are needed for juice, coffee, meal-size entrees, whole grain flours, and salads in the refrigerator or freezer. On the shelf, you can use tightly lidded quart containers for storing crackers, cookies, rice, cereals, dried beans, noodles, and many pastas.

• You can use even larger containers for flour, sugar, snack chips, and long, thin pasta.

The shape of a storage container has a bearing on its usefulness, too:

• Square containers take up the least amount of space and are easy to stack when they are in use and when they are themselves in storage.

• Straight-sided containers are easy to pack and to unpack, especially when the food inside is frozen.

For some, but not all, uses transparent containers have an advantage:

• Transparent storage containers help you to find foods in a hurry. When you can see what is inside it you are less likely to lose track of a package.

• Some food, such as flour, dried fruits, and dried herbs, need protection from light and are better off stored in opaque containers.

Plastic Bags

Plastic bags take up almost no space and conform readily to odd-shaped foods. There are several different types and they all come in various sizes. A securely sealed plastic bag can store all the same foods a rigid container can and also help to keep fresh fruits and vegetables longer in the refrigerator. Liquids, however, can be hard to handle in plastic bags and only those plastic bags specifically designed for the freezer are able to protect frozen foods. Putting very hot food into most plastic bags may melt them.

A big drawback to using plastic bags in the freezer is that, when packed, they don't hold a shape, so they are difficult to stack. There are paperboard cartons on the market that help you circumvent this problem. You line the stiff carton with a plastic bag, put in the food, then stack the carton in the freezer. These cartons are not airtight or moistureproof and are not suitable for storing food without plastic bag liners.

A number of brands of plastic bags are very soft and pliable, like Baggies and Glad Food Storage Bags. They adapt well to the shape of poultry, meat, and fish. They can be slipped over boxes of raisins, crackers, or other prepackaged foods that need extra protection. They are pretty sturdy but sharp bones or sharp corners on boxes may cause tearing. You can seal them tightly with the twist ties provided. Expel all the air from inside, twist the top of the bag closed, fold the twist over, and secure it with the tie or with a rubber band. These bags can be washed and reused. They also have the advantage of disposability, if leftovers should spoil.

Ziploc and A&P brand make stiffer plastic bags that hold their shape and are available in freezer-weight plastic. They are especially good for bunches of bite-size foods, cookies, crackers, chips. The top of the bag, when pressed firmly together, makes an airtight, watertight seal. These bags are reusable as well.

Boilable bags are specialized products more expensive and harder to find than other plastic bags. They are designed particularly, but not exclusively, to hold frozen foods. The bag of food can be taken right out of the freezer and heated in boiling water or in a microwave oven. They can be very time- and space-saving for freezing large quantities of fruits and vegetables and for freezing main dishes in serving sizes (homemade TV dinners). The heavy-duty plastic has been found safe by the FDA. A special heat sealer, that can cost anywhere from $10 to $70, closes the bag airtight and watertight. Some heat sealers have a device that draws excess air from the bag, creating a partial vacuum. Reducing the amount of air in the bag greatly enhances the quality of frozen food and extends storage life both in and out of the freezer. The vacuum- and heat-sealed bags help keep everything, from nuts, cookies, potato chips, flour, spices, and cereals to leftover creamed spinach, fresher on the shelf and in the refrigerator.

Many large plastic garbage bags are treated with chemicals that make them unfit for storing food. If the label does not say that you can use a bag for food, do not use it for food.

Wraps

Wraps are lightweight, flexible, and waste-free. You use only as much as you need.

• Aluminum foil can be folded into any shape. Using the drugstore wrap (p. 180), you can seal anything almost airtight. You can use foil to fashion a loose-fitting lid on open containers. Because it is slightly stiff, it provides some protection to fragile foods such as pies. Aluminum foil is apt to tear, however, and only heavy-duty aluminum should be used for freezing foods. Salty and high-acid food may cause the foil to darken and to develop tiny holes, but this can only happen when the food is in a non-aluminum metal container, such as stainless steel. Should this corrosion accidentally occur, the food is still safe to eat.

• Plastic wrap has all the advantages of aluminum foil and it is transparent, so you can see what is inside. Some brands stretch and stick, so they can form a tight seal over a bowl or other open container. Many brands are too flimsy to use as freezer wrap, tear easily, and get brittle at very low temperatures. Saran Wrap works best in the freezer. Read the label to see if your plastic wrap is meant for the freezer. You must use freezer tape to seal plastic-wrapped packages in the freezer. Extra-wide rolls are best for freezing meat, poultry, and other bulky foods.

If you are concerned about polyvinyl chlorides in plastic wrap, use Glad Cling Wrap or Handi-Wrap II instead of Saran Wrap or Reynolds. However, neither Glad nor Handi-Wrap brands are suitable for the freezer.

The PVC in Saran Wrap and Reynolds can leach into foods during microwaving. To prevent this, keep the plastic away from direct contact with the food and poke holes in the wrap to allow steam to escape.

• Freezer paper is thick enough to resist tearing and is suitable for wrapping meat, fish, poultry, and baked goods for the freezer. It must be plastic- rather than wax-coated and must be sealed with freezer tape. Freezer wrap does not mold as well as plastic or aluminum but, since it can be purchased in 18-inch widths, it is sometimes the best choice.

• Waxed paper is inexpensive but not as versatile as other wraps. In the refrigerator, it is ideal for wrapping fresh meat, fish, and poultry loosely and, sealed with tape, it protects other foods well enough. It should not be used to wrap foods for the freezer.

STORAGE CONTAINERS AND WRAPS COMPARED

Type	Benefits	Comments
RIGID CONTAINERS:		
All types	Reusable; good for liquids; prevent crushing	Unused containers take up storage space; need lids
Glass	Transparent	Must use annealed glass only in freezer: check label
Plastic	Unbreakable, lightweight	Will melt in high temperatures; some absorb color and odor; most nonflexible plastic not for freezer: check label
Metal	Unbreakable	May be corroded by salty and high-acid foods
Ceramic	Various sizes and shapes; some types go freezer to oven	Loose-fitting lids; some types crack in freezer: check label
PLASTIC BAGS	Lightweight, compact; fit odd-shaped foods; reusable (except boilable type)	Packed bags difficult to stack; some not for freezer: check label; high temperatures will melt nonboilable types.
WRAPS:		
All types	Lightweight, compact; fit odd-shaped foods; no waste: use just amount of material required; fit tight: leave no air space in package	Use drugstore wrap to seal food; secure with freezer tape
Aluminum foil	Holds shape well for stacking; lightproof; can write on package	Tears easily; use only heavy-duty type in freezer; can corrode; use drugstore wrap

Type	Benefits	Comments
Plastic wrap	Transparent	Some types not suitable for freezer
Freezer paper	Resists tears; can write on package	Use plastic-coated only: plastic side toward food
Waxed paper	Inexpensive	Not suitable for freezer

TRAVELING WITH FOOD

The same principles you apply to storing food safely in your kitchen extend to carrying it in your car, your boat, or on your bicycle. Keep hot foods hot. Keep cold foods cold. Keep the place where the food is stored clean and free from food spills.

Storage Equipment

The key to traveling with food is keeping it at the right temperature. You need some kind of insulation and, most often, ice for chilling. Having the right tool for the right job helps enormously.

• COOLERS. For many occasions, a Styrofoam cooler is quite adequate. These are the very inexpensive, lightweight coolers stacked in supermarkets and drugstores all summer long. In the house, they are useful to have on hand to keep food cold when you clean or defrost the refrigerator or freezer; or they can serve as extra refrigerator space during a party. In the trunk of your car, they make shopping trips easier in hot weather. Ask the grocery clerk to put all your cold foods in a separate bag, then pack them in the cooler on the way home.

For long-distance traveling, Styrofoam coolers have a few annoying defects. They have grips but no handles to make them easier to carry. The lids come off easily, so spills can occur. Worst of all, they squeak, a noise that is maddening inside a car.

You avoid these problems with a mid-priced molded plastic cooler. The one with handles on top can be carried and opened with one hand. The coolers with side handles have flat lids, making them easy to stack. The molded plastic containers costing between $10 and $20 have a 2- to 4-gallon capacity. These and the larger models are sometimes outfitted with beverage holders and trays to hold food above your ice.

Larger molded plastic coolers, from 4 to 9 gallons in capacity, cost between $20 and $30. Coleman and Thermos coolers of this size have metal bands for extra strength. Most brands have drains for emptying water that accumulates in the bottom of the cooler.

There are many larger coolers, but they are not necessarily better. Two 6-gallon coolers are more portable than one 12-gallon giant that gets very heavy when fully loaded with ice and food. But if you need one, there are 12- to 20-gallon molded plastic coolers available from $40 up. Because it is all plastic, the 86-quart Igloo cooler, fully loaded with 200 pounds of ice and fresh-caught fish, can stand up to the battering from a fishing boat's lurches and rolls without denting.

Plastic coolers often have a strange smell when they are new or absorb food odors as they are used. To deodorize one, wash it out with a solution of baking soda and water or soak a tiny piece of cloth in vanilla extract and leave it in the closed cooler overnight.

Many coolers are damaged by hot foods; some coolers cannot tolerate certain cleaning products. It is important to read the manufacturer's instructions to know just how to use and care for your cooler.

• INSULATED BAGS. For those times when a cooler is too cumbersome, you can use insulated bags. They are one step down from coolers in effectiveness and one step up in convenience. These soft, flexible bags that range in size from a large purse to a small suitcase are perfect for traveling on a plane or bus.

• ICE PACKS. Inside your cooler or insulated bag, you need ice or its equivalent to keep food chilled. There are two alternatives to plain ice cubes. Igloo manufactures bottles you can fill with water, then freeze at home. In your cooler, the ice-filled bottle chills food and, as the ice melts, provides you with fresh, cold drinking water.

Freeze paks are sealed plastic containers filled with a nontoxic gel. When frozen, the gel stays cold longer than regular ice. The containers come in many sizes to fit large or small coolers.

Both the bottles and the plastic containers eliminate the problem of water building up in your cooler as happens when loose ice cubes melt.

• THERMAL AND VACUUM BOTTLES. To keep coffee hot or lemonade cold when you are traveling, nothing works like a vacuum bottle. Thermos is the best-known manufacturer, but there are others who make these bottles that insulate by means of a vacuum between two sealed inner liners (usually glass) protected by a plastic or metal outer shell.

Thermal bottles are simply insulated with a thickness of plastic that is a poor conductor of heat. Thermal bottles do not work quite as effectively as vacuum bottles, but they come in a wider variety of sizes, including 3-gallon jugs and pint-size, wide-mouth jars suited to storing soups and stews.

Read all the instructions that come with your vacuum or thermal bottles so you know what they can and can't do. Vacuum bottles, for instance, may be damaged by carbonated liquids and the glass linings may break (they are usually replaceable).

• PLASTIC BAGS. If you had no other piece of equipment, you could successfully pack a picnic with nothing more than plastic bags. Even with a full supply of coolers and vacuum bottles, plastic bags are essential for traveling with food.

Heavy-duty Ziploc freezer bags and heat-sealed plastic bags are especially useful because they are completely leakproof. You can fill one with water and make your own ice pack. You can pack fruits in any brand of plastic bag to keep them from rolling around. You can pack cookies in plastic bags to keep them moist. When you have eaten all the cookies, the plastic bag can start holding trash. You can put damp cloths in a plastic bag and use them to wash your hands and face while you are in transit.

Packing Food to Travel

When you are preparing to travel with food, safety should be uppermost in your mind. Improper storage temperatures are a common cause of food poisoning.

• START WITH FOOD THAT IS THE RIGHT TEMPERATURE. Food should be very hot or very cold before you pack it. Salads with mayonnaise dressings, cold cooked meat, fish, and poultry, eggs, milk, custards, and cream pies, all must be chilled through before you pack them for travel. You can freeze sandwiches in advance, as described on p. 337.

Hot drinks, soups, and stews should be close to boiling when you seal them in insulated containers. For best results, warm vacuum or thermal bottles with a rinse of hot water before you pour in hot liquids. To keep cold drinks cold longest, refrigerate them overnight in their vacuum or thermal bottle.

• PACK IT FULL TO KEEP IT COLD. A cooler completely packed with ice and chilled food keeps colder longer than one that is only half full. Pick a cooler the right size for the occasion. Fill it about one quarter full of ice or gel-type freezer packs, then pack it to the top with cold food. After many hours, you may need to add more ice.

The ice does not have to stay at the bottom. You can pack around the sides of the food containers, particularly those that need the most chilling, such as salads, meats, and so on.

There are certain foods, like fresh fruit or vegetables, that you want cold but not touching the ice. Spread a flattened paper bag over the ice and stack these foods on top of it.

• PROTECT AGAINST BREAKAGE. Pack fragile foods in rigid containers. They take up more space than plastic bags but do protect the food. When you pack glass bottles and jars, apply strips of masking tape, vertically, on all sides of the container. If the glass should break, the tape prevents the pieces from scattering. Cushion glass containers with towel and paper napkins as you pack them.

9

Kitchen Crises

FOOD POISONING

WHAT IS FOOD POISONING?

There are three very important facts about food poisoning that every cook should know:

1. Food poisoning is not uncommon. Because the symptoms are usually mild and often mistaken for the flu, accurate statistics on the number of food-borne illnesses are nearly impossible to gather, but health experts estimate that one of every ten Americans suffers a bout of food-borne diarrhea each year.

2. The bacteria (called pathogens) that cause the vast majority of food-borne illnesses are virtually undetectable. Foods contaminated with dangerous bacteria can look, smell, and taste perfectly appetizing.

3. Improper storage and handling of foods *in home kitchens* cause the majority of food-borne illnesses.

There are three types of bacteria most often identified as the cause of food poisoning. When you understand how these pathogens operate—how they get into food and how they grow there—you can learn to control them. ("Ptomaine poisoning," by the way, is almost always a misnomer. Ptomaines exist. They are poisonous chemicals produced when plants or animals rot, but they are so obnoxious that no one would ever eat them by accident.)

Two sets of events must occur before these bacteria can make a person sick: *contamination* and *multiplication*.

Contamination occurs when the food comes in contact with the bacteria. Unfortunately these microbes are everywhere. They lurk on the skin and in the intestines of the animals we consume as meat; they cling to the surface of fruits and vegetables; they are on your hands and in your nasal passages; they even attach themselves to the dust floating through the air of your kitchen. Because they are so widespread, bacteria can easily contaminate food. But simple contact between food and pathogens is not enough to bring about the conditions that cause food poisoning.

The bacteria have to become active and multiply before they are dangerous. You can consume these pathogens in small numbers and they will have no effect on you at all. They make you sick only when you eat a lot of them at once. In order to grow and multiply, these pathogenic bacteria require moisture, warmth, a neutral environment (meaning neither too acid nor too alkaline), and time.

Flour is too dry to support the growth of the pathogens. It has less than 15 percent moisture content. A frozen chicken may carry bacteria but they are inactive at freezer temperatures. Fruits are generally too acidic and vegetables too alkaline to support the bacteria. An example of a hospitable setting for bacteria is ground beef set out to thaw at room temperature. The meat is moist and is neither too acid nor too alkaline. As it thaws, the temperature on the outer parts of the ground beef soon rises to the point that permits bacteria to grow. The three most common pathogens multiply at temperatures between 45° F. and 140° F. Room temperature is comfortably within this danger zone. But the warmth alone does not precipitate multiplication. The bacteria still require time to increase to dangerously high numbers.

Once the disease-causing bacteria find themselves in agreeable surroundings—nicely moist, neutral food at a comfortable temperature—they begin to grow furiously after about four hours. It is at this point that a single bacteria cell can turn into millions in a short space of time. When there are large numbers of bacteria in food, the food will make people sick.

SALMONELLA

Salmonella is one of the three types of bacteria that most often cause food poisoning.

Salmonellae come from the intestines of humans and animals. The bacteria are frequently present in raw poultry, meat, and eggs. Raw poultry, meat, seafood, and eggs can harbor salmonellae even after those foods have been frozen and, in the case of eggs, even after they have been dried.

In addition to being present in raw foods as they come from the market, salmonellae can be transferred to cooked foods by human hands and insect pests.

Salmonellae are readily killed by high heat, so proper cooking is essential to controlling the kind of food poisoning they cause.

Here is a sequence of events that would lead to salmonella food poisoning: a cook sets a raw chicken on a kitchen chopping board and cuts it up into parts for stewing. The chicken has come from the market with a few salmonellae microbes that end up on the chopping board. After the chicken has simmered for a couple of hours any salmonellae on the meat itself are destroyed by the heat. But the cook places the chicken parts back on the unwashed chopping board to slice the meat from the bone. The bacteria on the board reattach themselves to the warm meat. If the meat were served immediately, there would be little danger of illness because the salmonellae on the board would have had no more than two hours at room temperature, not enough time to multiply into millions. But in this case the meat is left on the board to cool for an hour, then made into chicken salad. The salad eventually goes into the refrigerator but does not cool down to below 40° F. for another two hours. Enough time has passed to allow the salmonellae a period of rapid growth.

The cook eats the cold chicken salad that night. The salmonellae cause what is called "food infection." The bacteria enter the body's alimentary canal and grow. During their life cycle inside the body the salmonellae produce toxins, substances that are poisonous and cause symptoms such as abdominal pain, nausea, diarrhea, headache, and fever. These symptoms develop twelve to thirty-six hours after the food is eaten and continue from one day to one week.

CLOSTRIDIUM PERFRINGENS

C. perfringens is a type of bacteria found in soil, dust, and human and animal intestines. Flies frequently carry large doses of *C. perfringens*.

These particular bacteria form heat-resistant spores that can survive most cooking processes. When conditions are right, the spores develop into active bacteria that readily multiply.

A case of *C. perfringens* food poisoning might happen this way: the cook prepares a large vat of beef stew to serve the next day. The spoon used to stir the food is resting on the stove top and a contaminated fly lands there, transferring *C. perfringens* to the spoon. The spoon dips back into the stew and now the stew contains a few spores. When the stew is finished, the cook leaves the vat out to cool for a couple of hours. The big pot of stew finally goes into the refrigerator, but a few more hours pass before the densely packed food cools sufficiently. During these hours the bacteria have had the time and the warmth to become active and to increase in number. Reheating the stew before serving may not destroy the bacteria that have grown—some types of *C. perfringens* survive for hours at the boiling point.

The cook eats the stew and the *C. perfringens* establish themselves in the alimentary canal. After eight to twenty-two hours the bacteria produce toxins that cause abdominal pain, nausea, and diarrhea. The symptoms may continue from twelve to forty-eight hours.

STAPHYLOCOCCUS AUREUS

About half the population carry staphylococcus organisms in the nose and throat and on the skin and hair. Skin infections usually contain staph.

Certain foods are more likely than others to support the growth of staph. They are cooked meats, cream-filled pastries, custards, egg dishes, stuffing, and gravy.

The scenario for staph food poisoning might be something like this: the cook slices warm ham and prepares sandwiches for the day's outing. A couple of sneezes is enough to contaminate the meat. The wrapped sandwiches sit inside a paper bag at room temperature half the day. By the time the cook eats the sandwich the staph has multiplied and formed toxins in the ham that quickly produce symptoms of illness. Within two to six hours staph toxins cause severe vomiting, abdominal pain, and diarrhea. The illness can persist for up to twenty-four hours.

CLOSTRIDIUM BOTULINUM

Clostridium botulinum causes the deadly disease botulism. Other types of food poisoning are very uncomfortable but rarely dangerous. Botulism has a fatality rate of about 65 percent.

The organism is found in soil and water. It is harmless until it encounters oxygen-free conditions, such as those inside canned, bottled, and vacuum-packed foods, or in the center of densely packed sausages, pastes, and cheeses. In these environments the bacteria form a toxin so virulent that even the smallest dose can be fatal. One milligram (1/28,000th of an ounce) is sufficient to cause the death of 8,000.

Botulism is rare, and when it does occur it is most often traceable to improperly processed home-canned vegetables. Unlike other agents of food poisoning, *C. botulinum* makes its presence known. Watch for warning signs in home and commercially canned foods, especially mushrooms, spinach, peas, beets, green beans, corn, meat, and fish. A can with bulging ends or a jar with a bulging lid should be thrown out—don't even open it. Discard badly misshapen cans of food. If a can or jar is leaking or if it spurts liquid when it is opened, throw it out. Canned food with a strange odor or an off color should be thrown away immediately. If you open a can of food that looks or smells suspicious, *do not taste any of it.* Throw it away.

Symptoms of botulism usually begin within twelve to thirty-six hours after the toxic food is eaten. They include vomiting, diarrhea, blurred vision, and difficulty with breathing and swallowing. Immediately contact a doctor if these symptoms appear. Early treatment of botulism increases the chances of survival.

OTHER KINDS OF FOOD-BORNE ILLNESS

Not all food-borne illnesses are caused by bacteria. The disease trichinosis comes from pork infected with a parasitic roundworm. Heat will destroy the parasite, which is why every recipe for pork calls for thorough cooking. Pork should be heated to at least 160° F. before serving.

Noxious chemicals may contaminate foods stored in containers made of unlined copper, lead, or zinc. Improperly glazed ceramic dishes can impart metallic poisons to food as well. Storing fruit punch in a galvanized metal bucket, for instance, can cause illness. Heavy-metal poisoning is often signaled by stomach upset and a metallic taste in the mouth.

Monosodium glutamate in large doses (1.5 grams or more) can cause headache and a burning sensation on skin of the chest, neck, and abdomen.

How to Prevent Food Poisoning

According to a study done by the Centers for Disease Control, the five most common factors leading to cases of food poisoning are:

1. Inadequate cooling.
2. The lapse of a day or more between preparing and serving food.
3. Infected persons handling foods which are not subsequently heat-processed.
4. Inadequate time or temperature or both during heat processing.
5. Insufficiently high temperatures during storage of hot foods on steam tables, buffets, etc.

There are a number of methods of handling and storing food that, if practiced routinely in your kitchen, can greatly reduce the chances of food poisoning.

Keeping dangerous bacteria out of our food is the first line of defense against food poisoning.

Here is a list of ways to keep large numbers of contaminants from getting into your food in the first place:

Wash your hands frequently while you are cooking.

• Wash them before you start cooking and wash them after you have handled raw meat, poultry, seafood, or eggshells.

• Wash your hands after using the toilet or diapering the baby.

• Wash your hands after contact with the family pet.

• Cover your face with a tissue when you cough or sneeze, discard the tissue, then wash your hands.

• Use paper towels or clean cloth towels to dry your hands.

Keep surfaces clean.

• Wash cutting boards and counter tops immediately after they have been in contact with raw meat or poultry.

• Every few days, wash wood cutting surfaces with bleach diluted in water or scrub the entire surface with a wire brush.

• Store raw meat, fish, and poultry in the refrigerator wrapped in such a way that juices cannot leak onto other foods. When defrosting meat, fish, or poultry, place it on a plate that catches all the drippings. Wipe up raw meat juices that spill in the refrigerator.

• Wash food particles from counter tops, shelves, stove top, and refrigerator racks with detergent and warm water.

• Take measures to eliminate flies and crawling insects from the kitchen. (See pp. 370–75 for specific information on pest control.)

Use kitchen utensils for only one job at a time.

• Wash all knives used to cut raw meat and poultry before they are used again.

• Use different spoons to stir raw and cooked foods.

• Taste foods with a different spoon than you are using to cook and use a clean spoon for each taste.

• Use different containers for storing and serving food. For instance, don't feed your child baby food directly out of the jar, then store the leftovers in the same jar. Use a clean spoon to measure servings into a dish.

Sanitize all your dishes and other kitchen equipment.

• Use detergent and hot water to wash dishes, pots, and pans. Rinse them in extremely hot water. A dishwasher uses water hot enough to sanitize the dishes.

• Wash every kind of kitchen gadget—slicers, grinders, processors, blenders, mixers, can openers, and so on—with detergent and hot water after every use.

• Use freshly laundered kitchen towels. A towel used to wipe up spills or food particles can harbor bacteria.
• Regularly clean kitchen sponges and dishrags in a solution of bleach and water.

Keep food clean.

• All poultry should be presumed guilty of carrying salmonella. Rinsing any bird inside and out will help reduce the number.
• The soil in which vegetables grow is laden with microbes, both safe and unsafe. Scrub or wash raw vegetables scrupulously.
• Before you eat fresh fruit and vegetables, rinse them in water to wash off dirt and pesticide residues.
• Store all foods in clean containers on clean shelves, both in the pantry and in the refrigerator. Never store food near drainpipes.
• Wipe dust and dirt from cans of food before opening.
• Keep prepared foods stored in the refrigerator in covered containers.

Even if your kitchen is immaculate, there is still a chance that contaminants will find their way into your food. The only danger this presents arises when pathogens are given the chance to multiply and to produce the toxins that cause illness.

Controlling the temperature of foods is the single most effective method of preventing food-borne illness. Bacteria that cause food poisoning are most active in food between the temperatures of 45° F. and 140° F. When food is held for longer than four hours, including preparation, storage, and serving time, it must be kept either colder than 45° F. or hotter than 140° F.

Heating and reheating foods to sufficiently high temperatures is another important preventative. Temperatures above 165° F. destroy most bacteria.

Keep hot foods hot.

• Hot foods held for later serving must be kept above 140° F. Warming ovens, hot plates, and chafing dishes may not be able to keep food hot enough, so use them with caution.

Use electric slow cookers only as recommended in the manufacturer's operating manual. Some electric slow cookers have temperature settings below 145° F. These settings are for special purposes, such as warming rolls. Do not cook foods at these low settings. If the cooker's markings are unclear, use a thermometer to make sure foods are heated to at least 150° F. within three hours and maintained at that temperature.

Heat foods to the proper temperature.

• Cook poultry to an internal temperature of at least 170° F. Use a meat thermometer to verify temperatures or use the maximum cooking time suggested in reliable recipes.
• Heat pork to an internal temperature of at least 160° F.
• Heat stuffing, either inside meat and poultry or cooked separately, to at

least 165° F. (Pack stuffing very loosely in the meat and poultry. This allows for more rapid heating.)
• Allow extra cooking time for frozen or partially frozen meat and poultry to be sure they reach the correct temperatures.
• Cook meat and poultry thoroughly, to the correct temperature, before storing them. Never partially cook meat or poultry, then finish cooking it later.
• To reheat broths, gravies, soups, and stews, bring them to a boil for several minutes.
• Carefully follow instructions for time and temperature when heating commercially prepared foods.

Chill foods rapidly.

• Put foods to be chilled in the refrigerator immediately. Leaving hot foods out to cool at room temperature only increases the time the food remains in the temperature danger zone (45° F. to 140° F.). Hot foods will not raise the temperature of your refrigerator significantly.
• When possible, store leftovers to be chilled in the smallest container that is practical. Hot food chills more quickly when it is stored loosely packed.
• Before storing, transfer leftovers from hot cooking pans to cooler storage containers.
• When you have an extremely large amount of very hot food, like a vat of stew, that must be chilled, first partially submerge the pot in a sinkful of ice water, then stir the hot food for several minutes. When it stops steaming, refrigerate the food in small, cool containers.
• Always store poultry and stuffing separately. A bird packed with stuffing takes a very long time to chill. The stuffing may stay warm for hours, even in the refrigerator. Wait to stuff the bird until the moment it is ready to go into the oven. After roasting, remove all the stuffing from the bird for separate storage.

Keep cold foods cold.

• Make sure your refrigerator maintains a temperature of 40° F. or lower. Buy a thermometer and use it to help you regulate your refrigerator's thermostat.
• Do not thaw food at room temperature. Freezing does not destroy all bacteria. If the food was contaminated going into the freezer, it will be contaminated coming out.
Frozen foods can be either cooked directly from the frozen state or thawed in the refrigerator. The only other safe way to thaw frozen food is to seal it inside a plastic bag and submerge the bag in cold water, no warmer than 50° F.
• Thoroughly chill prepared foods that can't be held in the refrigerator, such as potato salad for a picnic, or a sandwich in a bag lunch, before they

leave the house. If it is practical, keep prepared foods on ice while they are away from the kitchen. Never let foods sit at room temperature for more than four hours, including preparation time. (See p. 337 for information on freezing sandwiches and pp. 354–57 for suggestions on traveling with food.)

• When you go to the grocery store, plan your trip so that cold and frozen foods make it back home quickly.

Handle home-canned foods with care.

• Obtain reliable canning instructions and follow them explicitly as to proper equipment and processing times and temperatures. Good sources include "Home Canning of Fruits and Vegetables," a pamphlet available from the Superintendent of Documents, U. S. Government Printing Office, Washington, D.C. 20402; and Ruth Hertzberg et al., *Putting Food By,* 4th ed. (Lexington: Stephen Greene Press, 1988).

• Discard cans and bottles that leak or bulge.

• Discard cans or bottles with mold on the seal or on the contents.

• Discard cans or bottles with tiny bubbles in the contents.

• Discard cans and bottles of food that have an off odor, an odd color, cloudy liquid, or a strange texture. *Do not taste suspicious foods.* Discard the container and contents in such a way that no traces of the food are left on your hands, on kitchen surfaces, or any place that is accessible to pets or children.

• High heat completely destroys the toxin that causes botulism. Boiling home-canned meats, poultry, fish, corn, spinach, and mushrooms for 20 minutes and other vegetables for 10 minutes will eliminate the toxins. Fruits are usually, but not always, too acidic to support growth of the pathogen.

It is impossible to track down and to kill every microorganism in your food and in your kitchen. It is possible to hold them at bay with the methods described here. Remember it this way: keep foods clean; keep hot foods hot; keep cold foods cold.

CONTROLLING HOUSEHOLD PESTS

Insects outnumber people by immeasurably large numbers. Many of them eat the same food we do. It is hardly surprising that they manage from time to time to slip into our homes through some neglected crack in the wall or hidden inside a bag of flour. Don't be paralyzed with embarrassment the first time you find something black and crawly in the cupboard. There are things to do to get rid of cockroaches, ants, weevils, silverfish, and even the common housefly.

You want to get rid of these insects quickly, not only because they are repulsive, but also because they carry disease-causing bacteria. Once they have had time to lay eggs in hard-to-reach places, to reproduce, and generally to make themselves at home, it is doubly hard to evict them.

The first step toward getting rid of insect pests is to decide which insecticides you will need. Insecticides are chemicals that come in various forms for use in the home. Surface sprays are designed to coat walls, shelves, drawers, floors, and other surfaces with a liquid chemical that will stay in place for weeks, killing any insect that crawls there. Surface sprays come in pressurized containers or as liquids that can be used with a hand sprayer. Only a coarse, continuous spray action will work to coat a surface densely enough. The finer spray of aerosol insecticides will not effectively cover a surface.

Aerosol insecticides, or space sprays, work best on mosquitoes, houseflies, and other flying insects. They won't keep working for more than a few moments after they leave the can, so they are good for zapping an insect you can see but not for long-term protection. Spraying aerosols into tight corners and hidden cracks can drive out cockroaches, but you will need a surface spray (or a mallet) to actually kill them.

Pesticidal dust, applied with a hand-held duster, does two jobs. It will coat surfaces and it can be blown into hard-to-reach places.

Insecticides also come in cream and paste form. You can use a paintbrush to apply this type very accurately in places where spot treatment is needed.

Pesticides are poison. Before you use an insecticide, read the label from top to bottom, every word. These chemicals can be poisonous to your family and your pets if they are mishandled. They can contaminate the food you are trying to protect. Some damage plastics and flooring materials. Use them with the greatest care.

The different insects you are likely to find in your home each have different habits and hiding places. They are susceptible to different insecticides. You must recognize which pest you are dealing with before you can hunt down every last one and kill it.

Cockroaches. Cockroaches seem to have been designed specifically to elude man's every effort to fend them off. They will eat almost anything— starch, glue, paper, fabric—but they wait until dark to do it. They have hard, crushproof bodies that can survive all but the most aggressive blows. They are fast runners, too. If you find a flat bullet-shaped insect, brown or black, from 1/2 to 2 inches long, with long whiplike antennae, you are probably looking at a cockroach. They stay hidden during the day, but you may be able to catch them out by tiptoeing into a dark kitchen and flipping on the lights.

You will need a surface spray or dust insecticide. Buy one that states clearly on the label that it works against cockroaches. An inexpensive home remedy is to apply boric acid in place of a commercial product, but treat it with as much caution as you would stronger chemicals.

You don't want the poison to get into your food, food containers, dishes, or silverware, so clear out cupboard shelves and drawers before you begin to spray. Carefully vacuum or brush out shelves and cupboards to remove all food particles from the corners and cracks. Examine the food packages to see if any insects are lurking there. Put any infested food into a plastic bag, wrap it tightly, and throw it away.

Apply enough insecticide to coat all the places where the cockroaches may be found:

- Beneath the kitchen sink and drainboard.
- In cracks around cupboards and cabinets.
- On inside and outside surfaces of cupboards and cabinets, particularly in upper corners.

- Along exposed pipes or conduits and in cracks where they pass through the wall.
- Behind window and door frames, where they meet the wall.
- Along loose baseboards and molding strips.
- On the undersides of tables and chairs.

Cockroaches can make themselves comfortable in all the rooms of the house. If you suspect they have spread, spray carefully in the bathroom, especially around the pipes. Spray closets, bookshelves, and other areas where they may be feeding on paper or fabric.

Wait for the insecticide to dry before you put your kitchen back together. Line the shelves and drawers with paper so the food and dishes won't come into direct contact with the residual spray. Transfer foods that attract cockroaches, such as flour, cereal, crackers, and pasta, to glass jars with screw-top lids.

If the cockroaches have really gotten a foothold in your home, you may have to repeat these procedures a number of times. These insects are adept at hiding eggs in inaccessible places, so you may kill off one generation of pests only to be faced with another.

If repeated attempts fail to rid your home of cockroaches, call a professional exterminator.

Ants. Ants behave quite differently from cockroaches. When they find a ready food supply, such as a tempting pile of cake crumbs, on the counter, they notify their central nest and soon a long wavering line of marchers stretches from nest to crumbs, each ant shouldering a morsel for the trip back. They don't hide in the dark and they don't run from people, they just single-mindedly carry off all the food they can reach.

When you spot a column of ants, don't follow your impulses and start swatting. If you sneak up behind them and follow the ants to where they are hauling the food, you will have a good idea where to find the nest. The nest may be outdoors, beneath the floor, under a pile of papers, or just tucked into a corner. Destroying the nest will get rid of the ants. If you can't locate the nest, you can still spray all the surfaces where you have found the ants crawling.

Use a pesticide with a label that specifies it is effective against ants. Apply it to all surfaces where you have actually seen the ants. If you have really pinpointed the ants' line of march, you can apply liquid or paste insecticide with a brush in just that area. This will mean less upheaval in the kitchen, because you won't have to clear every last pantry shelf in advance. With a surface spray you must remove food, dishes, and utensils from the kitchen before you apply it (then put down paper before you reshelve things). Spray the nest if you can. Otherwise, spray openings that are likely entryways, such as:

- Cracks around baseboards, sinks, bathtubs, toilets, and cupboards.
- The lower edges of windows and door frames.
- Posts, pillars, and pipes under the house that might act as runways for the ants.
- Openings in the wall around electrical outlets and around plumbing and heating pipes.

After a few days the ants should be gone. More ants mean you need to search out more places to apply the insecticide.

Weevils. The tiny, wiggly creatures that infest flour, cereal, spices, and other dry foods in your pantry are usually called weevils. Weevils smuggle themselves into your house inside packages of crackers and other dry foods. If you spot them before they migrate to the next package on the shelf, it's easy to dispose of them. Just drop the one infested product into a plastic bag, seal it tight, and throw it away.

If the weevils have spread in the cupboard, sterner measures are called for. Take everything off the cupboard shelf and inspect it. If it is a can or bottle, wipe it off—any little dusting of flour or crumbs will attract weevils. If it is a dry food in an opened container—flour, noodles, baking mix, rice, crackers, nuts, cereal, oatmeal, cornmeal, nuts, dried meat, dried fruit, popcorn, spices, and so on—look for crawling insects inside and out. Discard infested food.

If you can't see the bugs, it is still possible that the product contains tiny eggs or larvae. To be on the safe side, you can sanitize foods with heat or cold. Prepared foods like crackers and breakfast cereal can go on a pan in the oven for 20 minutes at 150° F. Other products, especially baking mixes, can lose quality when heated, so put these into the freezer for three to seven days, depending on the size of the container.

Meanwhile, scrub down your pantry shelves with detergent and hot water. Be sure to remove any food particles jammed into cracks in the cupboard.

If you have done all this and the weevils reappear, repeat the whole procedure and add two steps. After you have washed the shelves, spray them with an insecticide designed for weevils. Put paper down before you return food to the cupboard.

Flies. Flies breed and feed on every kind of filth and garbage. When they crawl on the food, tables, and counter tops in your home, they leave behind microorganisms that can cause disease.

Screens on your windows and doors keep out most flies. Keeping garbage cans tightly lidded and cleaning up pet droppings from the yard will eliminate likely breeding spots.

A few flies can be swatted, but flies in greater numbers can be treated with insecticide. Use a space or aerosol spray designed for flying insects. Follow directions for use on the label. Do not use the spray around uncovered food, including pet food. Do not use No-Pest strip in the kitchen, but old-fashioned fly-paper works and it's safe.

Silverfish. Silverfish are shiny gray crawling insects that feed on sugar, flour, and other starches. They can be controlled by the same methods described for cockroaches.

Once you have triumphed over an invasion of household insects, there are many steps you can take to prevent another battle with the bugs.

- Use soap and water. Frequently clean all the hidden corners where insects like to hide—pantry shelves, baseboards, water pipes, kitchen drawers, and cracks between cupboards and walls. Don't wait to wipe up crumbs and spills in any food storage area.
- Carry out the garbage as often as possible. Outdoors, keep it in tightly lidded containers.
- Store food on pantry shelves in closed containers. Keep containers wiped clean on the outside.
- Before you buy dry foods at the market, check for signs of infestation. Be suspicious of any breaks or tears in the packaging.
- If you bring groceries home in a cardboard box, throw out the box right away. Cockroaches find these cartons very attractive; they may be hiding inside.
- Caulk all the cracks and crevices in the walls and cupboards where insects have been most numerous. Even fat little cockroaches can squeeze into a space no greater than the thickness of several pieces of paper.

Insects are not the only creatures that will invade your house in search of food. Rodents are bold and hungry gate-crashers who slip into your home through small, unseen holes in the floor and walls and foundation. Like insects, they spoil your food and carry disease.

Mice. If you spot a mouse or its droppings, go out and get a mousetrap. Bait it with peanut butter, chocolate candy, cake, nuts or, yes, cheese. Put the trap on the floor at a right angle to the wall, directly in the suspected path of the animal. You are very likely to catch it. There are humane traps designed to capture mice so you can simply carry them out of the house. These will not injure the mouse but may teach him a lesson.

In cases where there are many mice in residence, you may have to use poison bait. This is an unpleasant alternative because poison bait left lying around is potentially dangerous to children and pets. As with other pesticides, poison bait should be handled precisely as the label describes.

Rats. Rats are grim enemies that not only damage property and bear disease but also bite people.

Poison bait is the most effective way to kill rats in the home. Rats are a bit too large and wily for most traps. If your own attempts to destroy invading rats don't work, you should engage a professional exterminator without delay or apply to your local board of health for help.

Once they are gone, there are a number of things you can do to keep mice and rats from coming back:

- Close up the holes in your house. Check the basement, the outside foundation, interior walls and floors for gaps. Seal up any holes larger than 1/4 inch. Examine doors and windows to see that they shut tight and that the frames fit snugly.
- Deposit all garbage into tightly covered metal cans.
- Keep basements and storerooms neat and clean. Stack boxes, paper, lumber, and other supplies on shelving at least 1 foot above the floor.

All these pests, insects and rodents alike, may persist despite your best efforts. This is the time to call on outside help. A professional exterminator can supply more tools to kill off the invaders as well as specific advice about where in the house your particular problem might lie.

POWER FAILURE

When your home suffers a power failure, one of the first problems you confront is what to do with all the food in the refrigerator and freezer. If the food is allowed to thaw and to warm up above 45° F., it will begin to spoil and, worse, it will begin to sustain the growth of bacteria that can cause illness.

What to Do with the Refrigerator and the Freezer in a Power Failure. You have little to be concerned about when the power failure is brief. There might be a blown fuse in the house that can be repaired right away. A thunderstorm might knock out service for no more than an hour before the lights go back on. A fully loaded freezer stays cold up to two days. Half loaded, it can stay cold about one day. The larger and colder the freezer the longer it will keep. You need to take only one precaution. *Don't open the door.* Foods in the refrigerator stay chilled during a power failure up to about six hours, as long as you leave the door closed. Don't open it even for an instant. To keep foods cold longer, place bags of regular ice on the upper shelves. You should put some pans beneath the bags to hold the melting ice. The more ice you put in the longer the temperature will stay cold. Open the door only when you are adding more ice. This is the time you will congratulate yourself for owning a thermometer for your refrigerator. When you are adding ice or when the crisis is finally over, you can check the temperature in various parts of the refrigerator to make sure that the most perishable items—milk products, meat, fish, poultry, mayonnaise, and cooked foods—have been kept below 45° F.

To ensure that there will be no damage to the appliance's electrical system, you may want to pull the plug until power is fully restored. Uneven surges of power can cause burnout.

A freezer failure that you know will last longer than one or two days demands action. There are two ways to go. You can transport your food to a working refrigerator/freezer or you can get hold of some dry ice to cool down your own.

When you stand to lose a lot, you will probably want to find a place to store your food safely until the emergency is over. Try to find space for your frozen foods at a friend's house first. If that fails, look for meat and frozen food warehousers in your area. Look in the Yellow Pages under Warehouses-Cold Storage. Even if the commercial cold storage company is affected by the power failure, their freezers are large enough to maintain low temperatures for days. To move the frozen food, bundle it with newspaper or pack it in insulated cardboard boxes.

The alternative is to keep your food where it is and to keep it cold with dry ice. The ice company, listed in the Yellow Pages under Dry Ice, can advise you about the quantity of dry ice you will need for your freezer. The maximum you can expect is that 25 pounds of dry ice in a fully loaded 10-cubic-foot freezer will keep food frozen for three days.

Before putting dry ice in the freezer, cover the food with heavy cardboard. Put the ice on the cardboard directly on top of the food. Always wear gloves while you are handling the ice. Once the ice is in place, don't open the freezer again unless it is to add more ice.

What to Do with the Food After a Power Failure. When the crisis is over, you should examine all the food stored in the freezer. In general, foods that still contain ice crystals are safe to refreeze and foods that have defrosted completely should not be refrozen.

Squeeze packages wrapped in plastic or cardboard without opening them. If you can feel lots of hard crystals of ice and hear them crunch together, the food is only partially thawed. Slabs of meat should still have a hard central core of ice.

You will have to open rigid containers to inspect the contents. Again, look for evidence of lots of firm ice crystals throughout the food.

Vegetables. Partially thawed vegetables can be refrozen. Date the package so you use it within two weeks. Fully defrosted vegetables should be cooked immediately or discarded. Discard fully defrosted vegetables packed in butter or cream sauce. Fully cooked vegetables may be refrozen.

Fruits and Fruit Juices. Highly acidic fruit juices such as citrus and berry can be safely refrozen even if they are fully defrosted. Other fruits should be treated the same as vegetables.

Meat. Beef, veal, lamb, and pork that are still firm can be refrozen. Date the packages and use them within two weeks. An alternative is to cook the meat, preferably in a casserole or stew, and refreeze it that way. Meat that is fully defrosted but still cold should be cooked immediately. If the meat has been fully defrosted for more than twenty-four hours, or if it has any odd odor or color, discard it.

Poultry. Whether partially or fully thawed, poultry should be cooked immediately and either served or refrozen. Discard any stuffed poultry.

Variety Meats. Liver, sweetbreads, and other organ meats should be cooked immediately and not refrozen.

Fish. Cook fish immediately, whether or not it is partially or fully defrosted.

Prepared Entrees. Cook partially or fully thawed entrees immediately and do not refreeze.

Dairy and Egg Dishes. Eggs, custards, cream pies, cream sauces, ice cream, and similar foods should be discarded. Milk, butter, and cream should not be refrozen. Use them within twenty-four hours if they are fully defrosted.

Baked Goods. Bread, unfrosted cakes, uncooked piecrusts, fruit pies, and cookies can be refrozen. Bake doughs immediately.

Use all refrozen foods within two weeks. Throw out any fully or partially thawed foods with suspicious odor or color.

INDEX

Abalone, 239
Acidophilus milk, 138
Acorn squash, 53, 54
Additives, chemical, 341–42
Allspice, 289
Almonds, 121–22, 127
Aluminum foil, 352, 353
Amaranth, 249
Angel food cake, 269, 270, 272
Annato, 289
Animal fat, 221, 280, 283
Anise, 289
Anjou pear, 106
Antibiotics
  in meat, 176
  in poultry, 216
Anticaking agents, 341
Antioxidants, 341
Ants, control of, 372–73
Appenzeller cheese, 159,
  163
Apples
  canned, 79–80
  cider, 308–9
  ethylene gas in, 10, 21,
    73
  freezing of, 74, 79
  juice, 308, 309
  ripening of, 77
  selection of, 78–79
  storage of, 79, 118
  varieties of, 77–78
Apricots
  canned, 80
  dried, 116
  fresh, 80, 118
Arrowroot, 275, 279
Artichokes, 40–41, 64
Arugula, 15, 16, 64
Asiago cheese, 161, 163
Asian pear, 81, 118
Asparagus, 5, 39–40, 64
  canned, 40, 66, 339
Aspartame, 288
Avocado oil, 281
Avocados, 81–82, 118

Baby lamb, 190
Bacon, 199–200
  drippings, 281, 283
  nitrite-free, 207

Bacteria
  and food poisoning, 361–65
  and food spoilage, 329
Baked desserts
  cake, 269–70, 272
  cookies, 270, 272
  frozen, 268, 269
    in power failure, 378
  pastries, 271–72
  pie, 271, 272
  prepacked, 268–69
  selection of, 268–69
  white flour in, 246–47
Baking powder, 275–76, 279
Baking soda, 276, 279
Bamboo shoots, 62–63, 64
Banana squash, 53, 54
Bananas, 73, 82–83, 118
Barbecue sauce, 298
Barley, 253, 254
Barley flour, 248, 250, 251
Bartlett pears, 106
Basil, 289–90
Batter breads, 263, 264
Bay leaves, 290
Bean curd (tofu), 49–50, 65
Bean threads (cellophane noodles),
  255
Beans
  broad, 48, 64
  dried, 50–52, 66
  and flatulence, 52
  green, 9, 47–48, 64
    canned, 66, 339
  lectin content of, 52
  lima, 48, 51, 64
  long, 48, 65
  storage timetable for, 64, 65
  wax, 47–48, 64
Beef
  baby, 191–92
  corned, 205, 207
  freshness of, 186
  grade of, 185–86
  ground, 187–88. *See also* Ground
    meat
  jerky, 206, 207
  liver, 194, 196
  stock, 192
  storage of, 186, 196
  See also Meat

Beer cheese, 158
Beets, 5, 22–23, 64
    canned, 22, 66, 339
    greens, 14, 22
Bel Paese cheese, 158, 163
Belgian endive, 11, 12
Bermuda onions, 60
Biscuit dough, 263, 264
Bitter melon, 34, 64
Black beans, 50
Black-eyed peas, 50, 64
Black tea, 305
Black walnuts, 126
Blackberries, 83–84, 118
Blanching vegetables, 8
Blood orange, 102
Blue cheese, 151–52, 160–61, 163
Blueberries, 84–85, 118
Boilable bags, 351
Bok choy, 43, 64
Bologna, 207
Bonbel cheese, 157, 163
Bones
    meat, 192–93, 197
    poultry, 214–15
Bosc pear, 106
Botulism, 341, 354–65
Bouillon cubes and crystals, 276, 279
Bouquet garni, 290
Boursin cheese, 157
Brains, 194
Bran, 248, 250, 251
Brazil nuts, 122, 127
Breads
    crisp. See Crackers and crisp breads
    crumbs, 262
    dough, frozen, 263, 264
    hard-crusted, 262–63, 264
    selection of, 260–61
    soft-crusted, 261–62, 263–64
Brick cheese, 158, 163
Brie cheese, 153, 156, 157, 163
Broad beans, 48, 64
Broccoli, 5, 37–38, 64
Brown rice, 257, 258
Brown sugar
    reviving, 286
    storage of, 285, 288
Brussels sprouts, 38, 64
Buckwheat, 253
Buckwheat flour, 248, 250, 251
Buffalo, 195

Butcher wrap, for frozen meat,
    180
Butter
    clarified, 141, 149
    grading of, 140
    margarine combination, 142
    salted, 140
    storage of, 141, 149
    unsalted, 140
    whipped, 140
Butter cake, 269
Buttercup squash, 54
Butterhead lettuce, 10, 65
Buttermilk, 137, 140, 149
Butternut squash, 54

Cabbage, 34–35, 64
    Chinese, 35
    history of, 34
    sauerkraut, 35
    sulfur compounds in, 34–35
Cakes, 269–70, 272
Calf's liver, 194
Camembert cheese, 154, 156, 157,
    163
Canadian bacon, 200, 207
Candy, 320–21, 323
Canned foods
    botulism, 364–65, 369
    canning process, 338–39, 369
    fruit, 76, 119
    storage of, 338–39
    vegetables, 6, 7, 66
    See also names of food
Cantaloupes, 85–86, 118
Capers, 290
Capon, 215–16
Carambola (star fruit), 113, 119
Caraway, 290
Cardamom, 290
Carrots, 5, 20–21, 64
    cake, 270
    juice, 21
Casabas, 86–87, 118
Cashews, 122, 127
Cassava, 24–25, 64
Cassia, 290
Cauliflower, 36–37, 61, 64
Caviar, 231–32, 239
Cayenne, 290
Celeriac (celery root), 25, 64
Celery, 5, 41–42, 64

Celery cabbage, 35
Celery root (celeriac), 25, 64
Celery salt, 290
Celery seed, 290
Cellophane noodles, 255
Cèpes, 55
Cereals
  instant hot, 253, 254
  ready-to-eat, 252–53, 254
  storage timetable for, 254
  whole grain, 253, 254
Chanterelles, 55
Chard, 5, 42–43, 65
Chayote, 53
Cheddar-type cheese, 154, 158–59,
    163
Cheese
  blue, 151–52, 160–61, 163
  cheesemaking process, 151–53
  crèmes, 156–57, 163
  firm, 158–60, 163
    Cheddar-type, 158–59
    Dutch, 159–60
    Swiss-type, 159
  food, 162
  freezing of, 154, 155, 157, 158,
    160, 161, 162
  fresh, 154–55
  hard, 161
  mold on, 153–54
  processed, 161–62, 164
    French, 162
  puffs, 322–23, 324
  selection of, 152–53
  semisoft, 157–58
  soft, 156–57
  spread, 162
  storage of, 153, 163–64
  See also names of cheese
Cheesecake, 271, 272
Chemical additives, 341–42
Cherimoya, 87, 118
Cherries, 87–88, 118
  sour, 88
Chervil, 290
Cheshire cheese, 158, 163
Chestnuts, 122–23, 127
Chicken
  contamination of, 363
  fat, 280, 283
  freezing of, 217
  selection of, 215–16

  smoked, 224
  storage of, 216–17, 223
  types of, 215–16
  See also Poultry
Chickpeas (garbanzos), 50, 64
Chicory, 11–12, 64
Chiffon cake, 269, 270, 272
Chiffon pie, 271, 272
Chili (hot) peppers, 30–31, 61, 65,
    66
Chili powder, 290
Chinese cabbage, 35, 64
Chips
  chocolate, 318
  snack, 322–23
Chives, 290
Chocolate
  chips, 318
  commercial coating, 318
  milk, 134, 318, 319
  shavings, 319
  storage timetable for, 323
  substitutes, 319
  sweetened, 317–18, 319
  syrup, 318, 319
  unsweetened, 317
Chops
  storage of, 181, 197
  thawing of, 182
Chutney, 61
Cinnamon sticks, 290
Clams, 232–33, 240
Clarified butter, 141, 149
Clementine oranges, 102–3
Clostridium botulinum, 364–65
Clostridium perfringens, 363–64
Cloves, 290–91
Cockroaches, control of, 371–72
Cocoa, 318, 319
Coconuts
  fresh, 123, 127
  shredded, 123, 127
Coffee
  beans
    fresh-ground, 302, 303, 304
    green, 303, 304
    selection of, 301–2
    whole roasted, 303, 304
  brewed, 304
  canned, 300, 302–3, 304
  decaffeinated, 303
  flavored, 303–4

Coffee *(cont.)*
  freeze-dried, 303, 304
  instant, 303, 304
  reheating, 304
Colby cheese, 158, 163
Cold cuts, 204–5, 207
Collard greens, 14
Comice pear, 106
Conch, 239
Condensed milk, 135–36, 150
Condiments, 297–99
Confectioners' sugar, 285
Conserve, fruit, 117
Containers
  freezer bags, 9, 180, 181, 333, 351–52, 353
  for fruit, 76
  for meat, 180–81
  metallic poisons in, 365
  rigid, 9, 333, 349, 350, 353
  in travel, 354–57
  for vegetables, 9
  See also Wraps
Converted rice, 258
Cooked food, freezing of, 335–37
  commercially prepared, 184
  meat, 183, 197
  in power failure, 378
  recipe alteration for, 335
  reheating, 336
  thawing, 336–37
Cookies, 270, 272
  dough, 270, 272
Cooking staples, 275–79
Coolers
  packing of, 356–57
  types of, 354–55
Coriander, 291
Corn
  canned and frozen, 46
  cream-style, 47
  freezing of, 46
  freshness of, 45–46
  history of, 45
  storage timetable for, 64
Corn chips, 322–23
Corn oil, 142, 281, 283
Corn syrup, 286, 288
Corned beef, 205, 207
Cornmeal, 248, 250
Cornstarch, 276, 279
Cortland apple, 77

Cottage cheese, 154–55, 163
Crab, 234, 240
Crab apple, 77
Cracked wheat, 248
Crackers and crisp breads
  crumbs, 266
  packaging of, 265
  storage of, 266–67
Cranberry (ies), 88–89, 118
  juice, 309
Cranberry beans, 50, 64
Crayfish, 235, 240
Cream, 138–39, 149
Cream cheese, 155, 163
Cream puffs, 269
Cream of tartar, 276, 279
Crème fraîche, 145, 149
Crèmes (cheese), 156–57, 163
Crenshaw melons, 89, 118
Crepes, 263, 264
Criminis, 55
Crookneck squash, 52
Cucumbers, 31–32, 64
Cumin, 291
Cured meat. *See* Processed and cured meat
Curly endive, 12, 64
Currants
  dried, 116
  fresh, 89–90, 118
Curry powder, 291
Custard
  frozen, 146
  pie, 269, 271, 272
  storage timetable for, 149

Daikon (Japanese radish), 26
Dairy products, *See* Cheese; Milk; Milk products
Dairy substitutes
  contents of, 147
  creamers, 147–48
  dessert toppings, 148, 150
Dandelion greens, 14
Dasheen, 26
Dates, 90, 118
Decaffeinated coffee, 303
Derby cheese, 158, 163
Desserts
  baked. *See* Baked desserts
  frozen, 146, 149
  nondairy toppings for, 148, 150

Devil's food cake, 270
Dill, 291
Dill pickle, 61
Double Gloucester cheese, 158,
    163
Dough
    bread, 263, 264
    cookie, 270, 272
Doughnuts, 261, 264
Dressings, 297
Dried food
    fish, 230
    fruit, 115–17, 119, 339, 340
    herbs, 293–95
    home drying, 339
    legumes, 50–52
    meat, 206
    milk and milk products, 132, 136–
        37, 150
    peppers, 30
    potatoes, mashed, 19
    storage of, 340
    tomatoes, 29
Drugstore wrap, for frozen meat,
    180–81
Duck, 220–21, 223
    fat, 221, 280, 283
    See also Poultry
Dutch cheeses, 159–60, 163

Eclairs, 269
Edam cheese, 154, 159–60,
    163
Eels, 231, 240
Egg noodles, 255, 256
Egg whites
    freezing of, 170
    storage of, 168–69, 171
Egg yolks
    freezing of, 169–70
    storage of, 169, 171
Eggnog, 137, 149
Eggplant, 32–33, 64
Eggs
    freezing of, 169–70
    freshness of, 165, 166, 167
    hard-cooked, 169, 170, 171
    selection of, 166–67
    storage of, 168–69, 171
    substitutes, 170
Elderberries, 85, 118
Emmenthal cheese, 159, 163

Empire apple, 77
Endive, 11–12, 64
English peas, 44, 45
English walnuts, 126
Enok mushrooms, 55
Enzymes
    in food spoilage, 327
    and vegetable storage, 3–4
Escargots, 239
Escarole, 11–12, 64
Ethylene gas, in fruit, 10, 21, 71,
    73
Evaporated milk, 135, 150
Extracts, 276–77, 279

Farmer's cheese, 155, 163
Fats and oils
    animal, 221, 280, 283
    rendered, 221, 224, 281
    reusing, 282–83
    selection of, 280
    storage timetable for, 283
    types of, 281–82
Feijoa, 90–91, 118
Fennel
    fresh, 43, 64
    seed, 291
Fenugreek, 291
Feta cheese, 157, 163
Fiddlehead ferns, 14–15, 64
Figs
    dried, 116
    fresh, 91, 118
Filberts, 124, 127
Filé, 291
Filled milk, 148
Firelli pear, 106
Fish
    canned, 230
    caviar, 231–32, 239
    dried, 230
    fat, 239, 240 n.
    fillets, 226–27
    freezing of, 228–29, 239–40
    fresh-caught, 227
    freshness of, 225
    frozen, 225–26, 227
        in power failure, 378
        storage timetable for, 239–40
        thawing, 229
    lean, 239, 241 n.
    pickled, 230

Fish (cont.)
  refrigeration of, 227–28, 239–
      40
    cooked fish, 229, 240
    smoked, 229–30, 240
  steaks, 227
  variety fish, 231
  whole, 226
  See also Shellfish
Flies, control of, 373–74
Flour
  processing of, 245
  storage of, 249–51
  types of, 246–49
Fontina cheese, 159, 163
Food
  chilling of, 368
  cleaning of, 367
  heating and reheating, 367–68
  spoilage, 327–29
  staples, 275–79
  See also names of foods
Food poisoning
  botulism, 341, 364–65
  causes of, 365
  Clostridium perfringens, 363–64
  contamination–multiplication
      process in, 361–62
  metal poison, 365
  prevention of, 211–12, 365–69
  salmonella, 362–63
  Staphylococcus aureus, 364
  trichinosis, 365
Food preservation
  canning, 338–39. See also Canned
      foods
  commercial methods of, 340–41
    additives in, 341–42
    irradiation in, 342
  drying, 339. See also Dried food
  in power failure, 376–78
  refrigeration, 330–31, 368–69
  in root cellar conditions, 331–32
  in travel, 354–57
  See also Freezing; Frozen food
Food storage
  of canned foods, 338–39
  containers and wraps for. See
      Containers; Wraps
  shelves, cupboards, drawers in,
      343–44
Frankfurters, 207

Freeze packs, 355
Freezing
  advantages of, 332
  containers and wraps in, 9, 75–76,
      180, 333, 349–54
  cooked food, 335–37
    commercially prepared, 184
    meat, 183, 196–97
    recipe alteration in, 335
    reheating, 336
    thawing, 336–37
  fish, 228–29
  food preparation for, 332
  freezer selection and operation,
      347–49
    defrosting, 349
    of refrigerator/freezer, 345–47
  fruit, 74–76
  herbs, 295
  labels, 333–34
  meat, 179–81
  rapid, 334
  sandwiches, 337
  sugar and syrup packs in, 75
  temperature for, 334
  tray freezing, 9, 74–75
  vegetables, 8–9
  wrapping techniques in, 180–81
  See also Frozen food; names of food
French custard ice cream, 146
French endive, 11, 12
French ice cream, 146
Frogs' legs, 231, 240
Frosted cakes, 269–70, 272
Frozen desserts, 146, 149
Frozen food
  fish, 225–26, 227, 229
  fruit, 76
  in power failure, 377–78
  prepared, 184
  thawing of, 334–35, 368
    fish, 229
    meat, 181–83, 362
  vegetables, 7, 8–9, 65, 332
  See also Freezing; names of foods
Fructose, 284
Fruit
  butter, 117
  canned, 76, 119, 339
  dried, 29, 115–17, 119
    drying process, 339, 340
  drinks, 309–10

ethylene gas in, 10, 21, 71, 73
freezing of, 74–76
frozen, 76
  in power failure, 377
jellies, jams, and preserves, 117–18
juice, 74, 308–11, 377
maturity of, 69–70
out-of-season, 72
pickled, 61–62, 66
pies, 271, 272
poached, 73
ripening of, 70–71, 73
selection of, 71–72
storage of, 10, 21, 73–74, 118–19
  See also names of fruit
Fruit ripeners, 73
Fruitcake, 270, 272

Game birds, 222, 224
Game meat, 195–96, 197
Garbanzos (chickpeas), 50, 64
Garlic
  fresh, 57–58, 66
  peeling and crushing, 58
  spices, 291
Geese, 220–21, 223
  fat, 221, 280, 283
  See also Poultry
Gelatin, 277, 279
Giblets, 223, 224
Ginger (dried and preserved), 291
Ginger root (fresh), 25, 64
Gluten flour, 248, 251
Goat
  cheese, 157, 163
  milk, 134
  yogurt, 143
Gobo root, 25, 64
Golden Delicious apple, 77
Gooseberries, 92, 118
Gorgonzola cheese, 160, 163
Gouda cheese, 159–60, 163
Grains, 245
  cereal, 252, 254
  See also Breads, Flour; Pasta; Rice
Grana Padano cheese, 161
Granadilla (passion fruit), 92, 118
Granny Smith apple, 77
Grapefruit, 92–93, 118
  juice, 309

Grapes, 94–95, 118
  juice, 309
Grapeseed oil, 281, 283
Gravenstein apple, 77
Gravy, 183, 197, 224
Great Northern beans, 50
Green beans, 9, 47–48, 64
  canned, 66, 339
Green onions, 58–59, 64
Green peppers, 29–30, 65
Green tea, 305–6
Green tomatoes, 27, 29, 61
Greens
  native, 14–15
  pungent, 11–12
  spinach, 12–13
  storage timetable for, 64
Grits, 253, 254
Groats, 252, 253, 254
Ground meat
  beef, 187–88
  pork, 189–90
  storage of, 181, 196, 197
  thawing of, 182, 362
Gruyère cheese, 159, 163
Guavas, 95, 118

Half-and-half (dairy), 138
Ham
  canned, 201, 202, 207
  country, 201–2, 207
  curing, 200
  raw, 202
  selection of, 201–2
  storage of, 202–3
Hamlin orange, 102
Havarti cheese, 157, 163
Hearts, 194
Heavy cream, 138
Herbal tea, 307
Herbs and spices
  dried, 293–94
    storage of, 294, 296
  fresh, 294
    drying of, 295
    freezing of, 295
    storage of, 294–95, 296
  types of, 289–93
Herkimer cheese, 158, 163
Herring, pickled, 230
Hominy, 253, 254
Hominy grits, 253, 254

Honey, 287, 288
Honeydew melons, 95–96, 118
Hormones
  in meat, 176–77
  in poultry, 216
Horseradish
  root, 25, 64
  prepared, 25, 66
Hot (chili) peppers, 30–31, 61, 65, 66
Hubbard squash, 54
Huckleberries, 85, 118
Hydroponic lettuce, 10
Hydroponic tomatoes, 27

Ice cream
  air content of, 146
  labeling of, 145–46
  storage of, 146–47, 149
Ice milk, 146
Ice packs, 355
Iceberg lettuce, 10, 11, 65
Idaho potato, 16
Idared apple, 77
Imitation milk, 148
Indian nuts (pine nuts), 125–26, 128
Infant formula, 148, 150
Insect pests, control of, 370–74
Instant coffee, 303, 304
Instant rice, 258
Insulated bags, 355
Irradiation, 342

Jaffa orange, 103
Jams, 117–18, 119
Japanese radish (Daikon), 26
Jarlsberg cheese, 159, 163
Jellies, 117–18, 119
Jerky, 206, 207
Jerusalem artichoke, 25–26, 64
Jicama, 25, 65
Jonathan apple, 77
Juice
  aerating, 310
  bottled and canned, 310
  cartons, 310
  frozen, 210
  storage timetable, 311
  varieties of, 308–10
Juniper berries, 291

Kale, 5, 14, 65
Kasha, 253
Ketchup, 298
Kidney beans, 50
Kidneys, 194
Kiwifruit, 96, 118
Kohlrabi, 38–39, 65
Kosher salt, 277
Kumquats, 96–97, 118

Lactose-reduced milk, 138
Lamb, 190–91, 197
  liver, 194, 197
  See also Meat
Lard, 281, 283
Leaf lard, 281
Leaf lettuce, 10, 65
Leeks, 58–59, 65
Lemonade, 309
Lemons, 97–98, 118
  juice, 309
Lentils, 50–52, 66
Lettuce
  preparation for salad, 11
  selection and storage of, 5, 9–10, 65
Liederkranz cheese, 156, 163
Light cream, 138
Light whipping cream, 138
Lignin, 5
Lima beans, 48, 64
  dried, 51
Limburger cheese, 156, 163
Limes, 98, 118
Litchis, 98, 118
Liver, 194, 197
Liverwurst, 208
Lobsters, 235, 240
Long beans, 48, 65
Lox, 230

Macadamia nuts, 124, 127
Mace, 291
McIntosh apple, 78
Macoun apple, 77
Mandarin oranges, 103
Mangoes, 99, 118
Manioc, 24–25
Maple syrup, 287–88
Maraschino cherries, 88
Margarine, 142–43, 149

Marinade
  fish, 228
  meat, 184
Marjoram, 291
Marmalade, 117
Marrow, 193
Mayonnaise, 297–98
Meat
  bones and stock, 192–93, 197
  canned, 206, 207
  contamination of, 176, 362
  cured. See Processed and cured
    meat
  freezing of, 178–81
    cooked meat, 183
    power failure and, 377–78
    refreezing, 183
    timetable for, 196–97
  freshness of, 178
  game, 195–96, 197
  ground. See Ground meat
  industry regulation, 176–77
  lamb, 190–91, 197
  marinated, 184
  precautions against food poisoning,
    365–69
  processed. See Processed and cured
    meat
  rabbit, 195, 197
  refrigeration of, 178–79, 181
    cooked meat, 183
    timetable for, 196–97
  as staple food, 175
  thawing of, 181–83, 362
  uniform labeling and grading, 177–
    78
  variety meat, 193–94, 196–97
  veal, 191–92, 196
  wraps for, 180–81, 333
  See also Beef; Ham; Pork
Meatballs, freezing of, 188
Mellorine, 146
Melons
  cantaloupes, 85–86, 118
  casabas, 86–87, 118
  crenshaws, 89, 118
  honeydews, 95–96, 118
  Persian, 107, 118
Meringue pie, 271
Metal poisoning, 365
Mexican custard apple (sapote), 113,
  119

Mice, control of, 374
Microorganisms, 328–29
Microwave oven
  blanching vegetables in, 8
  thawing meat in, 182–83
  thawing vegetables in, 9
Milk
  acidophilus, 138
  buttermilk, 137, 149
  certified, 137–38
  chocolate, 134, 318, 319
  condensed, 135–36, 150
  contamination prevention, 133
  cream, 138–39, 149
  dry, 136–37, 150
  eggnog, 137
  evaporated, 135
  fat content of, 134
  filled and imitation, 148
  freezing of, 135, 149
  freshness of, 134
  goat, 134
  homogenized, 133–34
  lactose-reduced, 138
  low-sodium, 138
  multimineral and multivitamin, 138
  pasteurized, 131–32, 133
  refrigeration of, 134–35, 149
  UHT (ultra-high temperature), 136
Milk products
  butter, 140–41, 149
  crème fraîche, 145, 149
  dating code on, 132
  dried and canned, 132, 150
  ice cream, 145–47, 149
  infant formula, 148, 150
  margarine–butter combinations,
    142, 149
  prepared foods, 133
  refrigeration of, 132–33, 149–50
    in power failure, 378
  sour cream, 144–45, 150
  substitutes for. See Dairy substitutes
  yogurt, 143–44, 150
  See also Cheese
Millet, 253
Mineola orange, 103
Mint, 291
Mixes, packaged, 269
Molasses, 286–87, 288
Molds (microorganisms), 328–29
Monastery-type cheese, 157, 163

Monosodium glutamate, 365
Monterey Jack cheese, 158, 163
Morels, 55
Mozzarella cheese, 154, 155, 163
Muenster cheese, 154, 157, 158, 163
Muffin dough, 263, 264
Mung beans, 51
Mushrooms, 55–56, 65
Mussels, 236–37, 240
Mustard
  homemade, 291–92
  prepared, 298
  seeds, 292
Mustard greens, 14
Mutton, 190

Napoleons, 269
Nappa cabbage, 35
Navel orange, 102
Navy beans, 51
Nectar, 309
Nectarines, 100–1, 118
Neufchâtel cheese, 155, 163
New potato, 16–17
Newtown Pippin apple, 78
Nitrites and nitrates, 206–7
Nopalitos, 14, 65
Northern Spy apple, 78
Nutmeg, 292
NutraSweet, 288
Nuts
  canned, 121, 128
  freezing of, 121
  in shell, 120–21
  shelled, 121
  storage of, 121, 127–28
  See also names of nuts

Oat bran, 248, 251
Oat flour, 248, 251
Oats, 253
Octopus, 231
Oils. See Fats and oils
Okra, 33–34, 65
Olive oil, 281, 283
Olives, 62
Onions
  enzyme action in, 59–60
  freezing of, 61
  green, 58–59, 64
  pickled, 61

processed, 292
selection of, 60
storage of, 60–61, 65, 66
  separation from potatoes, 18
  sliced, 65
  varieties of, 60
Oolong tea, 305, 306
Oranges
  canned, 103
  color of, 101–2
  freezing of, 103
  juice, 103, 309
  storage of, 103, 118
  varieties of, 102–3
Oregano, 292
Organ meats, 193–94, 196–97
Oyster mushrooms, 55
Oyster plant (salsify), 26, 65
Oysters, 236, 240

Packaging. See Containers; Wraps
Pancakes, 263, 264
Papayas, 104, 118
Paprika, 292
Parmesan cheese, 154, 161, 163–64
Parmigiano Reggiano (Parmesan) cheese, 161
Parsley, 292
Parsnips, 5, 24, 65
Parson Brown orange, 102
Passion fruit (granadilla), 92, 118
Pasta, 255–56
Pastries, 271–72
Pea beans, 51
Peaches, 104–6, 118
  canned, 105
Peanut butter, 124–25, 128
Peanut oil, 281, 283
Peanuts, 124, 125, 127
Pearl barley, 253, 254
Pears, 106–7, 118
Peas
  black-eyed, 50
  dried, 51–52, 66
  English, 44, 45
  freezing of, 9, 45
  lectin content of, 52
  selection and storage of, 44, 65
Pecan pie, 271, 272
Pecans, 125, 128
Pecorino Romano cheese, 161
Pekin ducks, 220

Pepper, 292
Peppercorns, 292
Peppers
    canned, 30, 66, 339
    chili, 30–31, 61, 65, 66
    dried, 30–31
    freezing of, 31
    selection and storage of, 30, 65
    varieties of, 29–30
Pepper sauce, 298
Periwinkle, 239
Persian melons, 107, 118
Persimmons, 107–8, 118
Pest control
    insect pests, 370–74
    rodents, 374–75
Petits pois, 44
Pickled fruits and vegetables, 66
Pickles, 31, 61–62, 339
Pickling spice, 292
Pies, 271, 272
    shells, 271, 272
Pignolias (pine nuts), 125–26, 128
Pineapple
    canned, 109
    fresh, 108–9, 118
    juice, 309
Pineapple orange, 102
Pink beans, 51
Pink lentils, 50–51
Pinto beans, 51
Pistachios, 126, 128
Pita bread, 260, 261, 262, 263
Plain pack, in fruit freezing, 74–75
Plantains, 109, 119
Plastic bags in food storage, 6, 9, 76,
        180, 181, 350–51, 353, 356
Plastic coolers, 354–55
Plastic wrap, 352, 354
Plum pudding, 270
Plums, 110–11, 119
Poi, 26
Polyvinyl chloride (PVC), 352
Pomegranates, 111, 119
Popcorn, 322, 323, 324
Poppy seeds, 292
Pork
    bacon, 199–200, 207
    cooking, temperature for, 189
    ground, 189–90
    ham, 200–3, 207
    liver, 194, 197

salt pork, 206, 208
sausage, 189–90, 203–4, 208
selection and storage of, 189,
        197
and trichinosis, 189, 365
See also Meat
Porridge, 252
Port du Salut cheese, 154, 157, 164
Pot cheese, 164
Potato(es)
    canned and frozen, 18
    chips, 322–23
    dried mashed, 19
    freezing of, 19
    history of, 16
    medium-weight, 18
    new, 16–17
    selection and storage, 17–18, 66
    varieties of, 16–17
Potato flour, 248, 251
Poultry
    barn-raised, 209
    chicken, 215–17
    cooked, 214–15, 217
    duck, 220–21, 223
    freezing of, 212–13, 214, 217,
        219
    frozen, 210
        in power failure, 378
        thawing, 213, 219–20
    game birds, 222, 224
    geese, 220–21, 223
    giblets, 223, 224
    precautions against food poisoning,
        211–12, 363, 365–69
    seasoning, 292
    selection of, 209–10
    servings per pound, 211
    smoked, 215, 224
    squab, 221, 223
    storage timetable, 223–24
    turkey, 217–20
Pound cake, 269, 270
Powdered sugar, 285
Power failure, food preservation in,
        376–78
Pré-salé lamb, 191
Preserves, 117–18, 119
Prickly pears, 111, 119
Processed cheeses, 161–62, 164
Processed and cured meat, 198–
        99

Processed and cured meat (cont.)
  bacon, 199–200
  canned, 206
  cold cuts, 204–5
  corned beef, 205
  dried, 206
  ham, 200–3
  nitrite-free, 207
  sausage, 203–4
  storage of, 199, 206–8
Prosciutto, 202
Provolone, 154, 160, 164
Prunes
  dried, 116
  fresh, 110–11, 119
  juice, 309
Ptomaine poisoning, 361
Pudding, 146, 149
Pullet, 215
Pumpkin, 54
  pie, 271, 272
  seeds, 126, 128
Purees, fruit, 75

Quinces, 112, 119
Quinoa, 253, 254

Rabbit, 195, 197
Radicchio, 12
Radishes, 26, 65
Raisins, 116
Rape, 15, 65
Raspberries, 74–75, 83–84, 119
Rats, control of, 374–75
Raw sugar, 285
Red beans, 51
Red cabbage, 34
Red Delicious apple, 78
Red peppers, 30, 65
Refrigeration
  commercial, 341
  and food preservation, 330–31,
    368–69
  in power failure, 376–77
  refrigerator/freezer selection and
    operation, 345–47
Rhode Island Greening apple, 78
Rhubarb, 112–13, 119
Rice
  and beans, 50
  length of grain, 257
  milling of, 257

  storage of, 258, 259
    cooked, 258–59
    types of, 258
Rice flour, 249, 251
Ricotta cheese, 155, 164
Ripening bowl, 73
Roasts, thawing of, 182
Rock Cornish game hen, 216, 217,
    223
Rocket, 15
Rolled oats, 253, 254
Romaine lettuce, 10, 65
Romano cheese, 154
Rome Beauty apple, 78
Root cellars, 331
Root and tuber vegetables, 3–4, 24–
    26
Roquefort, 151–52, 160, 164
Rosemary, 292
Russet potato, 16
Rutabagas, 23–24, 65
Rye, 253
  flour, 249, 251

Safflower oil, 142, 281–82, 283
Saffron, 292
Sage, 293
Salad dressings, 299
Salami, 208
Salmon, smoked, 230
Salmonella, 211, 362–63
Salsify (oyster plant), 26, 65
Salt, 277, 279
Salt pork, 206, 208
Sandwiches, freezing of, 337
Sapote (Mexican custard apple), 113,
    119
Sauces, seasoning, 297–99
Sauerkraut
  canned, 35, 66, 339
  fresh, 35, 65
Sausage
  fresh, 189–90, 197, 203–4, 208
  smoked, cooked and dried, 204,
    208
Savory, 293
Savoy cabbage, 34
Scallions, 58–59
Scallop squash, 52
Scallops, 237, 240
Sea salt, 277
Sea urchins, 237

Seasoned salt, 277
Seaweed, 62, 66
Seckel pears, 106
Seeds, 121, 126, 128
Semolina, 249, 251
Sesame oil, 282, 283
Sesame seeds, 293
Shallots, 60
Shellfish
  clams, 232–33, 240
  crab, 234, 240
  crayfish, 235, 240
  lobsters, 235, 240
  mussels, 236–37, 240
  oysters, 236, 240
  scallops, 237, 240
  sea urchins, 237
  shrimp, 238, 240
  storage timetable for, 239–40
  univalves, 239, 240
Sherbet, 146
Shitakes, 55
Shortening, vegetable, 282
Shrimp, 238, 240
Silverfish, control of, 374
Smithfield ham, 201
Snack foods, 322–24
Snails, 239
Snow peas, 44, 45
Sodium nitrate, 341
Soft drinks, 312–13
Sorrel, 14, 65
Soup barley, 253
Sour cream, 144, 150
Soy flour, 249, 251
Soy sauce, 298
Soybeans, 51, 65
  oil, 142, 282, 283
  tofu, 49–50
Spaghetti squash, 54
Spanish onion, 60
Spices. See Herbs and spices
Spinach, 5, 12–13, 65
Sponge cake, 269, 270, 272
Sprouts, 49, 65
Squab, 221, 223
Squash
  summer, 52–53, 65
  winter, 53–55, 66
Squid, 231, 240
Staphylococcus aureus, 364
Staples, 275–79

Star fruit (carambola), 113, 119
Steaks
  storage of, 181, 196
  thawing of, 182
Stew meat
  storage of, 181, 196
  thawing of, 182
Stewing chicken, 216
Stilton cheese, 160, 164
Stock
  bones for, 192–93, 197
  poultry, 224
  storage of, 181, 193, 197
Straightneck squash, 52
Strawberries, 113–14, 119
Stuffing, poultry, 212–13, 214, 224
Styrofoam coolers, 354
Suet, 194, 280, 283
Sugar
  storage of, 288
  substitutes, 288
  types of, 284–86
  See also Sweeteners
Sugar Baby watermelon, 114
Sugar cane, 63, 66
Sugar pack, in fruit freezing, 75
Sugar peas, 44, 45
Sugar snap peas, 44
Summer squash, 52–53, 65
Sunflower oil, 282
Sunflower seeds, 126, 128
Sweet potatoes, 7, 19–20, 66
Sweetbreads, 194
Sweeteners, 284, 286–88
Swiss chard, 5, 42–43, 65
Swiss-type cheese, 154, 159, 164
Syrup
  corn, 286, 288
  maple, 287–88
Syrup pack, in fruit freezing, 75

Tamarillo, 31, 65
Tamarinds, 114, 119
Tangelo, 102
Tangerine, 102
Tapioca, 277, 279
Taro root, 26, 65
Tarragon, 293
Tea, 305–7
Teff, 253, 254
Temple orange, 102

Thawing, 334–35, 368
 fish, 229
 meat, 181–83, 362
Thermal bottles, 355–56
Thyme, 293
Tillamook cheese, 158, 164
Tilsiter cheese, 157, 164
Tofu (bean curd), 49–50, 65
Tomatillos (tomatoes verdes), 31, 66
Tomatoes
 canned, 7, 29, 66, 339
 freezing of, 29
 green, 27, 29, 61
 greenhouse, 27, 28
 history of, 26–27
 hydroponic, 27
 juice, 310
 selection and storage of, 6, 28–29,
   65, 66
 sun-dried, 29
 vine-ripened, 27, 28
Tongue, 194
Tortillas, 261, 262, 263
Travel with food
 packing for, 356–57
 storage equipment in, 354–56
Tray freezing, 9, 74–75
Trichinosis, 189, 365
Tripe, 194
Triticale flour, 249, 251
Truffles, 57, 65
Tuber and root vegetables, 3–4, 24–
 26
Turkey, 217–20, 223
 See also Poultry
Turmeric, 293
Turnips, 23–24, 65
 greens, 14, 23

Uglifruit, 114, 119
UHT (ultra-high temperature) milk,
 136
Univalves, 239, 240

Vacuum bottles, 355
Vacuum packs, 341
Velencia oranges, 102
Variety meats, 193–94, 196–97
Veal, 191–92, 197
 See also Meat
Vegetable oil, 282, 283
Vegetable shortening, 282

Vegetables
 canned, 7, 66
 canned vs. frozen, 6–7
 cooked, 9, 64
 cutting and soaking of, 6
 enzyme action on, 3–4
 freezing of, 8–9
 frozen, 7, 9, 64–65, 332
  in power failure, 377
 grading of, 6
 humidity requirements of, 4, 5,
   6
 pickled, 61–62, 66
 refrigeration of, 5–6
 selection of, 5
 separation from fruit, 10, 21
 storage timetable for, 64–66
 See also names of vegetables
Venison, 195
Vinegar, 277–78, 279
 homemade, 278

Waffles, 263, 264
Walnut oil, 282, 283
Walnuts, 126–27, 128
Water chestnuts, 63, 65
Water ice, 146
Watercress, 15, 65
Waterglassing, of eggs, 169
Watermelon, 114–15, 119
Watermelon rinds, 61
Wax beans, 47–48, 64
Waxed paper, 352, 354
Weevils, control of, 373
Westphalian ham, 202
Wheat berries, 248, 253
Wheat germ, 249, 251
Whelks, 239
Whipped butter, 140
 filling, 269, 272
 freezing of, 139
 prepared, 138, 139
 storage of, 139, 149
White flour, 246–47, 250
White rice, 258
Whole wheat flour, 247–48, 250
Wild rice, 258
Winesap apples, 78
Winter squash, 53–55, 66
Worcestershire sauce, 298
Wrapping methods, for meat, 180–
 81, 333

Wraps
    meat, 180–81, 333
    types of, 352, 353–54

Yam, 19
Yearling lamb, 190
Yeast, 278, 279

Yeasts (microorganisms), 329
Yellow peppers, 30, 65
Yogurt, 143–44, 150
    frozen, 146
York imperial apple, 78

Zucchini, 52, 53